Major Constrains and Verdict of Crop Productivity

The Editor

Dr. U.N. Bhale, presently Assistant Professor, Department of Botany, Arts, Science and Commerce College Naldurg Tq. Tuljapur, Dist. Osmanabad affiliated to Dr.Babasaheb Ambedkar Marathwada University Aurangabad (MS). He has completed his Ph.D. in Botany (Plant Pathology) from Dr.Babasaheb Ambedkar Marathwada University Aurangabad (MS). He has 12 years teaching and 15 years research experience. He has contributed more than 90 papers in National and International journals of repute. He has published 3 books, guided 06 Ph.D. students and attended 37 National and International conferences. He has completed 03 Minor and 01 Major research projects. He has successfully organized a National Conference as Organizing Secretary funded by University Grants Commission (UGC). His areas of interest are Mycology, Plant pathology and VAM technology. He is an Associate Editor of International Journal of Bioscience Discovery and Life Member of Academy of Plant Sciences, IPS, ISMPP, BBRC, MSI, IMS, JBB and IJPS.

Major Constrains and Verdict of Crop Productivity

Volume 1

Editor

Dr. U.N. Bhale

Department of Botany,
Arts, Science and Commerce College,
Naldurg – 413 602
Tal. Tuljapur, Dist. Osmanabad,
Maharashtra (India)

2015

Daya Publishing House®

A Division of

Astral International Pvt. Ltd.

New Delhi – 110 002

Cataloging in Publication Data--DK
Courtesy: D.K. Agencies (P) Ltd. <docinfo@dkagencies.com>
Major constrains and verdict of crop productivity / editor, Dr. U.N. Bhale.
pages cm
Includes bibliographical references and index.
ISBN 9789351243489 (Set)
ISBN 9789351243694 (Vol. 1)
ISBN 9789351243700 (Vol. 2)
ISBN 9789351305545 (International Edition)
1. Crops and soils--India. 2. Soil productivity--India. I. Bhale, U. N., editor.
DDC 631.4954 23

Published by : **Daya Publishing House®**
A Division of
Astral International Pvt. Ltd.
– ISO 9001:2008 Certified Company –
4760-61/23, Ansari Road, Darya Ganj
New Delhi-110 002
Ph. 011-43549197, 23278134
E-mail: info@astralint.com
Website: www.astralint.com

Laser Typesetting : **Classic Computer Services**, Delhi - 110 035

Printed at : **Sanat Printers**

PRINTED IN INDIA

Prof. C. MANOHARACHARY
M.Sc, Ph.D, D.Sc., F.N.A.Sc., F.N.A.A.S., F.B.S., F.P.S.I., F.A.P.A.S.
Professor Emeritus (CSIR)
Former Head, Chairman, Principal,
Co-ordinator (UGC SAP)
Dean, Development & UGC Affairs
NASI Senior Scientist

DPARTMENT OF BOTANY
Osmania University,
Hyderabad – 500 007, A.P. India.
Off: 91-40-2768 22 44
Mob:9391164243

Foreword

I have great pleasure in introducing the academic venture entitled " Major constrains and verdict of crop productivity" an edited volume of 54 scientific articles by Dr. U.N. Bhale, Department of Botnay, Arts, Science and Commerce College, Naldurg, Maharashtra state, India. India is an agricultutally dependent country and the growing population needs food and nutritional security.

Though India reached targeted productivity both in wheat and rice during first green revolution. Huge losses have incurred in crop prouductivity due to increase in pests and diseases, and also loss of soil fertility because of non judicious usage or overuse of pesticides, fungicides and fertilizers.

Further there has been repeated drought periods because of failure of monsoon and less precipitation. All these factors advocate that we need a second green revolution where in drought and disease resistant varieties need to be produced besides increasing the production of pulses, oil seeds and fertile land. All the 56 scientific contributions deal with biotechnology of crop production, usage of biofertilizers and biopesticides, management of pests and diseases, biodiversity, phytotoxicity, plant and microbial metabolites and related issues. The contributed authors have necessary expertise. I hope that this book will add its strengths to the existing knowledge and furtherance of new knowledge besides bringing an awareness on issues related to Agriculture. I hope that all categories of readers and libraries will find it useful. Finally I compliment the editor Dr. U.N. Bhale and authors for their concerted efforts.

1/7/2014.

Prof. C. Manohara Chary
NASI Senior Scientist

Preface

India is an Agricultural country and reached targeted productivity especially rice, wheat and others crops in green revolution. Now it is huge losses in crop productivity due to pest, diseases and loss of soil fertility because of overuse of chemical pesticides. Therefore number of constrains are found in crop productivity. Drought conditions are thoroughly increased due to failure of monsoon in time.A constraint free soil environment is very important for achieving higher food production. Abiotic stresses like nutrient-poor, degraded and often acidic, soils limit crop production in many tropical regions. Good agronomic practices help minimize the water and other physiological stresses imposed by subsoil constraints.When coupled with the high cost of inorganic fertilizer, much small-scale agriculture occurs under conditions of nutrient scarcity and/or metal ion toxicity. Drought is perhaps the biggest constraint to agricultural productivity worldwide. Biotic stresses like diseases are a huge problem worldwide for crop productivity.

This book is an outcome of to create awareness about the emerging trends in agriculture designing, developing new economic process to farmers and provide a platform for forthcoming researchers. I have attempted to compile the invaluable and carefully selected contributions of plant biologists in this consolidation of this book with all originality and alertness.The book considers more specifically the particular implications of different approaches to technical improvement of productivity in farming. The book concludes by outlining agricultural areas in which research needs to take a new approach to investigating the constraints to crop productivity improvement and the relationship of these to rural poverty. The expertise has in the fields of plant protection, plant genetic resources, forestry, soil science, and extension and is well resourced to address crop production problems in the region. To my mind, the best approach to identifying key constraints is to establish a much better dialogue with between bench scientists, researchers and those key breeders in the country who actually talk to farmers and understand local agricultural issues.

I am sincerely thanks to all contributors and those directly or indirectly have rendered their help to bring out this book. My special genuine gratitude to Prof. C. Manoharachary, Professor Emeritus (CSIR),NASI Senior Scientist, Osmania University, Hyderabad for encouraging and giving Foreword letter. Lastly, my sincere thanks to publisher Astral International (P) Limited, New Delhi who provides an opportunity to publish this book.

Dr. U.N. Bhale

Contents

Volume-2

2015, Major Constrains and Verdict of Crop Productivity Pages 1–20
Editor: Dr. U.N. Bhale
Published by: ASTRAL INTERNATIONAL (P) LIMITED, NEW DELHI

Chapter 1

Constraints to Cowpea Production in Africa

Oigiangbe and Osawe Nathaniel*

Department of Zoology, Ambrose Alli University, Ekpoma, Edo State, Nigeria

Introduction

There are about 170 species of the genus *Vigna* out of which only three are important grain legumes in the world (Onwueme and Sinha, 1991). These are cowpea (*Vigna unguiculata*), green gram (*V. radiata*) and black gram (*V. mungo*). Five inter-fertile subspecies of *V. unguiculata* are recognized by the lengths of their calyx lobes and by the length and dehinscence of their fruits (Steele *et al.*, 1985). Two of these subspecies are wild types with short scabrous, dehiscent fruits and small seeds which show some dormancy because of their relatively impermeable testas. The wild types are found only in Africa and are the ancestors of the cultivated subspecies. *Vigna unquiculata* ssp. *dekindtiana* is the common wild and weedy cowpea of the African savannas. It has both perennial and annual forms in West Africa. The second wild cowpea, *V. ungniculata* ssp. *mensensis* differs a little from the subspecies *dekindtiana* and extends more into the wetter forest regions.

Cowpeas were first domesticated in Ethiopia or West Africa more than 4000 years ago (Onwueme and Sinha, 1991). The earliest cultivars were prostrate or spreading, photosensitive, short-day types of the subspecies *unguiculata*. It is believed that this subspecies reached India with sorghum and bulrush millet from East Africa around 1,300 B.C. Cowpea probably reached southern Europe from Asia before 300 B.C., and the New World in the seventeenth century from West Africa and Europe (Steele, 1972). Although cowpea moved to India and Asia much earlier than America,

* E-mail: osawe_oigiangbe@hotmail.com

it has formed an integral part of the cropping systems of both continents despite the fact that it is less important than in Africa (Ng and Marechal, 1985).

Cowpea is usually grown as the cover crop in a system based on cereals or tuber crops (Okigbo and Greenland, 1976, Blade *et al.*, 1997). Their advantages in the system include rapid establishment, reduction in soil erosion and temperature, as well as suppression of weeds (Zuofa *et al.*, 1992). The crop is adapted to warm weather and can withstand drought and heat much better than other legumes (Onwueme and Sinha, 1991). Cowpea has well developed tap root system and like other legumes can fix atmospheric nitrogen in the soil by mycorrhiza association between their roots and soil fungi (IITA, 1977). Cereals intercropped with cowpea is believed to derive some nitrogen from the root nodules of the cowpeas, especially towards the end of the growing period. In the Savanna and Sahel regimes of West Africa, the roots of cowpea releases some substances which provokes the germination of the parasitic witchweed, *Striga hermontica* which is a serious problem on cereals. By causing such suicidal germination of the weed, which cannot grow to the reproductive phase, the seed bank of the weed is reduced in the soil. Thus, cowpea act as a trap crop for this weed, which is often very difficult to control (Singh and Emechebe, 1991).

Nigeria is the largest producer and consumer of cowpea in the world. Data by the Food and Agricultural Organization of the United Nations (2001-2010) show that Nigeria produces an average of 2.58 ± 0.31 million metric tonnes of the crop per annum. The country's demand deficit of the commodity is met by imports from other West African countries notably Burkina Faso and Niger. All parts of cowpeas plant are consumed in one way or the other by inhabitants of the regions where the crop is grown. Different dishes are prepared from the leaves, green pods, green peas and the dry seeds in different cultures. These parts provide proteins, vitamins, minerals and micronutrients. The seeds are richer in the amino acids lysine and tryptophan making it a very good supplement to cereals. The economic value of the crop is derived from the food value of the grain.

Production Constraints

Cowpea production is affected by a wide range of biotic and abiotic factors. The biotic factors include pests, diseases, and parasitic weed attacks. The abiotic factors include rainfall, temperature, soil quality and texture, drought, poor crop management, wrong planting dates, low plant population and mixed cropping.

I. Biotic Factors

Pests

Cowpea is a highly nutritious plant for both man and other animals. All parts of the plant are exploited by one or more types of insects, rodents or nematodes. Even lizards are known to feed on germinating seeds in the soil (Oigiangbe, unpublished). Insects and some vertebrates form the bulk of primary consumers of cowpeas. Birds are not often a problem on cowpea. Vertebrate pests include rats, squirrels, hares and some ungulates. These are usually controlled by trapping, snares, barriers or food poisons. Much of the cowpea plant is consumed by a wild array of insect pests which have generated extensive studies. Although all parts of the plant are attacked, the flowers and the pods are the most important and constitute the greatest hazards in cowpea production (Onwueme and Sinha, 1991). Insects from the major phytophagous taxa attack the crop from seedling to harvest and can cause economic damage at all stages of plant growth (Taylor, 1967; Singh and Jackai, 1985; Jackai and Daoust, 1986; Singh *et al.,* 1990). For the purpose of this review, only insects damaging cowpea in Africa will be considered. These include:

a. Aphids

Two types of aphids infest cowpea in Africa. The most important and widely distributed one is *Aphis craccivora* Kock or the bean aphid. The second one is *A. fabae* (Scopoli) or the black bean aphid which has been reported as a minor pest in East Africa (Singh and van Emden, 1979). Females are usually found and they reproduce parthenogenetically. They are usually without wings when abundant food and optimal environmental conditions prevail. Under reduced food supply and hash environmental conditions such as crowding, alates, which have wings, are produced. These alates leave the old habitat for new ones where there is more food supply. Aphids usually infest young seedlings, but their population can build up again when the pods on which they also feed start appearing. They inflict direct damage to the plant by removal of its sap leading to distortion of leaves, stunting of the plants and poor nodulation of the root systems (Onwueme and Sinha, 1991). This reduces yield and in extreme cases the plant dies (Singh and Van Emden, 1979). Apart from the direct damage, aphids also cause indirect damage which can sometimes be more serious, by transmitting viruses to the plant (Bock and Conti, 1974). Naturally, aphid populations are kepted in check by natural enemies such as parasites and predators. Secondly, some cultivars with acceptable levels of resistance have been developed.

b. Leafhoppers

These are made up of the *Empoasca* complex (Homoptera: Cicadellidae) which include species found all over tropical Africa. One of the commonest species but of little economic importance is *E. dolichi* Haust and some other species have been reported (Singh, 1980). Generally, leafhoppers are considered minor pests and most cowpea varieties grown in Africa appear to be resistant to them (Singh *et al.,* 1990) However, they cause serious damage to cowpea seedlings during drought (Raman *et al.,* 1978). Both adults and nymphs suck sap from the underside of the leaves.

c. Bean Fly

The Bean fly, *Ophiomyia phaseoli* (Tryon), synonym: *Melanogromyza phaseoli*, is a pest of cowpeas in East Africa. Eggs laid on leaves of cowpeas hatch into maggots which behave like leaf miners, penetrating underneath the epidermis of the leaf until it reaches the vein. (Singh and Jackai, 1985). From here it tunnels to the midrib and to the leaf stalk, finally reaching the stems near the ground surface or in the petiole in older plants. Infested plants have punctured leaves, stunted growth, yellowing and can eventually die. Most often the stem is swollen at the base, cracked and black brown due to rotting of the tissues. Population build up is encouraged by drought, leading to heavy reductions in yield.

d. Leaf Worm

The Egyptian leaf worm, *Spodoptera littoralis* (Boisduval), synonym: *Prodenia litura* (Fabricius) has been reported as a serious pest of cowpea in Egypt (Singh and van Emden, 1979). Older larvae usually hide in the soil during the day, but feeds voraciously at night. This is a polyphagous insect which behaves like cutworms attacking a wide variety of crops. *Spodoptera littoralis* is essentially a defoliator which can sometimes bore into green pods or feed inside the pod from the outside. Large populations on a field can lead to total yield loss. Natural enemies of this insect include parasites and a polyhedrosis virus.

e. Foliage Beetles

Several beetles feed on the leaves of cowpea and some are vectors of important viral diseases (Singh and Taylor, 1978). Foliage beetles are sporadic pests whose direct damage is often minimal, when populations are low. The most important of these insects is *Ootheca mutabilis* (Sahlberg) (Coleoptera: Chrysomelidae) which attacks cowpea in the East and West Africa (Booker, 1965; Halleren, 1971). Another species, *O. bennigseni* Weise is reported to be important in East Africa (Le Pelley, 1959). Adults of these beetles feed on the surface of the leaves between the veins, while the larvae feed on the cowpea roots (Singh and Taylor, 1978). Indirect damage is done by the adults by transmitting cowpea yellow mosaic virus (Bock, 1971).

The striped foliage beetle, *Medythia quaterna* (Fairmaire), synonym: *Lasperodes lineata* (Kara) (Coleoptera; Chrysomelidae), is found mainly in the forest and derived savanna zones of West Africa (Singh and Jackai, 1985). Adults feed on newly emerged leaves, mostly at the margins, and can completely defoliate or kill the plants. They are also vectors of the cowpea yellow mosaic virus.

Some other beetles of the Family Lagriidae also feed on the leaves of cowpea but their populations are not high enough to cause economic damage. These include *Lagria villosa* Fabricius and *Chrysolagria* species.

Similarly, the striped bean weevil, *Alcidodes leucogrammus* (Erickson) (Coleoptera; Curculionidae) has been reported as a sporadic pest of cowpea in East and West Africa (Singh and Jackai, 1985). The adults feed on the leaves, but females lay eggs on the stem. After hatching, the larvae tunnel and feed inside the stem, causing stunted growth. The stem later splits and the plant die.

f. African Bolloworm

The African bollworm, *Heliothis armigera* (Hubner) (Lepidoptera: Noctuidae) is an occasional pest of cowpea mainly in areas where the crop is grown with maize (Singh and van Emden, 1979). It is a polyphagous pest, which attacks other cultivated crops. The larvae feed in the leaves, flower buds, flowers and green pods of cowpea. Young larvae enter the pods and feed on the seeds inside while older larvae feed on the seeds from outside the pod. Robertson (1963) reported some parasites of *H. armigera.*

g. The Llegume Pod Borer

The Legume Pod Borer, *Maruca vitrata* Fabricius Synonym: *Maruca testulalis* (Geyer) (Lepidoptera: Pyralidae) is one of the most important pests of cowpea in Africa (Singh and ran Emden, 1979; Singh and Jackai, 1985, Jackai and Daoust, 1986, Singh *et al.*, 1990). It causes 20 – 80 per cent yield loss if adequate control measures are not put in place. The early larvae feed on young tender shoots and peduncles before the appearance of flower buds and flowers. As soon as the flower buds and flowers starts forming, the larvae move to these floral parts. They also feed on the green pods when these are produced. During the day, they hide in the flowers or pods, but move around the host plants at night and infest new flowers or pods. They usually web leaves, flowers and pods together (Taylor, 1967).

h. Thrips

The legume bud thrips, *Megalurothrips sjostedti* (Trybom), Synonym: *Taeniothrips sjostedi* (Trybom) (Thysanoptera: Thripidae), is a major pest of cowpea (Singh and Taylor, 1978, Singh *et al.*, 1990, Jakcai and Adalla, 1997). The nymphs and adults feed on racemes and flower buds which can result in total loss of flowers. Under heavy infestation, the racemes produce no flower buds and flowers are diseased when they are formed (Singh, 1978).

In the semi-arid regions, where drought is often a problem, the foliage thrips, *Sericothrips occipitalis* Hood (Thysanoptera: Thripidae) is a minor pest of cowpea seedlings. As the rains establish, the infestation reduces leading to a recovery of the plants.

i. Blister Beetles

Several species of blister beetles belonging to the Family Meloidae are known to damage flowers of cowpea. The most important species are *Mylabris farquharsoni* Blair and *M. amplecteus* Gerstaecker. Some species of *Coryna*, especially *C. apicicorus* (Guenn), are also important pests. These beetles are more important when cowpea is planted with maize or close to maize fields (Singh and Jackai; 1985). The adults feed on flowers and pollen.

j. Cowpea Seed Moth

The cowpea seed moth, *Cydia ptychora* (Meyrick), Synonym: *Laspeyressia ptychora* (Meyrick) (Lepidoptera: Tortricidae), feeds on the green pods of cowpea in the humid regions of Africa. The first instar larvae enter the pods and feed on the developing seeds causing extensive damage (Jackai and Daoust, 1986). In some cases, infestation continues after harvest.

k. Lima Been Pod Borer

The Lima bean pod borer, *Etiella zinckerella* (Treitschke) (Lepidoptera: Pyralidae) infest cowpea pods in East Africa. The larvae feed inside the pods on developing seeds (Singh *et al.*, 1990).

l. Pod Bugs

Many species of bugs do considerable damage to pods of cowpea. Their populations are usually high as the adults constantly move from wild host plants to cowpea. Under favourable climatic conditions and abundant food supply, these bugs breed throughout the year. Adults and nymphs cause considerable damage as they suck the sap from developing pods and seeds. Infested pods dry up prematurely as the seeds remain poorly formed. Four species of these bugs in the family Coreidae (Heteroptera) are important in Africa. *Clavigralla tomentosicollis* (Stal), Synonym: *Acanthomia tomentosicollis* Stal, are found in West and East Africa, *C. shadabi* Dolling, Synonym. *A. horrida* (Germer), found in East Africa, *C. elongata* Signoret in East Africa and *Anoplocnemis curvipes* (Fabricius), in East, West and Central Africa. The populations of these bugs are often kept in check by several species of parasites and predators (Singh and Jackai, 1985, Singh *et al.*, 1990, Jackai and Adalla, 1997).

Riptortus dentipes (Heteroptera; Alydidae) is commonly found on cowpea feeding on pods or resting on foliage (Singh and Jackai, 1985).

Nezara viridula (Linnaeus) (Heteroptera: Pentatomidae) is found throughout Africa and is a minor pest of cowpea. *Aspavia armigera* (Fabricius), another pentatomid, also feeds on pods of cowpea.

m. The Pod Weevil

The pod weevil, *Piezotrachelus varius* Wagner, Synonym: *Apion varius* Wangner (Coleoptera: Curculionidae), feeds in cowpea pods. Eggs are laid on the green pods and the larvae feed inside the seeds (Singh and Jackai, 1985). The population is usually low and is regarded as a minor pest.

n. Storage Weevils

The cowpea bruchid, *Callosobruchus maculatus* Fabricius, is the principal post harvest pest of cowpea and occurs wherever the crop is grown (Jackai and Daoust, 1986). This bruchid frequently infests up to 100 per cent of the stored seeds within 3 to 5 months under ordinary storage conditions (Booker 1967; Singh 1977, 1980). Often, farm storage for 6 months is accompanied by about 30 per cent loss in weight with up to 70 per cent of the seeds being infested and virtually unfit for consumption (Singh and Jackai, 1985). In Egypt, losses in seed weight commonly reach 50 per cent after only 3 months of storage (Hussain and Abdel-Aal, 1982). The damage pattern caused by *C. maculatus* to stored cowpea was studied in northern Nigeria in an open market survey over an 8 year period by Caswell (1981). This study showed that each *C. maculatus* larva consumes an average of 10 per cent of one seed and that an estimated 30 tons was lost to bruchids annually.

The reduction in seed weight due to bruchid attacks was found to be directly proportional to the number of holes or "windows" produced in seeds (Singh *et al.*, 1983). Thus seed losses can be easily predicted for different levels of attack by *C. maculatus*. *Callosobruchus rhodesianus* and *Bruchidius atrolineatus*, also infest cowpea under field conditions. Generally they develop poorly on stored grains (Booker, 1967), although the latter species has been reported as an important storage pest of cowpea in northern Nigeria (Raheja, 1973).

For large scale storage, efficient methods to control bruchids have been developed, and these involve the use of proper storage facilities and various fumigants. On subsistence farms, treatments with pirimiphos-methyl are suitable. Wherever the produce can be fumigated under air tight conditions, Phostoxin (Aluminum

Phosphide) has been found effective. Phostoxin is effective against the eggs and larvae inside the seeds and does not have any residue. Therefore, produce has to be protected from re-infestation in storage (Singh and Jackai, 1985). For subsistence farmers, who produce less than 10 kg seeds for their own consumption, mixing the seeds with groundnut oil (about 5ml/kg of seed) were found practical and effective (Singh *et al.*, 1979).

Diseases

Cowpea is susceptible to a number of fungal, bacterial, and viral diseases such as Cercospora leaf spot, ashy stem blight, bacterial blight, blackeye cowpea mosaic potyvirus (BICMV), cowpea aphid-borne mosaic potyvirus (CABMV), and cowpea mosaic comovirus (CPMV).

The root knot nematode, *Meloidogyne incognita,* is a serious pest of cowpeas in most growing areas of the world (Sikora *et al.*, 2005). This nematode constitutes a major constraint to cowpea production (Sikora and Greco, 1990). Symptoms of damage induced by root-knot nematode include patches of stunted and yellowed plants, presence of root galls, excessive branching of roots, and reduced root systems (Adegbite, 2011). Poor germination and death of seedlings may be observed in cases of heavy infestations (Mishra, 1992).The root knot nematode, *Meloidogyne incognita* was first reported in Nigeria on cowpea in 1958 and documented in 1960 (Anon, 1961). Bridge (1972) reported a yield loss of 40 per cent and Olowe (1976) demonstrated a yield reduction of 25 per cent at inocula level of 133 eggs of *M. incognita* per kg of soil or a 91 per cent reduction with 13,300 eggs per kg soil. Similarly, Caveness (1973) and Ogunfowora (1976) reported yield losses of 20 per cent and 50 per cent respectively, due to infestation by *M. incognita*. A cowpea grain yield loss of 69 per cent caused by root knot nematode was also reported (Babatola and Omotade, 1981). Heavy infestation of cowpea by *M. incognita* led to early senescence of the crop (Olowe, 1978).

Parasitic Weeds

Cowpea is attacked by two parasitic plants, *Striga gesnerioides* (Wild) Vatke and *Alectra vogelii* (L.) Benth, both belonging to the Family Scrophulariaceae (Kuijt, 1969). Of the two, *Striga* is more widely distributed covering sub-Saharan Africa, and parts of Asia and USA, whereas *Alectra* is presently restricted to Africa, (Musselman *et al.*, 1991, Parker and Riches, 1993). Both parasitic plants cause considerable damage to cowpea, with substantial yield reduction, especially in Africa (Emechebe *et al.*, 1991; Lagoke *et al.*, 1994). At present, the Sudano-Sahelian belt of Africa is more affected by *Striga* and the Guinea-Sudan savanna belt by *Alectra*, but both parasites are fast spreading beyond these limits (Singh and Emechebe, 1997). According to Singh and Emechebe (1997) the *Striga-Alectra* problem is becoming even more acute with more monocropping under growing population pressure. This problem is further made worse in areas with sandy soils, poor fertility and low rainfall. *Alectra* is more prevalent in the northern Guinea savanna and southern Sudan savannas of West Africa as well as in East and southern Africa, whereas *Striga* is mostly found in West and Central Africa (Singh, 2002). The most seriously infested areas are in the semi-arid drought striken belts of Nigeria, Niger, Burkina Faso, Mali and Senegal (Aggarwal and Haley, 1987). Cowpea parasitized by these weeds are usually destroyed or the yield reduced to almost zero (Leleji, 1987).

Although, both parasites constitute severe constraints to cowpea production in most of the growing regions in sub-Sahara Africa, *Striga, gesnerioides* is considered the more important of the two (Emechebe *et al.*, 1991; Lagoke *et al.*, 1994; Lane *et al.*, 1997). The cowpea strain of *S. gesnerioides* attacks only cowpea but there are wild strains that attack tobacco and sweet potato in certain parts of Africa. Populations of *A. vogelii* on cowpeas cause serious damage to groundnut, and occasional infection of soybean. The effect of *Alectra* is equally devastating in heavily infected cowpea fields, as has been observed in Botswana (Riches, 1989; Parker and Riches, 1993), Kenya (Bagnall-Oakeley *et al.*, 1991), and Nigeria (Singh and Emechebe, 1991). Singh and Emechebe (1991) observed that in fields that are heavily infected with both *Striga* and *Alectra*, the crop is mostly infected by *Striga* if the host cultivar is equally susceptible to both parasites. However, if the host cultivar is resistant, or moderately resistant to *Striga* and susceptible to *Alectra* (as in the case of IT82D-849 and Suvita-2), *Alectra* infection is very severe, causing substantial yield losses. In West and Central Africa, where *Striga* and *Alectra* are present, control measures must be effective for both. Singh (2002) reported that five strains of *S. gesnerioides* have been identified from collaborative work with some national and regional programs, Strain 1 is found in Burkina Faso, Strain 2 in Mali, Strain 3 in Nigeria and Niger, Strain 4 in Benin Republic and Strain 5 in Cameroon. These strains have serious implications in breeding for resistance against this parasitic weed.

Striga and *Alectra* seeds germinate when exposed to root exudates from cowpea, other hosts and a few nonhosts (Singh and Emechebe, 1997). Radicles elongate, showing a chemotropic response to a concentration gradient of root exudates. Once it contacts with host roots, the radicle apex develops numerous hairs, which attach to host roots. The *Striga* radicle penetrates and stimulates cell division in the host roots. The new host cells, together with growing parasitic tissues, form a large haustorium, uniting the parasite with tissue in the host's stele which permits transfer of water and nutrients from host to parasite. *Alectra* radicle, stimulates profuse root formation by the host and form a larger haustorium than *Striga*. Both *Striga* and *Alectra* shoots emerge from the haustorium about 2 weeks after infection and grow into plants which may be 15-25 cm tall (Singh and Emechebe, 1997).

The common symptoms of *Striga* infections are interveinal chlorosis, general stunting and smaller leaves (Singh and Emechebe, 1997). In the case of severe infection, straw coloured necrotic spots develop on the lamina, followed by complete desiccation of the leaves (Emechebe *et al.*, 1991). Plants infected by *Alectra* have symptoms similar to, but less drastic than plants infected by *Striga*. Infected plants usually show general stunting and wilting with reduced numbers of flowers and pods, but plants may be completely wilted if there is acute moisture deficit (Mugabe, 1983).

The extent of yield reduction depends upon the time and level of infection (Singh and Emechebe 1997). Aggarwal and Ouedraogo (1989) reported a mean yield loss of 30 per cent due to *Striga* infection in Burkina Faso, with 50 per cent loss in the most susceptible local variety. In Nigeria, total destruction of cowpea has been reported from Sokoto, Kano, Katsina, Borno and Bauchi States due to *Striga / Alectra* complex (Emchebe *et al.*, 1991).

An estimate of the yield loss due to *Striga* infection of cowpea under natural conditions was provided by Muleba *et al.* (1997). They studied yield losses in cowpea genotypes of varying susceptibility to *Striga* under *Striga*-infested and *Striga*-free conditions. They found that yield losses in *Striga* – infested plots varied from 3.1 per cent at the experiment station to 44.2 per cent in farmers' fields. Also, depending on the susceptibility of the cowpea genotypes to *Striga*, the yield loss varied from 3.1 per cent to 36.5 per cent. The effect of *Striga* infection on cowpea photosynthesis was reported by Hibberd *et al.* (1996). They showed that the allometric relationship between shoot and root dry weight was similar in parasitized plants relative to the control plants as was the proportion of the dry matter partitioned into leaf, stem and root tissues. However, infected plants failed to make any significant investment of dry matter in pods.

Both plant parasites are rather difficult to control because they produce large amounts of seeds, and their adaptation/dormancy mechanism permit the seeds to stay alive in the soil for several years (Saunders 1933, Kust, 1963). Each *Striga* plant can produce up to 90,000 seeds, which may remain viable for 15-20 years (Parker, 1991). Control of *Striga* is difficult for subsistence farmers as up to 75 per cent of the crop damage is done before the seedlings emerge from the ground (Singh, 2002). Crop resistance offers an appropriate and sustainable control method (Lane *et al.*, 1997). A local landrace, B301 from Botswana confers complete resistance to *Striga* and *Alectra* in Burkina Faso, Cameroon, Mali, Niger and Nigeria (Singh, 2002). However, it has moderate levels of resistance to the strain from Benin Republic. Other lines such as IT8ID-994, IT89KD-288, 58-57 and Gorom local confer complete resistance to strains from Benin Republic and Burkina Faso. A number of new varieties have been developed from crosses between these complementary parents with combined resistance to *Alectra* as well as all the five strains of *Striga* (Atokple *et al.*, 1995; Berner *et al.*, 1995; Singh and Emechebe 1997; Singh *et al.*, 1997; Singh, 2000).

Another strategy for *Striga* control is the reduction of the population of *Striga* seeds in the soil. One way of accomplishing this is to induce *Striga* seeds to germinate in the absence of the host with the subsequent death of the *Striga* seedling (Emechebe and Lagoke, 2002). It has been suggested by Berner and Williams (1998) that *S. gesnerioides* control involving rotation with nonhost cultivars has potential for success only if these cultivars are selected with the *Striga* isolate(s) from the locality of intended deployment of the nonhost cultivar.

II. Abiotics Factors

Rainfall

Cowpea is sown as rain fed crop all over Africa (Onwueme ad Sinha, 1991) with the crop surviving on stored water in the soil after the cessation of rains. Rainfall decreases with increasing northern latitude and the region between 12-13°N latitude representing the Sudano-Sahelian belt receives reasonably reliable rainfall between

July and August (Singh, 1987). Further north, the rainfall reduces gradually and becomes too short and unreliable for crop production beyond 16°N latitude. The availability of adequate soil moisture for crop growth depends not only on the amount and distribution and rainfall, but also on the water holding capacity, soil type, depts. and evaporation rates (Omanga *et al.*, 1987).

Cowpea has a wide range of adaptation in the tropics from short duration cultivars in semi-arid regions (<600 mm annual rainfall) through the sub-humid to the humid forest belt (1000-1500 mm of annual rainfall) (Summerfield *et al.*, 1985). Summerfield *et al.* (1985) reports that the crop is said to need adequate water for only 55 days (65 per cent) of its life. The question is not therefore, only how much is adequate, but also on which days should it be available. Ziska and Hall (1983) and Ziska *et al.* (1985) conducted studies on the response of cowpea to drought during the vegetative and reproductive phases of growth. The results showed that grain yield of cowpea is strongly dependent upon the water supply during the reproductive stage, with relatively little influence of vegetative stage drought on grain yield.

Drought spells of 7-10 days or more are fairly common and can occur any time during the crop growth period in the Sudan savanna and the Sahel area (Aggarwal and Haley, 1987). The effects of a 6 days dry spell on a 60 day cowpea are much more devastating than the same on a 90 day cowpea (Leleji, 1987). This is because the 60 day variety is more likely to be a relatively small plant with corresponding small root system which may not be efficient in tapping the soil moisture. According Hall *et al.* (1997), early erect cowpea cultivars, which begin flowering about 30 days after sowing in the tropics, have proved to be useful in some dry environments and years because of their ability to escape drought (Hall and Patel, 1985). These cultivars, however, may be more sensitive to mid season drought than medium cycle spreading cultivars (Thiaw *et al.*, 1993). Gwathmey *et al.* (1992) reported some indeterminate cowpea that begin flowering early, but have delayed leaf senescence (DLS) after producing the first flush of pods, which enables them to produce a second flush of pods. The DLS was shown to enhance adaptation to midseason drought by permitting substantial recovery in pod production after the drought (Gwathmey and Hall, 1992). The combination of early flowering and DLS may increase adaptation of cowpea to environments which frequently experience either terminal or midseason droughts (Hall *et al.*, 1997). Thus, an early cultivar would be effective when the rainy season is short but without drought at midseason and the DLS would enhance adaptation when midseason droughts occur, and increase forage production and quality. Early flowering cultivars with DLS should therefore, be adapted to the broad range of conditions that can occur in the Sahelian zone of Africa, where the annual rainfall may vary from 200 mm, with a short growing season of less than 60 days, to 400 mm in wetter years, but with the possibility of midseason droughts (Dancette and Hall, 1979; Khalfaoui, 1991). The plants avoid drought by reducing in leaf areas, decreases in stomatal conductance and changes in leaflet orientation (Shackel and Hall, 1979).

Temperature

In addition to low and erratic rainfall, high atmospheric and soil temperatures and sand blast are quite common in the Sahel (Aggarwal and Haley, 1987). The damage by sand blasts is particularly high when the crop is at the seedling stage. High air temperature, particularly a combination of effective day and night temperature, respectively, above 33 °C and 26 °C have been reported to adversely affect cowpea seed yield if they occur during flowering even with well watered cowpea crop (Warrag and Hall, 1983; Turk *et al.*, 1980). The observed yield reduction was mainly due to male sterility induced by high day and night temperatures and to embryo abortion induced by high day temperatures (Warrag and Hall, 1983).

High soil temperatures do not appear to affect flower abscission if they were not combined with high night temperatures (Warrag and Hall, 1983). However, soil temperatures above the optimal level adversely affect nutrient uptake, root biosynthesis, plant growth and development (Muleba, 1987). According to Muleba (1987) hot soil temperatures result in low seedling dry weight 15 days after planting, delayed flowering or bud formation and flowering (as a result of flower bud abscission), few number of branches and flowers per plants, and increased length of the main stem and the longest branch. The overall effect of these is a reinforcement of apical dominance and increased vegetative growth by hot soil temperatures. Hall *et al.* (2002) reported that cowpea exhibits incomplete emergence when soil temperatures are below 19 °C. When night temperatures are above 17 °C cowpea can exhibit floral bud suppression and low pod set (Hall *et al.*, 2002). Night temperatures of 19 °C compared to 24 °C can delay flowering in some lines by as much as 28 days (Huxley and Summerfield 1974 cited by Wien and Summerfield, 1980). Similarly, day temperatures of 33°C compared to 27°C reduced seed yield by an average of 45 per cent in 22 cowpea cultivars tested in 1976 in controlled glasshouses (Summerfield *et al.*, 1978). The IITA lines yielded an average of 29 per cent less at 33°C than at 27°C day temperature while the traditional accessions from more northernly locations yielded an average of 3 per cent more (Wein and Summerfield, 1980). These workers observed that this trend was particularly apparent when nights were also warm (24°C) and was more marked for those lines that originated from the northernmost locations which appear to be well adapted to warmer days of the early part of the dry season. Singh *et al.* (1999) grew 102 cowpea breeding lines at IITA Kano Station from March to May when the temperatures ranged from 24 to 27°C in the night and from 38 to 42°C during the day. Most of the lines showed severe flower abortion with little or no pods and these were rated as heat susceptible. The most susceptible lines (IT97K-461-2 and IT97K-461-4) showed complete sterility with no development of pollen beyond the microspore stage. These lines are otherwise normal and very high yielding in the regular crop season (July-October) when day temperatures are below 35°C and night temperatures are below 24°C.

The heat tolerant lines, however, had normal pollen, good pod set, and normal grain yield (Singh *et al.*, 2002).

III. Other Factors

Other abiotic factors contributing to low yield are problem soils, especially low pH, high Aluminum (Al), fertility depleted or high pH, saline soils, excessive moisture, photosensitivity, intercropping and inadequate management (Rachie, 1985). Even in the semi-arid regions, intense rain storms can lead to ephemeral water logging of soils, especially during periods of vegetative growth. Research in controlled environments showed that after 16 days of water logging, plants were 60 per cent smaller than controls (Minchin and Summerfield, 1976). Recovery of the plants afterward was dominated by root and nodule growth, perhaps at the expense of initiation and growth of branches (Hong *et al.*, 1977). Water logging during the vegetative stage of short duration genotypes had the most severe effects on seed yield (reduction of about 48 per cent) (Minchin *et al.*, 1978) and both nitrogen fixing and non-nodulated plants suffered similar yield reductions. When cowpea is intercropped with tall cereal crops, lights is an important limiting factor (Terao *et al.*, 1997). From studies in different intercropping systems, these workers observed that the cereal canopy should not be too dense, so as to allow for the better growth of legume crops and thus keep total utilization of light high in the system. Terao *et al.* (1997) found that at least about 40 per cent of incident light is necessary to grow healthy cowpea plants. They also recommended that the type of cowpea adapted to intercropping is the spreading type, improved to retain substantial root system, and high translocation efficiency. Photoperiod has little or no direct effect on leaf appearance and its effects on other aspects of vegetative growth have not been reported (Craufurd *et al.*, 1997). In contrast, photoperiod can have great effects on reproductive developments, although some genotypes are insensitive (Ellis *et al.*, 1994).

References

Adegbite, A. A. (2011). Assessment of yield loss of cowpea (*Vigna unguiculata* L.) due to root knot nematode, *Meloidogyne incognita* under field condtions. *American Journal of Experimental Agriculture*, **1**(3): 79-85.

Aggarwal, V. D. and Haley, S. D. (1987). Breeding for dought and *Striga* resistance in cowpea. In pp. 307-319 J. M. Menyounga, T. Bezuneh and A. Youdeowei (Eds.) *Food grain production in semi-arid Africa. Proccedings of an International Drought Symposium held at the Kenyatta Conference Centr, Nairobi, Kenya 19th to 23rd May, 1986*. Ouagadougou, Semi-Arid Food Grain Research and Development.

Aggrawal, V. D., and Ouedraogo, J. T. (1989). Estimation of cowpea yield loss from *Striga gesnerioides* infestation. *Tropical Agriculture*, **66**; 91-92.

Anon. (1961). *Report of the Department of Agricultural Research for the year 1959/1960.* Federal Printing Division, Lagos, p. 22.

Atokple, I. D. K., Singh, B. B., and Emechebe, A. M. (1995). Genetics of resistance to *Striga* and *Alectra* in cowpea. *Journal of Heredity*, **86**: 45-49.

Babatola, J. O. and Omolade, O. (1981). Chemical control of the nematode pests of cowpea, Vigna unguiculata (L.) Walp. *Crop Prtection*, **10**: 131-134

Bagnall-Oakeley, H., Gilbberd, V., and Nyongesa, T. E. (1991). The incidence and control of *Alectra vogelii* in Embu District Kenya. In pp. 340-345 J. K. Ransom, L. J. Musselman, A. D. Worsham, and C. Parker (Eds.) *Proceedings, 5th International Symposium on Parasitic Weeds*, 24-30 June 1991, Nairobi, Kenya. The International Maize and Wheat Improvement Center (CIMMYT), Mexico, D. F. Mexico.

Berner, D. K., Kling J. G. and Singh, B. B. (1995). *Striga* research and control: a perspective from Africa. *Plant Disease,* **79**: 652-660.

Berner, D. K. and Williams, O. A. (1998). Germination stimulation of *Striga gesnerioides* seeds by hosts and nonhosts. *Plant Disease,* **82**(11): 1242-1249.

Blade, S. F., S. V. R. Shetty, T. Terao and B. B. Singh (1997). Recent developments in cowpea cropping systems research. In pp. 114-128 B. B. Singh, D. R. Mohan Raj, K. E. Dashiell and L. E. N. Jackai (Eds.) *Advances in cowpea research.* Ibadan. Co publication of International Institute of Tropical Agriculture (IITA) and Japan International Research Center for Agricultural Sciences (JIRCAS).

Bock, K. R. (1971). Notes on East African plant virus diseases. I: cowpea mosaic virus. *East African Agriculture and Forestry Journal,* **37**: 60-62.

Bock, K. R. and M. Conti (1974). *Cowpea aphid-borne mosaic virus.* Kew, UK, CMI (Commonwealth Mycological Institute). CMI/AAB Description of plant viruses 134.

Booker, R. H. (1965). Pests of cowpea and their control in northern Nigeria. *Bulletin of Entomological Research,* **55**: 663-672.

Booker, R. H. (1967). Observation on three bruchids associated with cowpea in northern Nigeria. *Journal of Stored Product Research,* **3**: 1-15.

Bridge J. (1972). Plant parasitic nematodes of irrigated crops in the northern States of Nigeria. *Samaru Miscellaneous Papers* No. 42.

Caswell, G. H. (1981). Damage to stored cowpea in the northern part of Nigeria. *Samaru Journal of Agricultural Research,* **1**: 11-19.

Caveness, F. E. (1973). Nematode attack on cowpea in Nigeria. In pp. 213-218 *Proceedings of the First International Institute of Tropical Agriculture (IITA) Grain Legume Improvement Workshop,* October 29 to November 2, 1972, IITA, Ibadan.

Craufurd, P. Q., Summerfield, R. J., Ellis, R. H. and Roberts, E. H. (1997). Photoperiod, temperature, and the growth and development of cowpea. In pp. 75 – 86 B. B. Singh, D. R. Mohan Raj, K. E. Dashiell and L. E. N. Jackai (Eds.) *Advances in cowpea research.* Ibadan. Co publication of International Institute of Tropical Agriculture (IITA) and Japan International Research Center for Agricultural Sciences (JIRCAS).

Dancette, C. and Hall, A. E. (1979). Agroclimatology applied to water management in the Sudanian and Sahelian zones of Africa. In pp. 98-118 A. E. Hall, G. H. Cannell, and H. W. Lawton (Eds.) *Agriculture in semi-arid environments. Ecological studies.* Springer-Verlag, Berlin.

Ellis, R.H., Lawn, R. J., Summerfield, R. J., Qi. A., Roberts, E. H., Chay, P. M., Brouwer, J. B., Rose, J. L. and Yeates, S. J. (1994). Towards the relative prediction of time of flowering in six annual crops. III. Cowpea (*Vigna unguiculata*). *Experimental Agriculture,* **30**: 17-29.

Emechebe, A. M., and Lagoke, S. T. O. (2002). Recent advances in research on cowpea diseases. In pp. 94-123 C. A. Fatokun, S. A. Tarawali, B. B. Singh, P. M. Kormawa and M. Tamo (Eds.) *Challenges and opportunities for enhancing suitable cowpea production.* Proceedings of the World Cowpea Conference III held at the International Institute of Tropical Agriculture (IITA), Ibadan, Nigeria, 4-8 September 2000. Ibadan. IITA.

Emechebe, A. M., Singh, B. B., Leleji, O. I., Atokple, I. D. K., and Adu, J. K. (1991). Cowpea *Striga* problems and research in Nigeria. In pp. 18-28 S. K. Kim (Ed.) *Combating Striga in Africa*. IITA, Ibadan, Nigeria.

Gwathmey, C. O. and Hall, A. E. (1992). Adaptation to midseason drought of cowpea genotypes with contrasting senescence traits. *Crop Science,* **32**: 773-778.

Gwathmey, C. O., Hall, A. E., and Madore, M. A. (1992). Adaptive attributes of cowpea genotypes with delayed monocarpic leaf senescence. *Crop Science,* **32**: 765-772.

Hall, A. E., and Patel, P. N. (1985). Breeding for resistance to drought and heat. In pp. 137-151 S. R. Singh and K. O. Rachie (Eds.) *Cowpea: research, production and utilization*. Chichester. John Wiley and Sons Ltd.

Hall, A. E., Thiaw, S., Ismail, A. M., and Ehlers, J. D. (1997). Water-use efficiency and drought adaptation of cowpea. In pp. 87-98 B. B. Singh, D. R. Mohan Raj, K. E. Dashiell and L. E. N. Jackai (Eds.) *Advances in cowpea research.* Ibadan. Co publication of International Institute of Tropical Agriculture (IITA) and Japan International Research Center for Agricultural Sciences (JIRCAS).

Hall, A. E., Ismail, A. M., Ehlers, J. D., Marfo, K. O, Cisse, N., Thiaw, S., and Close T. J., (2002). Breeding cowpea for tolerance to temperature extremes and adaptation to drought. In pp. 14-21 C. A. Fatokun, S. A. Tarawali, B. B. Singh, P. M. Kormawa and M. Tamo (Eds.) *Challenges and opportunities for enhancing suitable cowpea production.* Proceedings of the World Cowpea Conference III held at the International Institute of Tropical Agriculture (IITA), Ibadan, Nigeria, 4-8 September 2000. Ibadan. IITA.

Halleren, P. van (1971). Insect pests of cowpea *Vigna unguiculata* (L.) Walp. in the Accra plains. *Ghana Journal of Agricultural Science,* **4**: 121-123.

Hibberd, J. M., Quick, W. P., Press, M. C., and Scholes, J. D. (1996). The influence of the parasitic angiosperm *Striga gesnerioides* on the growth and photosynthesis of its host, *Vigna unguiculata*. *Journal of Experimental Botany,* **47**(297): 507-512.

Hong, T. D., Minchin, F. R., and Summerfield, R. J. (1977). Recovery of nodulated cowpea plants [*Vigna unguiculata* (L.) Walp.] from water logging during vegetative growth. *Plant and Soil,* **48**: 661-672.

Hussain, M. H. and Abdel-Aal, Y. A. I. (1982). Toxicity of some compounds against the cowpea seed beetle *Callosobruchus maculatus* (Fab.) (Coleoptera: Bruchidae). *International Journal of Pest Control,* **24**: 12, 13, 16, 17.

IITA (International Institute of Tropical Agriculture) (1977). *Annual Report*. Ibadan. IITA.

Jackai, L. E. N. and. Daoust, R. A. (1986). Insect pests of cowpea. *Annual Review of Entomology,* **31:** 95-119.

Jackai, L. E. N. and Adalla, C. B. (1997). Pest management practices in cowpea: a review. In pp. 240-258 B. B. Singh, D. R. Mohan Raj, K. E. Dashiell and L. E. N. Jackai (Eds.) *Advances in cowpea research.* Ibadan. Co publication of of International Institute of Tropical Agriculture (IITA) and Japan International Research Center for Agricultural Sciences (JIRCAS).

Khalfaoui, J. L. B. (1991). Determination of potential lengths of the crop growing in semiarid regions of Senegal. *Agricultural Forest Meteorology,* **55**: 251-263.

Kuijt, J. (1969). *The biology of parasitic flowering plants*. University of California Press. Berrkeley, CA, USA., pp.346.

Kust, C. A. (1963). Dormancy and viability of witchweed seeds as affected by temperature and relative humidity during storage. *Weeds,* **11**: 247.

Lagoke, S. T. O., Shebayan, J. Y., Weber, G., Olufajo, O., Elemo K., Adu, J. K., Emechebe, A. M., Singh, B. B., Zaria, A., Awad, A., Ngawa, L., Olaniyan, G. O., Olafare, S. O., and Adeoti, A. A. (1994). Survey of *Striga* problems and evaluation of *Striga* control methods and packages in crops in the Nigerian savanna. In pp. 91-120 S. T. O. Lagoke, R. Hoevers, S. S. M'boob and R. Traboulsi (Eds.) *Improving Striga Management in Africa. Proceedings, 2nd General Workshop on Pan-African Striga Control Network* (*PASCON*). 23-29 June 1991. Nairobi, Kenya. FAO/PASCON. Accra, Ghana.

Le Pelley, R. H. (1959). Index of insect hosts parasites. In pp. 1-296 T. J. Aderson (Ed.) *Agricultural Insects of East Africa*. Nairobi, Kenya, East Africa High Commission.

Leleji, O. (1987). Cowpea improvement and production in the semi-arid regions of Nigeria. In pp. 321-330 J. M. Menyounga, T. Bezuneh and A. Youdeowei (Eds.) *Food grain production in semi-arid Africa. Proccedings of an International Drought Symposium held at the Kenyatta Conference Centr, Nairobi, Kenya 19th to 23rd May, 1986*. Ouagadougou, Semi-Arid Food Grain Research and Development.

Lane, J. A., Moore, T. H. M., Child, D. V., and Bailey J. A. (1997). Variation in virulence of *Striga gesnerioides* on cowpea: new sources of crop resistance. In pp. 225-230 B. B. Singh, D. R. Mohan Raj, K. E. Dashiell and L. E. N. Jackai (Eds.) *Advances in cowpea research.* Ibadan. Co publication of International Institute of Tropical Agriculture (IITA) and Japan International Research Center for Agricultural Sciences (JIRCAS).

Minchin, F. R., and Summerfield, R. J. (1976). Symbiotic nitrogen fixation and vegetative growth of cowpea [*Vigna unguiculata* (L.) Walp.] in water logged conditions. *Plant and Soil,* **25**: 113-127.

Minchin, F. R., Summerfield, R. J., Eaglesham, A. R. J., and Stewart, K. A. (1978). Effects of short term water logging on growth and yield of cowpea (*Vigna unguiculata*). *Journal of Agricultural Science Cambridge,* **90**: 335-336.

Mishra, S. D. (1992). Nematode pests of pulse crops. In: pp. 97-106 D. S. Bhati and R. K. Walia (Eds.). *Nematode pests of vegetable crops.* CBS Publishers and Distributors, Delhi, Indian.

Mugabe, N. R. (1983). Effect of *Alectra vogelii* Benth. On cowpea (*Vigna unguiculata* [L.] Walp.) I. Some aspects of reproduction of cowpea. *Zimbabwe Journal of Agricultural Research,* **21**: 135-147.

Muleba, N. (1987). Response of cowpeas to high soil temperatures and drought. In pp. 331-349 J. M. Menyounga, T. Bezuneh and A. Youdeowei (Eds.) *Food grain production in semi-arid Africa. Proccedings of an International Drought Symposium held at the Kenyatta Conference Centr, Nairobi, Kenya 19th to 23rd May, 1986.* Ouagadougou, Semi-Arid Food Grain Research and Development.

Muleba, N., J. T. Ouedraogo, and J. B. Tingnere (1997). Crop yield loss attributed to *Striga* infestations. *Journal of Agricultural Science,* **129**(1): 43-48.

Musselman, L., Bharthalakshmi, J., Safa, S. B., Knepper, D. A., Mohamed, K. I., and White, C. L. (1991). Recent research on the biology of *Striga asiatica, S. gesneriodes* and *S. hermonthica*. In pp. 31-41 S. K. Kim (Ed.) *Combating Striga in Africa*. IITA, Ibadan, Nigeria.

Ng, N. Q. and Maréchal, R. (1985). Cowpea taxonomy, origin and germplasm. In S. R. Singh and K. O. Rachie (Eds.) *Cowpea: research, production and utilization*. Chichester. John Wiley and Sons Ltd., pp. 11-21.

S. R. Singh and K. O. Rachie (Eds.) *Cowpea: research, production and utilization*. Chichester. John Wiley and Sons Ltd.

Ogunfowora, A. O. (1976). Research on Meloidogyne at the Institute of Agricultural research and Training, University of Ife, Moor Plantation, Ibadanm In: pp. 9-14 *Proceedings of the First IMP Resaerch Planiing Conference on Root knot Nematodes, Meloidogyne spp., Region IV*, IITA, Ibadan, Nigerian, June 7-11, 1976.

Okigbo, B. N. and D. J. Greenland (1976). Intercropping systems in tropical Africa. In pp. 63-10 R. I. Papendick. P. A. Sanchez and G.B. Triplett (Eds.). *Multiple Cropping*. American Society of Agronomy, Special Publication No. 27, Madison, Wisconsin, USA.

Olowe, T. (1976). Research work on root knot nematodes at the National Cereals Research Institute, Ibadan. In: pp. 15-19 *Proceedings of the First IMP Resaerch Planiing Conference on Root knot Nematodes, Meloidogyne spp., Region IV,* IITA, Ibadan, Nigerian, June 7-11, 1976.

Olowe, T. (1978). International Meloidogyne Project progress report, 1978. In: pp. 58-69 *Proceedings of the Second IMP Resaerch Planiing Conference on Root knot Nematodes, Meloidogyne spp.,* February 20-24, 1978, Abijan, Ivory Coast.

Omanga, P. G. A., Shakoor, A. and Ngugi, E. C. K. (1987). Grain legume production strategies in the semi-arid and arid areas of Kenya. In pp. 351-360 J. M. Menyounga, T. Bezuneh and A. Youdeowei (Eds.) *Food grain production in semi-arid Africa. Proccedings of an International Drought Symposium held at the Kenyatta Conference Centr, Nairobi, Kenya 19th to 23rd May, 1986*. Ouagadougou, Semi-Arid Food Grain Research and Development.

Onwueme, I. C. and T. D. Sinha (1991). *Field crop production in tropical Africa.* The Netherlands. Technical Centre for Agricultural and Rural Cooperation

(CTA)., p. 400.

Parker, C. (1991). Protection of crops against parasitic weeds. *Crop Protection*, **10**: 6-22.

Parker, C. and Riches, C. R. (1993). *Parasitic weeds of the world: biology and control*. CAB International, Wallingford, Oxon, UK.

Rachie, K. O. (1985). Introduction. In pp. xxi-xxviii S. R. Singh and K. O. Rachie (Eds.) *Cowpea: research, production and utilization*. Chichester. John Wiley and Sons Ltd.

Raheja, A. K. (1973). A report on the insect pest complex of grain legumes in Northern Nigeria. pp. 295-299. *Proceedings of the 1st IITA Grain Legumes Improvement Workshop*. Ibadan, Nigeria.

Raman, K. V., S. R. Singh and H. F. van Emden (1978). yield losses in cowpeas following leafhopper damage. *Journal of Economic Entomology*, **71**: 836-838.

Riches, C. R. (1989). *The biology and control of Alectra vogelii* Benth. (Scrophulariaceae) in Botswana. Ph. D. thesis, University of Reading, UK., p. 208.

Robertson, D. G. (1963). Cowpea virus research in Nigeria. In pp. 75-76 *IAR, Proceedings of the first Nigerian Grain Legume Conference*, Samaru, Zaria, Nigeria. IAR, Ahmadu Bello University.

Saunders, A. R. (1933). Studies on phanerogramic parasitism with particular reference to Striga lutea (Lour). *South African Department of Agricultural Science Bulletin*, **128**: 56.

Sikora, R. A. and Greco, N. (1990). Nematode parasites of food legumes. In pp. 197-198 M. Luc, R. A. Sikora and J. Bridge (Eds.) *Plant parasitic nematodes in Subtropical and Tropical Africa*. 1st Edition, Willingford, UK, CABI Publishing.

Sikora, R. A., Greco, N., and J. F. V. Silva (2005). Nematode parasites of food legumes. In pp. 259-318 M. Luc, R. A. Sikora and J. Bridge (Eds.) *Plant parasitic nematodes in Subtropical and Tropical Africa*. 2nd Edition, Willingford, UK, CABI Publishing.

Singh, B. B. (1987). Breeding cowpea varieties for drought escape. In pp. 291-306 J. M. Menyounga, T. Bezuneh and A. Youdeowei (Eds.) *Food grain production in semi-arid Africa. Proccedings of an International Drought Symposium held at the Kenyatta Conference Centr, Nairobi, Kenya 19th to 23rd May, 1986*. Ouagadougou, Semi-Arid Food Grain Research and Development.

Singh, B. B. (2000). Breeding cowpea varieties for wide adaptation by minimizing genotype x environment interactions. In pp. 173-181 I. J. Ekanayake and R. Ortiz (Eds.) *Gentype x environment interactions analysis of IITA mandate crops in sub-Saharan Africa*. International Institute of Tropical Agriculture, Ibadan, Nigeria.

Singh, B. B. (2002). Breeding cowpea varieties for resistance to *Striga gesneroides* and *Alectra vogelii*. In pp. 154-163 C. A. Fatokun, S. A. Tarawali, B. B. Singh, P. M. Kormawa and M. Tamo (Eds.) *Challenges and opportunities for enhancing suitable cowpea production*. Proceedings of the World Cowpea Conference III held at the International Institute of Tropical Agriculture (IITA), Ibadan, Nigeria, 4-8

September 2000. Ibadan. IITA.

Singh, B. B., and Emechebe, A. M. (1991). Breeding for resistance to *Striga* and *Alectra* in cowpea. In pp. 303-305 J. K. Ransom, L. J. Musselman, A. D. Worsham, and C. Parker (Eds.) *Proceedings, 5th International Symposium on Parasitic Weeds,* 24-30 June 1991, Nairobi, Kenya. The International Maize and Wheat Improvement Center (CIMMYT), Mexico, D. F. Mexico.

Singh, B. B. and Emechebe, A. M. (1997). Advances in research on cowpea *Striga* and *Alectra*. In pp. 215-224 B. B. Singh, D. R. Mohan Raj, K. E. Dashiell and L. E. N. Jackai (Eds.) *Advances in cowpea research.* Ibadan. Co publication of International Institute of Tropical Agriculture (IITA) and Japan International Research Center for Agricultural Sciences (JIRCAS).

Singh, B. B., S. K. Asante, D. Florini, L. E. N. Jackai, C. Fatokun, and Wydra, K. (1997). Breeding for multiple disease and insect resistance. *IITA Annual Report 1997.* Project, 11., p. 24.

Singh, B. B., Mai-Kodomi, Y., and Terao, T. (1999). relative drought tolerance of major rainfed crops of the semi-arid tropics. *Indian Journal of Genetics,* **59**: 1-8.

Singh, B. B., Ehlers, J. D., Sharma, B., and Freire Filho, F. R. (2002). Recent progress in cowpea breeding. In pp. 22-40 C. A. Fatokun, S. A. Tarawali, B. B. Singh, P. M. Kormawa and M. Tamo (Eds.) *Challenges and opportunities for enhancing suitable cowpea production.* Proceedings of the World Cowpea Conference III held at the International Institute of Tropical Agriculture (IITA), Ibadan, Nigeria, 4-8 September 2000. Ibadan. IITA.

Singh, S. R. (1977). Cowpea cultivars resistant to insect pests in world germplasm collection. *Tropical Grain Legume Bulletin,* **9**: 3-7.

Singh, S. R. (1978). Resistance to pests of cowpea in Nigeria. In pp. 267-279 S. R. Singh, H. F. van Emden and T. A. Taylor (Eds.) *Pests of grain legumes: ecology and control.* London, Academic Press.

Singh, S. R. (1980). Biology of cowpea pests and potential for host plant resistance. In pp. 389-421 M. K. Harris (Ed.) *Biology and breeding for resistance to arthropods and pathogens in agricultural plants.* College Station, Texas A and M University Bulletin MP1451.

Singh, S. R and Taylor, T. A. (1978). Pests of grain legumes and their control in Nigeria. In pp. 99-111 S. R. Singh, H. F. van Emden and T. A. Taylor (Eds.) *Pests of grain legumes: ecology and control.* London, Academic Press.

Singh, S. R and van H. F. Emden (1979). Insect pests of grain legumes. *Annual Review of Entomology,* **24**: 255-278.

Singh, S. R. and Jackai, l. E. N. (1985). Insect pests of cowpeas in Africa: Their life cycle, economic importance and potential control. In pp. 217-231 S. R. Singh and K. O. Rachie (Eds.) *Cowpea: research, production and utilization*. Chichester. John Wiley and Sons Ltd.

Singh, S. R., Luse, R. A., Leuschner, K., and Nangju, D. (1979). Groundnut oil for control of cowpea weevil. *Journal of Stored Products Research,* **14**: 77-80.

Singh, S. R., Singh, B. B., Jackai, L. E. N., Ntare, B. R. (1983). *Cowpea Research at IITA. Information Service No. 14.,* p.20.

Singh, S. R., Jackai, L. E. N., Dos Santos, J. H. R., and Adalla C. B. (1990). Insect pests of cowpea. In pp. 43-89 S. R. Singh (Ed.) *Insect pests of tropical food legumes.* Chichester, John Wiley and Sons Ltd.

Steele, W. M. (1972). *Cowpeas in Africa.* Reading, UK. University of Reading. (Ph. D. thesis).

Steele, W. M., Allen, D. J., and Summerfield R. J. (1985). Cowpea (*Vigna unguiculata* (L.) Walp.). In: pp. 520-583 R. J. Summerfield and E. H. Roberts (Eds.) *Grain legume crops.* Collins. London.

Summerfield, R. J., Minchin, F. R. and Roberts, E. H. (1978). *Report 17, Reading University – IITA Tropical Grain Legume Physiology Project.*

Summerfield, R. J., J. S. Pate, E. H. Roberts and H. C. Wien (1985). The physiology of cowpea. In pp. 65-101 S. R. Singh and K. O. Rachie (Eds.) *Cowpea: research, production and utilization.* Chichester. John Wiley and Sons Ltd.

Taylor, T. A. (1967). The bionomics of *Maruca testulalis* Gey., a major pest of cowpeas in Nigeria. *Journal of the West African Science Association,* **12**: 112-129.

Terao, T., I. Watanabe, R. Matsunaga, S. Hakoyama and B. B. Singh (1997). Agro-physiological constraints in intercropped cowpea, an analysis. In pp. 129-140 B. B. Singh, D. R. Mohan Raj, K. E. Dashiell and L. E. N. Jackai (Eds.) *Advances in cowpea research.* Ibadan. Co publication of International Institute of Tropical Agriculture (IITA) and Japan International Research Center for Agricultural Sciences (JIRCAS).

Thiaw, S., Hall, A. E., and Parker, D. R. (1993). Varietal intercropping and the yield and stability of cowpea production in semiarid Senegal. *Field Crops Reaserch.* **33**: 217-233.

Turk, K. J., Hall, A. E., and Asbell C. W. (1980). Drought adaptation of cowpea I. Influence of drought on seed yield. *Agronomy Journal,* **72**: 413-420.

Warrag, M. O. A., and Hall, A. E. (1983). Reproductive response of cowpea to heat stress: Genotypic differences in tolerance to heat at flowering. *Crop Science,* **25**: 1084-1090.

Wien, H. C. and Summerfield, R. J. (1980). Adaptation of cowpeas in West Africa: Effects of photoperiod and temperature responses in cultivars of diverse origin. In pp. 405-417 R. J. Suand and A. H. Bunting (Eds.) *Advances in legume science Vol. 1 of the Proceedings of the International Legume Conference,* Kew, 31 July – 4 August 1978, held under the auspices of the Royal Botanic Gardens, Kew, the Missouri Botanic Garden, and the University of Reading.

Ziska, L. H. and Hall, A. E. (1983). Seed yields and water use of cowpeas *Vigna unguiculata* (L.) Walp. Subjected to planned-water-deficit irrigation. *Irrigation Science*, **3**: 237-245.

Ziska, L. H., Hall, A. E., and Hoover, R. M. (1985). Irrigation management methods for reducing water use of cowpea (*Vigna unguiculata* [L.] Walp.) and lima beans (*Phaseolus lunatus* L.) while maintaining seed yield at maximum levels. *Irrigation Science*, **6**: 223-239.

Zuofa, K., N. M. Tariah and Isirimah, N. O. (1992). Effects of groundnut, cowpea and melon on weed control and the yields of intercropped cassava and maize. *Field Crops Research*, **28**: 309-314.

2015, Major Constrains and Verdict of Crop Productivity Pages 21–50
Editor: Dr. U.N. Bhale
Published by: ASTRAL INTERNATIONAL (P) LIMITED, NEW DELHI

Chapter 2

Fungicide Resistance in Fungal Pathogens of Various Crops and its Integrated Management

N.S. Suryawanshi

Research Laboratory, Department of Botany,
D.S.P. Mandal's K.V. Pendharkar College of Arts, Science and Commerce,
Dombivli (E) Dist. Thane – 421 203, M.S., India

ABSTRACT

There are many fungal pathogens viz., Alternaria alternata, Colletotricum musae, Penicillium expansum, Botrytis cinerea, Aspergillus flavus and Fusarium oxysporum causing diseases to strawberry, Banana, Apple, Grape, Papaya and Onion respectively. India is the 2nd largest vegetables and 3rd largest fruits producer in the world. They are integrated management by using different fungicides, plant extracts; essential oils, homoeopathic medicines and various agrochemicals use individually or in mixture with chemical or medicinal plant extract or any form that is called integrated management of fruit as well as vegetable pathogens; however, there are reports that are fungicide resistance in different pathogens. The present investigation of fungicide resistance cases of plant pathogens in various countries is reviewed since 1980. Maximum cases of fungicide resistance have been reported from USA and followed by France, Japan, Netherland and India. Recently the alternative for fungicide farmar used biological origin compound. Different fungicides, Thiophanate methyl, and Carbendazim use for management of different crop pathogen in brief reviwed.

Keywords: *Plant pathogen, Fungi-fungicide resistance, Fruits and vegetables, Integrated management.*

* E-mail: suryawanshins002@gmail.com

Introduction

Plant disease continuous to threaten crop protection in Modern agriculture and plays a direct role in the destruction of pathogen. Particular fruits and vegetable pathogen especially fungi cause important losses are most aggressive. Some of the important fruit and vegetables are infested by various types of pathogen *viz. Alternaria alternata, Colletotricum musae, Penicillium expansum, Botrytis cinerea; Aspergillus flavus* and *Fusarium oxysporum* have detrimental effects on crop of economics importance. When the use of resistant varieties, cultural practices or alternation of inadequate to suppress these pathogens sufficiently. The use of chemical becomes essential. However the use of chemical also encounters certain problems like the development of resistance in pathogen. This is because of the organisms have the ability to adapt to changing and often adverse condition well known is the development of resistance to fungicide in fungi, antibiotics to bacteria and insecticide in insect which have the control of fungal diseases of man diseases and pest in agricultural crop plants (Dekker, 1982). Resistance in fungi to fungicides has been well known in agricultural field since 1970. But fungicide resistance in plant pathogenic fungi is also recorded by some researchers.

Use of Fungicides

The use of synthetic fungicides, such as Carbendazim, Benomyl Thiophanate methyl, mancozeb and bendimidazole in fruit and vegetable packaging are primary tools used for management of postharvest decay caused by *Alternaria alternata, Colletotrichum musae, Penicillium expansum, Botrytis cinerea, Aspergillus flavus* and *Fusarium oxysporum etc.* However, due to Problem released to fungicides application, such as fungicide toxicity, development of the resistance by pathogen, growing concern for human safety and protection of the environment. These problem call for research into new potential strategies for controlling postharvest decay of fruit and vegetable (Suryawanshi, 2005) one way to achieve this goal is the use of effective natural compound that have minimal adverse effect on environment and health. The minimum use of fungicide is not adverse effect on fruit and vegetable crops. But last seven decade farmar used excessive amount of any fungicide is an effect on crops and environment and to increased the resistance problem details reviewed given since 1970.

Benomyl was introduced in 1971 and maximum cases of resistance of fungal pathogens. It is followed Carbendazim, Thiophanate methyl, Copper oxychloride, Mancozeb etc. All these are the systemic and non systemic fungicides. Benomyl resistance was reported in *Botrytis cinerea* (Bollen and Scholten, 1971) *Erysiphe graminis* (Vargas, 1973), *Cercospora beticola* (Dovas *et al.*, 1976) and in *Fusarium moniliforme* (Wada *et al.*, 1970) Carbendazim resistance reported in *Verticillium fusicola* (Bollen and Zayen, 1975) and *Septoria tritici* (Fisher and Griffin 1984). In India attention towards fungicide problem was focused by Gangawane 1982, 1990, later studied on this aspect have been carried out by different workers (Table 2.1). There are nearly 18 pathogens showing resistance towards 24 fungicides from different part of the country Gangawane (1997). Field survey has been done for pathogens like *Aspergillus flavus, Venturia inaequalis, Gloeosporium amphelophagum* (Chaudhary and Pulto, 1984,

Table 2.1: Fungicide Resistance in Fungal Pathogens Reported by different Crops and Countries

Sl.No.	*Name of Pathogen*	*Fungicide*	*Author/Country*
1.	*Ceratocystis ulmi*	Carbendazim	Gibbs and Brasier UK 1980
2.	*Macrophomina phaseolina*	Carbendazim	Pan and San (1980) India
3.	*Cercoporidium personotum*	Carbendazim	Katan (1980) Isral
4.	*Pseudocercosperella horpotrichoides*	Carbendazim	Horsten and Fehrmann (1980) Netherland
5.	*Septoria nodorum*	Carbendazim	Horsten and Fehrmann (1980) Netherland
6	*Aspergillus flavus*	Carbendazim	Gangawane saler (1981) India
7.	*Fusarium oxysporum*	Carbendazim	Delen and Yildiz (1981) Turkey
8.	*Sclerotinia sclertiorum*	Carbendazim	Delen and Yildiz (1981) Turkey
9.	*Botrytis cinerea*	Carbendazim	Delen and Yildiz (1981) Turkey
10.	*Cladosporium* spp.	Carbendazim	Delen and Yildiz (1981) Turkey
11.	*Rhizoctonia solani*	Carbendazim	Delen and Yildiz (1981) Turkey
12.	*Cocomyces hiamalis*	Carbendazim	Jones and Ehert (1981) USA
13.	*Aspergillus nidulans*	Carbendazim	Georgopoulos (1981) Greece
14.	*Botrytis cinerea*	Carbendazim	Tripathi and Schlolesser (1982) Kenya
15.	*Venturia inaequalis*	Carbendazim	Chaudhary and Putto (1994) India
16.	*Septoria tritici*	Carbendazim	Fisher and Griffin (1984) U.K
17.	*Aspergillus flavus*	Carbendazim	Gangawane and Reddy (1985) India
18.	*Aspergillus flavus*	Carbendazim	Gangawane and Reddy (1986) India
19.	*Aspergillus flavus*	Carbendazim	Gangawane and Reddy (1987) India
20.	*Puccinia arachidis*	Carbendazim	Gangawane an (1988) India
21.	*Macrophomina phaseolina*	Carbendazim	Kamble (1991) India
22.	*Gloeosporium ampelophagum*	Carbendazim	Chander and Thind (1995) India
23.	*Alternaria tenuissima*	Carbendazim	Bhale (2002) India
24.	*Sclerotium rolfsii*	Carbendazim	Hiwale (2004) India
25.	*Sclerotium rolfsii*	Carbendazim	Solunke *et al.* (2001) India
26.	*Gloeosporium ampelophagum*	Carbendazim	Deokate *et al.*, 2002
27.	*Alternaria alternata*	Carbendazim	Nasreen, 1997
28.	*Alternaria tenuis*	Carbendazim	Kadam, 1997
29.	*Venturia inaequalis*	Carbendazim	Singh and Kumar, 1999
30.	*Cercospora arachidicola*	Carbendazim	Jadeja *et al.*, 1999
31.	*Botrytis cinerea*	Carbendazim	Giuseppe Lima *et al.*, 2003
32.	*Penicillium expansum*	Carbendazim	Giuseppe Lima *et al.*, 2003
33.	*Gloeosporium amphelopagum*	Carbendazim	Khillare *et al.*, 2004
34.	*Scleotinia sclerotium*	Carbendazim	Quis *et al.*, 2005

Contd...

Table 2.1–*Contd...*

Sl.No.	*Name of Pathogen*	*Fungicide*	*Author/Country*
35.	*Colletotrichum capsci*	Carbendazim	Quis *et al.*, 2005
36.	*Aspergillus* spp.	Carbendazim	Raju and Naik, 2006
37.	*Penicillium digitatum*	Carbendazim	Raju and Naik, 2006
38.	*Trichoderma koeningii*	Carbendazim	Kapoor *et al.*, 2006
39.	*Fusarium oxysporum*	Carbendazim	Kapoor *et al.*, 2006
40.	*Trichoderma harzianum*	Carbendazim	Pandey *et al.*, 2006
41.	*Colletotrichum capscii*	Carbendazim	Srinivas *et al.*, 2006
42.	*C. gloeosporioides*	Carbendazim	Patel, 2009
43.	*Alternaria alternata*	Carbendazim	Sharma and Gaur, 2009
44.	*Rhizopus stolonifer*	Carbendazim	Sharma and Gaur, 2009
45.	*Aspergillus flavus*	Carbendazim	Sharma and Gaur, 2009
46.	*Aspergillus sydowii*	Carbendazim	Sharma and Gaur, 2009
47.	*Aspergillus niger*	Carbendazim	Sharma and Gaur, 2009
48.	*Macrophomina phaseiolinia*	Carbendazim	Patil and Kamble, 2011
49.	*Alternaria alternata*	Carbendazim	Dahiwale and Suryawanshi, 2011
50.	*Penicillium expansum*	Carbendazim	Baviskar and Suryawanshi, 2012
51.	*Penicillium expansum*	Carbendazim	Baviskar and Suryawanshi, 2013
52.	*Penicillium expansum*	Carbendazim	Baviskar and Suryawanshi, 2014

Gangawane and Reddy, 1987, Annamalai and Lalithakumari, 1987, Chander and Thind, 1995, There is need of field survey of fungicides resistance in different pathogens in India. Fungicide is reported in *Aspergillus flavus* to Carbendazim, Benomyl, Thiophanate methyl and captain. Gangawane and Salver, 1981, Gangawane and Reddy, 1985) in *Macrophomina phaseolina* to Carbendazim Kamble, 1991, in *Rhizoctonia bataticola* to Maneb, Kocide and other fungicide (Anilkumar and Shashtry, 1979), Gangawane and Saler, 1988) in *Fusarium udam* to benomyl (Kamble and Gangawane, 1994) Fungicides have been to be very useful in controlling of fungal pathogen. However during last 30 years application of fungicides has plagued with several problem among which development of fungicide resistant in pathogenic organism (Dekker, 1982) Gangawane and Saler, 1981, Gangawane, 1990). Detail of fungicides resistance in plant pathogenic fungi are also illustrated by some worker reported in (Tables 2.1–2.4) this shows that maximum cases of resistance in plant pathogenic fungi are recorded against fungicides. This is because Carbendazim and other fungicides application for the management of fungal diseases have been revived. Recently studies on the use of plant extracts have opened a new path for the control of the plant pathogen. The plant extracts have been reported to be safe, non phototoxic to man but effective against pathogens.

Table 2.2: Use of Thiophanate Methyl against Various Diseases and Pathogens

Sl.No.	Disease	Pathogen	Reference
1.	Brown rot of Cherry	*Monilina* spp.	Ishii, 1970
2.	Apple scab	*Venturia inaequalis*	Cole *et al.*, 1971
3.	Grey mould of strawberries	*Botrytis cinerea*	Gilchrist and Cole, 1971
4.	Anthracnose of Banana	*Gloeosporium musarum*	Long, 1971
5.	Bitter rot	*Gloeosporium* spp.	Foschi *et al.*, 1972
6.	Aporicot blossom	*Monilina laxa*	Ogawa *et al.*, 1972
7.	Citrus decay	*Penicillium* spp.	Yamada *et al.*, 1972
8.	Cherry blossom blight	*M. fructicola*	Byrde and Willets, 1977
9.	Citrus decay	*Penicillium* spp.	Patel, 1988
10.	Grey mould of strawberries	*Botrytis cinerea*	Elad *et al.*, 1988
11.	Apple scab	*Venturia inaequalis*	Thakur *et al.*, 1995
12.	Grey mould of strawberries	*Botrytis cinerea*	Butters and Hollomon, 1999
13.	Grey mould of strawberries	*Botrytis cinerea*	Yourman *et al.*, 2000
14.	Grey mould of strawberries	*Botrytis cinerea*	Seiya and Shunji, 2009
15.	Grey mould of strawberries	*Botrytis cinerea*	Yang *et al.*, 2011
16.	Grey mould of grape	*Botrytis cinerea*	Dahiwale and Suryawanshi (2013)

Table 2.3: Use of different Fungicides for Management of different Pathogens

Sl.No.	Fungicide	References
1.	Organo mercurial	Noble *et al.*, 1966
2.	Carbendazim	Dekker,1977
3.	Triphenyltins	Giannopolities, 1978
4.	Carbendazim	Geeson,1978; Delan and Yildiz, 1981; Tripathi and Schloessor, 1982
5.	Aromatic hydrocarbon	Eckert, 1982
6.	Dodine	Gilpatrick, 1982
7.	Iprodione (Rovral 25 per cent)	Dennis, 1983a
8.	Phenylamides	Staub, 1994
9.	Carbendazim	Leroux and Clerjeau, 1985; Khoury and Saad, 1986
10.	Carboxanilides	Locke, 1986
11.	Procymidone	O' Brien *et al.*, 1986
12.	Diethofencarb (N-phenylcarbamate)	O' Brien *et al.*, 1986
13.	Dicarboximide (Vinclozolin)	Hunter *et al.*, 1987
14.	Iprodione (Rovral 25 per cent)	Aharoni and Barkai-Golan, 1987
15.	Dichlorofluanid Procymidone	Hunter *et al.*, 1987
16.	Benzimidazole (Thiophanate methyl) Carbendazim	Elad *et al.*, 1988

Contd...

Table 2.3–*Contd...*

Sl.No.	*Fungicide*	*References*
17.	Phosphorothiolates	Kato, 1988
18.	Dicarboximide	Lorenz, 1988; De Kock and Holz, 1991; Washington *et al.*, 1992
19.	Copper hydroxide Thiram (Pomarsol 75 per cent wp) Vinclozolin (Ronilan 50 per cent)	Washington *et al.*, 1992
20.	Dicarboximide (Vinclozolin)	Beresford, 1994; De Kock and Holz, 1994
21.	DMIs	DeWarrd, 1994
22.	Procymidone Diethofencarb (N-phenylcarbamate)	Beresford, 1994
23.	Folpet 50 per cent I prodione (Rovral 25 per cent) Mancozeb (Dithane M-45 80 per cent) Procymidone (Sumislex 25 per cent)	Milkota *et al.*, 1996; Harman *et al.*, 1996
24.	Anilinopyrimidine cyprodinil	Hilber and Schuepp, 1996 a,b
25.	Anilinopyrimidine cyprodinil	Suty *et al.*, 1997
26.	Pyrimethanil Hydroxyanilide fenhexamid Mepanipyrim Phenylpyrrol fludioxanil	Kuck *et al.*, 1997
27.	Anilinopyrimidine cyprodinil Hydroxyanilide fenhexamid Mepanipyrim Phenylpyrrol fludioxanil Pyrimethanil	Rueegg *et al.*, 1997; Suty *et al.*, 1997; Leroux *et al.*, 1998; Hilber and Hilber, 1998; Rosslenbroich *et al.*, 1998
28.	Benzimidazole (Thiophanate methyl)	Butter and Hollomon, 1999
29.	Hydroxyanilide fenhexamid Anilinopyrimidine cyprodinil	Rosslenbroich and Stuebler, 2000
30.	Procymidone	Kalamarakis *et al.*, 2000
31.	Dicarboximide (Vinclozolin) Benzimidazole (Thiophanate methyl)	Yourman *et al.*, 2000
32.	Cyprodinil	Kalamarakis *et al.*, 2000
33.	Iprodione (Rovral 25 per cent)	Pande *et al.*, 2001
34.	Thiram (Pomarsol 75 per cent wp) Captan, Thiobendazol	Blacharski *et al.*, 2001
35.	Dithane M-45 Triadimefon Vinclozolin (Ronilan 50 per cent) Diethofencarb(N-phenylcarbamate) Procymidone	Pande *et al.*, 2001
36.	Melanin Biosynthesis Inhibitors (Dehydratase) (MBI-D) SO_2	Leroux *et al.*, 2002
37.	Boscalid	Kaku *et al.*, 2003

Contd...

Table 2.3–*Contd...*

Sl.No.	*Fungicide*	*References*
38.	Benzimidazole (Thiophanate methyl)	Droby and Lichter, 2004
39.	Chlorothalonil (Bravo 50 per cent)	Serey *et al.*, 2007
40.	Eugenol, Benomyl (Benlate 50 per cent) Benzimidazole(Thiophanate methyl)	Seiya and Shunji, 2009
41.	Chlorothalonil (Bravo 50 per cent)	Wang *et al.*, 2010
42.	Diethofencarb(N-phenylcarbamate) Thiram (Pomarsol 75 per cent wp) Sterol derivatives	Yang *et al.*, 2011
43.	Benomyl	Telmore *et al.*, 2013

Table 2.4: Use of Carbendazim for the Management of Pathogens

Sl.No.	*Pathogen*	*References*
1.	*Ceratocystis ulmi*	Gibbs and Brasier (1980)
2.	*Septoria leucanthem*i	Paulus *et al.* (1976)
3.	*Colletotricum coffeanum*	Ogawa *et al.* (1976)
4.	*Venturia pirina*	Shabi and Ben (1976)
5.	*Aspergillus nidulans*	Devidse (1981), Georgopoulos (1981)
6.	*Botrytis cinerea*	Geeson (1978), Delen and Yildiz (1981)
7.	*Cercosporidium personotum*	Katan (1980)
8.	*Septoria nodorun, Pseudocercosporella herpotrichoides*	Horsten and Fehrmann (1980)
9.	*Fusarium oxysporum*	Delen and Yildiz (1981)
10.	*Sclerotinia sclerotirum*	Delen and Yildiz (1981)
11.	*Cladosporium* spp.	Delen and Yildiz (1981)
12.	*Coccomyces hiemalis*	Jones and Ehret (1981)
13.	*Aspergillus flavus*	Gangawane and Saler (1981)
14.	*Septoria tritici*	Fisher and Griffin (1984), Metcalfe (1985).
15.	*Aspergillus flavus*	Gangawane and Reddy (1986)
16.	*Aspergillus flavus*	Gangawane and Reddy (1987)
17.	*Alternaria alternata*	Madhukar and Reddy (1988)
18.	*Sclerotium* sp.	Solunke *et al.* (2001)
19.	*Gloeosporium amphelopagum*	Deokate *et al.* (2002)
20.	*Alternaria alternata*	Nasreen,(1992)
21.	*Alternaria alternata*	Kadam (1997)
22.	*Alternaria alternata*	Singh and Majumdar (2001)
23.	*Venturia inaequalis*	Singh (1980)

Contd...

Table 2.4–*Contd...*

Sl.No.	*Pathogen*	*References*
24.	*Cercospora arachidicola, Puccinia arachidis*	Jadeja *et al.* (1999)
25.	*Botrytis cinerea, Penicillium expansum*	Giuseppe Lima *et al.* (2003)
26.	*Gloeosporium amphelopagum*	Khilare *et al.* (2004)
27.	*Sclerotinia sclerotiorum*	Qais *et al.* (2005)
28.	*Colletotrichum capscii*	Srinivas *et al.* (2006)
29.	*Aspergillus niger,P. digitatum*	Raju and Naik (2006)
30.	*Trichoderma koeningii, Fusarium oxysporum*	Kapoor *et al.* (2006)
31.	*Trichoderma harzianum*	Pandey *et al.* (2006)
32.	*Taphrina deformans, Monilinia laxa, Cladosporium* spp., *Pythium, Fusarium, Rhizoctonia* spp.	Thind (2007)
33.	*Exserohilum turcicum*	Bunkar and Mathur (2008)
34.	*Colletotrichum gloeosporioides*	Patel (2009)
35.	*Alternaria alternata, Aspergillus tennuissima, A. flavus, A. fumigates, A. niger, A. sydowii, Rhizopus stolonifer*	Sharma and Gour (2009)

Use of Plant Extracts

Biological control of plant pathogens is preferred over the hazardous chemical based products. Botanicals like leaf, root, stem, rhizomes, bulbs and other plant parts are used as extracts to control pathogens by treatment. There is an urgent need to find out effective, alternative methods of diseases control, which are less harmful to human beings and environment. In view of these the present investigation was undertaken to screen some plant part extracts against fruits and vegetable pathogenic fungi and the data has been presented in this paper. Recently plant extracts are gaining importance in integrated disease management of various crops. *Punica granatum* and *Plumbago zeylanica* extracts were found to be inhibitory against *Alternaria tenuis, Curvularia lunata, Helminthosporium speciferum, F. moniliforme* (Bambode and Shukla, 1974). Upadhyay *et al.* (1980) reported that garlic juice possess inhibitory effect against fourteen fungal species. Forty wild plants were tested by Al-bed *et al.* (1993), out of these *Anagallis avensis* was most effective against plant pathogenic fungi such as *Botrytis cinerea, Fusarium oxysporum, Helminthosporium sativum* and *Cladosporium herbarum.* Prasad *et al.* (1993) concluded that neem extract exerted varying degrees of fungal toxicity which may be due to volatile sulfurous compounds or the limonoid gedunin.

Niaz *et al.* (1994) suggested that turmeric oil completely inhibited the growth of following fungal species *viz. Aspergillus niger, A. flavus, A. fumigatus* and *A. wentil.* Storage rot of apple caused *Alternaria alternata* was controlled by using pathogenic fungi such as *Botrydiplodia theobromae, Fusarium moniliforme* and *F. solani.* Mughal, *et al.* (1996) observed that aqueous leaf extracts of *Allium sativum, Datura alba* and *Withania somnifera* inhibited the growth of *Alternaria alternata, A. brassicola* and *Myrothecium roridum*. Khilare and Gangawane (1997) reported that use of twenty five plant extracts in the management of thiophanate methyl resistant of *Penicillium digitatum* causing green mould of sweet orange. Out of twenty five plants *Azadirachta indica, Curcuma longa, Terminalia chebula* and *Ocimum sanctum* are most effective in the management of Thiophanate methyl resistant in *P. digitatum*. Dixon (2001) reported that plant produce an enormous array of secondary metabolites and it is commonly accepted that a significant part of this chemical diversity serves to protect plants against microbial pathogen. These plant compounds are routinely classified as antimicrobial on the basis of susceptibility tests that produce MIC in the ranged between 100 to 1000 μg/ml (Singh and Majumdar, 2001; Tegos *et al.*, 2002; Paul and Sharma, 2002).

Deokate *et al.* (2004) reported that the potential of plant extracts against the Carbendazim resistant of *Gloeosporium ampleophagum* was observed individually and mixture. Out of 20 plant extracts, *Curcuma longa, Emblica officinalis, Terminalia bellerica* and *T.chebula* was highly effective as they gave (85.72) value to 100 per cent. Suryawanshi (2005) reported that leaf extract of tulsi and *Bougainvillea* was effective against *Colletotrichum vanilla* causing growth reduction of 74.07 and 57.40 per cent. Neem leaf extract reduced the growth of Exserohilum turcicum recorded by Bunker and Mathur (2008), Basha *et al.* (2009) summarized that the growth of *Colletotrichum musae* and *Fusarium moniliforme* also reduced. Yusuf *et al.* (2011) reported that antifungal activity of 26 plant extracts against *Phytophthora infestans* confirmed by radial growth technique. Out of *Xanthium strumarium, Lauris nobilis, Salvia officinale* and *Styrax officinale* were the most active plants that showed potent antifungal activity. List of pathogens *Monilinia fructicola, Drechslera graminea, Gloeosporium ampelophagum, Penicillium digitatum, Alternaria alternata, Fusarium solani* control using various plant extracts were recorded (Janisiewicz and Korsten, 2002; Deokate *et al.*, 2004; More, 2006; Khilare and Gangawane, 1997; Chakraborty *et al.*, 2007; Thind, 2009; Zaker and Mosallanijad, 2010; Sharma *et al.*, 2010; Saini *et al.*, 2011, Baviskar and Suryawanshi 2011, Dahiwale and Suryawanshi, 2011, 2012, 2013, Suryawanshi *et al.*, 2011 and also supported research work given in Table 2.5.

Use of Bio-Control Agents

There are some few reports available on biocontrol agents of various fruits and vegetable pathogens. Fruits and vegetables are the most important component of a balanced diet and we can now, grow varieties of different vegetables round the year. India is the world's second largest producer of vegetables next to China. Vegetables are more susceptible to insect's pests and diseases due to their tenderness and softness as compared to other crops and virtual absence of resistance characters because of intensive hybrid cultivation. The native value of healthy vegetables is altered because of fungal attack and sometimes fungi produce certain mycotoxin in them and make them unsuitable for human consumption. Therefore, biological control of plant pathogens has been considered as a potential control strategy in recent years and

Table 2.5: Pathogens Controlled using different Medicinal Plant Extracts

Pathogens	*Medicinal Plants*	*References*
Alternaria tenuis	*Punica granatum*	Bombade and Shukla (1973,1974)
Alternaria soloni	Wild plant	Al-bed *et al.* (1993)
Alternaria tenuis Rhizopus sp.	Termeric oil	Niaz *et al.* (1994)
Fusarium sp.	*Glosscardia adscendens*	Kareppa and Sangai (1996)
Fusarium monoliforme	*Azadirachta indica*	Bankole and Adebanjo (1995)
Xanthomonas campestris	*Datura metal*	Tiwari *et al.* (2004)
Alternaria brassicae	*Brassica juncea*	Patni *et al.* (2005)
Alternaria burnsii	*Cuminum cyminum*	Vihol *et al.* (2009).
Colletotrichum musae	*Azadirachta indica*	Basha *et al.* (2009)
Alternaria alternata	*Eucalypus globulus*	Zaker and Mosalaanejal (2010)
Alternaria alternata	*Terminalia alata*	Shinde *et al.* (2011)
	Terminalia arjuna	
Alternaria solani	*Pongamia pinnata*	Pattnaik *et al.* (2012)
Phytophthora capsici	*Aegel marmelos*	Pattnaik *et al.* (2012)
Pythium debaryanum	*Piper nigrum*	Pattnaik *et al.* (2012)
Botrytis cinerea	*Magnolia liliflora*	Bajpai and Kang (2012)
Penicillium expansum	*Azadirachta indica*	Baviskar and Suryawanshi (2012)
Penicillum expansum	Plants Leaf extract	Baviskar and Suryawanshi (2014)

search for these biological agents is increasing. *Trichoderma* is the most commonly used fungal biological control agent and have long been known as effective antagonists against plant pathogenic fungi (Chet *et al.*, 1997). During the last decade, species of *Trichoderma* have emerged as the most powerful bioprotectants for the management of a wide variety of plant diseases by virtue of their broad spectrum action against a number of plant diseases caused by fungi, bacteria, viruses and even nematodes (Mukhopadhyay, 1995). Thus the present study was aimed to evaluate the antagonistic activity of *Trichoderma* spp. in laboratory conditions. In the recent years, the biological control has received a worldwide attention and is being integrated effectively with other pesticides. Moreover, in the context of environmental pollution the use of biological agents is considered quite safe. Three *Trichoderma* species *viz. T. viride, T. harzianum, Trichoderma* sp. were evaluated against *Aspergillus niger, A. flavus, Phytophthora* sp., *Fusarium oxysporum, Rhizoctonia solani, Penicillium notatum* and *Alternaria solani*. The results were found all the *Trichoderma* sp. effective against above pathogens *Trichoderma* proved to be superior for inhibition of the above pathogen as compared to other *Trichoderma* species Fungal antagonists were the first to be identified as biocontrol agents of plant pathogens and have received much attention in the last three decades worldwide (Harman *et al.*, 2004; Cook and Baker, 2010). A wide range of fungal antagonists including species of *Trichoderma, Gliocladium virens, Aspergillus niger, A. flavus, Fusarium, Rhizoctonia, Agrobacterium radiobacter, Bacillus* spp., *Pseudomonas fluorescens* and *Pseudomonas* spp. have been researched and commercialized. Several species of different fungal antagonists have been effective against serious plant pathogens. Mostly *Trichoderma* spp. have been used through solid and liquid fermentation technologies to develop talc, gypsum based formulation for seed, soil, foliar and post harvest applications (Nakkeeran *et al.,* 2002). It has attracted attention of several workers of test fungal antagonists along with fungicides (Chahal, 2010).

Bio-control agents provide economical and relatively non-polluting delivery systems for protective material compared to other application system. The idea of bioagents against plant pathogen comes through the first experiment by Millard and Taylor (1927) who showed that the potato grown in sterilized soil and inoculated with *Streptomyces scabies* with *S. precox* (Gangawane and Deshpande, 1977). Before commencing the search for biological control agents agreement needs to be sought from the Natural Resource Management Standing Committee (NRMSC), with initial application through the Australian weeds committee (AWC). In addition to meet the requirement of the quarantine act 1908 and the Environment Pollution and Biodiversity Conservation Act, 1999 (EPBC), biological control must pass through assessment before approval can be given. D'Ercole *et al.* (1988) reported that very good results in controlling *B.cinerea* with *Trichoderma* species. Very limited research has been done on biological control of *B.cinerea* using *Trichoderma* in chickpea. Integrating a biocontrol agent with sub lethal doses of fungicide seems to be very promising in controlling plant pathogens without disturbing the biological equilibrium. Wilson *et al.* (1991) reveals that *Trichoderma* used for controlling *Penicillium digitatum* green mould on citrus. O'Neill *et al.* (1996) showed that *Trichoderma harzianum* effective in suppression of tomato stem rot caused by *B. cinerea.*

Agarwal and Tripathi (1999) reported that *Trichoderma viride* exhibited the antagonistic effect against chickpea grey mould pathogen *B.cinerea* mycoflora examination revealed that the hyphal tips of *B.cinerea* swelled and becomes curved. Importance of *Trichoderma* is so increased that 113 species are known all over the world. Of these 13 species are reported in India (Nagmani *et al.,* 2002). Nearly 32 crop diseases have been controlled by different species of *Trichoderma* reported by Ade *et al.,* 2005. Population dynamics of biocontrol agents and plant pathogens have very well been discussed by Paulitz and Belager (2001). Similarly Suryawanshi (1998) studies antibiotic resistance in *Xanthomonas oxonopodis pv. citri* managed with *Trichoderma viride* and *Bacillus subtilis.* Janisiewicz and Korsten (2002) reported that the antibiotics subtulin produced by *B. subtilis* and Pyrrolnitrin produced by *Pseudomonas cepacia* reduced *in vitro* growth and spore germination of *B. cinerea, P. expansum* and *M. fructicola.* Sharma and Trivedi (2005) observed that the isolate of *Trichoderma virens* and *Trichoderma viride* were the most effective in which maximum growth inhibition of mycelium of *Fusarium oxysporum* was recorded (77.7 per cent and 77.5 per cent). Mukhopadhyay (2011) revealed that *Trichoderma* emerged as most potential bioprotectants for management of wide variety of plant diseases. Cong *et al.* (2011) reported that 72 endophytic bacteria isolated from healthy tomato stems and leaves gave strongly inhibitory effect on *B.cinerea* Pers. Several fungal pathogens controlled using biocontrol agents. Baviskar and Suryawanshi (2012) recommended that 16 microorganisms against Carbendazim resistant of *Penicillium expansum* causing blue mould of apple. Out of which *Bacillius subtilis* more fruitful microorganisms use to control fruits and vegetable pathogens and supported researcher given in (Table 2.6).

Table 2.6: Various Pathogens Controlled Using Bio-control Agents

Sl.No.	Bio-agents	Host	Disease	References
1.	*Pseudomonas syringe*	Apple	Blue Mould	Pusey and Wilson, 1984.
2.	*Trichoderma harzianum*	Grape Strawberry	Grey mould Rhizopus rot Grey mould	Pusey and Wilson, 1984 Elad *et al.*, 1990; Wisniewski and Wilson, 1992; Elad *et al.*, 1993
3.	*Pseudomonas fluorescens* *Erwinia amylovora*	Pear	Pear blossom	Wilson and Lindow, 1993
4.	*Candida oleophila*	Orange	Orange rot	O' Neill *et al.*, 1996
5.	*Bacillus subtilis*	Citrus	Sour rot	Pusey and Wilson, 1994
6.	*Candida oleophila* *Candida famata*	Orange	Orange rot	Arras, 1996
7.	*Candida oleophila*	Orange	Orange rot	De Meyer *et al.*, 1998
8.	*Aureobasidium pullulans*	Tomato	Rhizopus rot	Ippolito *et al.*, 2000
9.	*A. pullulans*	Apple	Blue mould	Castoria *et al.*, 2001
10.	*Pseudomonas cepacia*	Pomegranate	Blue mould	Janisiewicz and Korsten, 2002
11.	*Aureobasidium. pullulans*	Grape	Grey mould	Schena *et al.*, 1999
12.	*Agrobacterium tumefaciens*	Grape Kiwifruit	Grey mould Bitter rot	Beever *et al.*, 2005
13.	*Trichoderma*	Eucalyptus	Grey mould	Salome and Sanfuentes, 2010
14.	*Bacillus subtilis*	Brown rot	Peach	Prusky and Gullino, 2010
15.	*Bacillus subtilis*	Peaches	Brown rot	Wilson *et al.*, 2011
16.	*Trichoderma harzianum* *Bacillus amyloliquefaciens*	Banana	Black sigatoka, Leaf spot of banana	Alvindia,2012
17.	*Pseudomonas syringae*	Potato	Potato rot	Ernest and Livingston, 2012

Use of Homoeopathic Medicines

There are few reports on the use of homoeopathic medicines in the management of various pathogens. Inhibitory effect of homoeopathic drugs such as *Lycopodium, Thuja, Arsenicum, Zincum* etc. against *Alternaria alternata, Fusarium moniliforme, Gloeosporium psidii, Colletotrichum gloeosporioides* and *Pestalotia sp*. and certain fruit rot pathogens have been reported by Khanna and Chandra, 1989 and 1992; Chandra *et al.*, 1981; Wilson *et al.*, 1991. Mohanty *et al.*, 1995 showed that five homoeopathic drugs *viz. Thuja occidentalis, Baptisia tinctoria, Arsenicum album* and *Sepia officinale* are inhibitory to brinjal fruit rot pathogen caused by *Phomopsis vexans.* Dahiwale and Suryawanshi (2010) observed that fruit rot of pomegranate caused by *Alternaria alternata* is one of the most important post harvest diseases. It was revealed that altogether 13 homoeopathic medicines were inhibitory against *A. alternata*. Out of which *Arsenicum album, Argentum metallicum* was highly effective fallowed by *Zincum metallicum, Baptisia tinctoria, Belladonna, Tabaccum, Lycopodium clavatum, Thuja occidentale, Cyanopodium, Ustilago maydis, Sepia officinale* and *Iris versicolar* in decreasing PCE values against *A. alternata*. Dahiwale and Suryawanshi (2014) observed that grey mould of grape caused by *Botrytis cinerea* control using homoeopathic medicine.

Use of Essential Oils

Application of chemical compounds is considered as the most inexpensive and common method in plant disease control. However, their adverse affects on human health and the environment, prompted man to produce natural pesticides. Biologically active compounds found in plants appear to be more adaptable, acceptable and safer than synthetic compounds and display a wealthy source of potential pathogen control agents. Extracts and essential oils of medicinal plants are effective against fungal and bacterial pathogens; meanwhile they are biodegradable compounds which have high potential for using in integrated pest management programs. The use of biological compounds extracted from plants may be an alternative to conventionally used fungicides to control phytopathogenic fungi, due to their being bioactive chemicals such as flavonoids, phenols, tannins, alkaloids, quinone, saponins and sterol.

During recent years, some commercial companies used the results of research findings about the pesticide properties of plant essential oil and presented new pesticide compounds to the market. For example Mycotech company making Cinnamite™ as a fungicide for use in the greenhouse and Valero™ as a fungicide and acaricide for use on crops such as grapes, citrus and nuts. There are many reports about the impact of medicinal plant essential oils and plant extracts against food and grain storage fungi, foliar pathogens, soil born fungi and nematodes. The response to the different plant essential oils and plant extract generally depends on the fungal species tested and may include ranges from resistant (inhibition of growth) to various degree of susceptibility (Hayes and Laws, 1991; Burt, 2004; Soylu *et al.*, 2006).

Essential oils used for the controlling of postharvest pathogens. It seems that the anti-fungal and anti-microbial effects are the result of many compounds acting synergistically. Few reserarchers has been done to demonstrate the potential of essential oils for use against postharvest pathogens. Work showed that essential oils from *Thymus zygis, Eugenia caryophylla* and *Cinnamomum zeylanicum* prevented the growth of *B.cinerea*. Other researchers have shown that the essential oil of *Monarda citrodora* and *Melaleuca alternifolia* also exhibit antifungal activity against a wide range of common postharvest pathogens. We have looked at the effect of tea tree oil as a vapour on the growth of *Botrytis* from grapes tea tree oil has antibacterial and antifungal properties that have secured it a place in the commercial pharmaceutical market. We wanted to determine if these properties may also be useful for the postharvest control of fungi on grapes. Our work showed that concentrations of between 50 to 100 were able to prevent the growth of fungi when it was grown in the laboratory (Mishra and Dubey, 1994; Wilson *et al.*, 1997; Jobling, 2000; Tripathi and Dubey, 2004; Khan *et al.*, 2007; Tripathi *et al.*, 2008; Singh *et al.*, 2008; Sirirat *et al.*, 2009;

Kumar *et al.*, 2010; Ines *et al.*, 2011; Samane *et al.*, 2011; Marandi *et al.*, 2011; Shinde and Wadje, 2011; Amini, 2012 simillarly some casses mentioned in (Table 2.7).

Table 2.7: Fruit Pathogens Controlled Using Essential Oils

Sl.No.	*Common Name*	*Essential Oils*	*Pathogen*	*References*
1.	Cinnamon oil	*Cinnamomum martini*	*H. oryzae*	Singh *et al.*, 1980
	Tea oil	*Trachyspermum ammi*		
2.	Lemon oil	*Citrus limonum*	*B. cinerea*	Wilson *et al.*, 1997
	Red thyme	*Thymus zygis*		
	Citronella oil	*Cymbopogon nardus*		
	Thymus lemon	*Thymus hiemalis*		
	Magjorum	*T.mastichiana*		
	Clove	*Eugenia caryophyllata*		
	Basil	*Ocimum basilicum*		
	Fir	*Abies balsamea*		
	Juniper	*Juniperus communis*		
	Orange	*Citrus aurantium*		
	Peppermint	*Mentha piperita*		
	Pine	*Pinus syvestris*		
	Nilgiri oil	*Eucalyptus globulus*		
	Carrot	*Daucous carrota*		
	Cedar wood oil	*Juniperus Mexicana*		
	Cypress	*Cupressus semiperviens*		
	Fennel	*Foeniculum vulgare*		
	Cinamomum leaf	*Cinnamomum zeylanicum*		
	Tea oil	*Melaleuca alternifolia*		
3.	Thyme	*Thymus vulgaris*	*B. cinerea*	Reddy *et al.*, 1997
		Thymus zygis	*R.stolonifer*	
4.	Red thyme	*Eugenia caryophyllata*	*B. cinerea*	Jobling, 2000
	Clove	*Cinnamomum zeylanicum*		
	Cinnamon leaf			
5.	Thymus lemon	*Thymus hiemalis*	*B. cinerea*	Samane *et al.*, 2011
	Magjorum	*T.mastichiana*		
6.	Ginger	*Zingiber offcinale*	*B. cinerea*	Tripathi *et al.*, 2008
	Tulsi oil	*Ocimum sanctum*		Sirirat *et al.*, 2009
	Almond oil	*Prunus amygdalus*		
	Cinnamon oil	*Cinnamomum verum*		
7.	Olive oil	*Olea europaea*	*B. cinerea*	Vagelas *et al.*, 2009

Contd...

Table 2.7–*Contd...*

Sl.No.	*Common Name*	*Essential Oils*	*Pathogen*	*References*
8.	Orange	*Citrus aurantium*	*B. cinerea*	Samane *et al.*, 2011
	Juniper	*Juniperus communis*		
	Fir	*Abies balsamea*		
	Fennal	*Foeniculum vulgare*		
9.	Triphala	*Terminalia arjuna*	*Aspergillus niger*	Shinde, 2011,
		T.bellerica	*Alternaria alternata*	
		T. catappa	*A.brassicola*	Shinde and Wadje, 2011
10.	Nilgiri oil	*Eucalyptus globulus*	*B. cinerea*	Amsalu *et al.*,2011
	Rose mary oil	*Lavendula officinalis*		
	Nut meg oil	*Myristica fragrans*		
	Olive oil			

Acknowlegement

Auther is grateful to the Emeritus Prof. Dr. L.V. Gangawane, Dr. Babasaheb Ambedkar Marathwada University, Aurangabad, India for encouragement and motivation of Research and U.G.C. (WRO), Pune for Financial Assistance under Minor Research Project. I have also grateful to Principal of my college for providing research Laoratory Facilities.

References

Ade, A.B., Gangawan, e L.V. and Khilare, V.C. (2005). Application of micro-organisms in the biological control of plant pathogens. In: Frontiers in Plant sciences (Mukerji K.G., Tilak K.V., Reddy B.R, Gangawane L.V., Prakash P. and Kunwar I. K. eds.) *I. K. International Public. New Delhi.*, pp. 533-548.

Agarwal, Abha and Tripathi, H.S. (1999). Biological and chemical control of *Botrytis* grey mould of Chickpea. *J.Mycol.Pl. Pathol.*, **29**(1): 52-56.

Aharoni, Y. and R. Barkai - Golan (1987). Pre harvest fungicide sprays and polyvinyl wraps to control *Botrytis* rot and prolong the post harvest life of strawberries. *J. Hortic. Sci.* **62**: 177-181.

Al-bed A.S., S.R. Dassem and H.A. Abu- Balan (1993). Antifungal effects of some common wild plants species on certain plants pathogenic fungi. *Dis. Ser. Appl. Sci.*, **20**(3): 149-198.

Alvindia, G. Dionisio (2012). Inhibitory influence of biocontrol agents, plant oils and an inorganic salt on *Mycosphaerella fijiensis* and *Cordana musae*, the causal pathogen of black sigatoka and leaf spot of banana. *African J. Microbiol. Research.*, **6**(19): 4179-4184.

Amini, M., N. Safaie, M. J. Salmani, Shams-Bakhsh, M. (2012). Antifungal activity of three medicinal plant essential oils against some phytopathogenic fungi. *Trakia J. Sciences.*, **10**(1): 1-8.

Amsalu, Abera, Fikre Lemessa and Diriba Muleta (2011). The antifungal activity of some medicinal plants against coffee berry disease caused by *Colletotrichum kahawae*. *Int. J. Agricultural Research*, **6**: 268-279.

Anilkumar, T.B. and Shashtry, M.N.L. (1979). Development of tolerance to fungicides in *Rhizoctonia bataticola*. *Phytopathology*., **91**: 126-131.

Annamalai, P. and Lalithakumari D. (1987). Cross resistant mutants of *Drechslera oryzae*, *Ind. Phytopath*., **40**: 296.

Arras, G. (1996). Mode of action of an isolate of *Candia famata* in biological control of *Pencillium digitatum* in orange fruits. *Post harvest Biol. Technol.*, **8**: 191-198.

Bajpai, V.K. and Kang, S.C. (2012). *In vitro* and *in vivo* inhibition of plant pathogenic fungi by essential oil and extract of *Magnolia liliflora*. *Desr. J. Agr. Sci. Tech.*, **14**: 845-856.

Bambode, R. S. and Shukla, V. N. (1974). Studies on fungitoxic properties of *Plumbago zeylanica* L. extract. *PKV Res. J.*, **2**(2): 110-112.

Bankole, S. A. and Adebanjo A. (1995). Inhibition of growth of some plant pathogenic fungi using some Nigerian plants. *Int. J.Tropic Pl. Dis*., **13**(1): 91-95.

Basha, R.A. Lal, A.A. Gupta, S. Srivatava, R. and Gupta P. (2009).Ecofriendly management of post harvest fungal pathogen causing crown rot and antracnose of Banana in Allahbad. *Myco.Plant Pathol.*, **39**: 14-18.

Baviskar, R. N. and Suryawanshi, N.S. (2011). Integrated management of *Penicillium expansum* causing blue mold disease of apple using plant extracts. *J. of Sci. and Tech. Bionanofrontier.*, **4**(1): 145-147.

Baviskar, R.N. and Suryawanshi, N.S. (2012). Management of *Penicillium expansum* Link causing blue mold of apple using fungicides. *Journal of Bionanofrontier. International Society of Science and Technology.*, **5**(1): 81-83.

Baviskar, R.N. and Suryawanshi, N.S. (2013). Application of microorganismsin the biological control of *Penicillium expansum. Bionano frontier.*, **6**(1): 60-62.

Baviskar, R.N. and Suryawanshi, N.S. (2013). Effect of passage on the development of cabendazim resistant in *Penicillium expansum* Link causing blue mold of apple. *Bionano frontier.*, **6**(2): 287-289.

Baviskar, R.N. and Suryawanshi, N.S. (2014). Inhibitory effect of leaf extracts on carbendazium resistant *Penicillium expansum* causing blue mold of Pear. *Bionano frontier.*, **7**(1): 140-142.

Baviskar, R.N.and Suryawanshi, N.S. (2012). "Management of *Penicillium expansum* Link. Using certain micronutrient. *Bionano frontiers.*, **5**(2) 226-228. :

Beever, R.E., K.M. Plummer and Wurms, K.V. (2005). Novel Approaches to controlling fruit Pathogens. *New Zealand Plant Protection.*, **58**: 68-73.

Beresford, Robert (1994). Understanding fungicides resistance. *The orchar dist.*, **67**(9): 24.

Bhale, U.N. (2002): Studies on management of some important diseases of spinach in Maharashtra.Ph.D. Thesis Dr. Babasaheb Ambedkar Marathwada University, Aurangabad.

Blacharski, R.W., Bartz J. A., Xiao C.L. and Legard, D.E. (2001). Control of post harvest *Botrytis* fruit rot with preharvest fungicide application in annual strawberry. *Plant Dis.*, **85**: 597-602.

Bollen, G. J. and Scholten (1971). Acquired resistance to benomyl and some other systemic fungicides in a strain of *Botrytis cinerea* in Cyclamen. *Neth. J. Plant Pathol*. **77**: 83-90.

Bollen, G. J. and Zayen Van A. (1975). Resistance to benzimidazole fungicide in pathogenic strain of Bollen G. J. and V*erticillium fungicola*. *Neth.J.Plantpathol.,* **81**: 157-167.

Bunker, R.N. and Mathur K. (2008). Evaluation of neem based formulations and chemical fungicides for the management of *Sorghum* leaf blight. *Ind. Phytopath.*, **61**(2): 192-196.

Burt, S. (2004). Essential Oils: their antibacterial properties and potential applications in foods - a review. *Int. J. Food Micribiol.*, **94**: 223- 253.

Butters, J. A. and Hollomon, D. W. (1999). Resistance to benzimidazole can be caused by changes in β-tubulin isoforms. *Pestic. Sci.*, **55**: 486-503.

Byrde, R. J. W. and Willets, H. J. (1977). The brown rot fungi of fruit: Their biology and Control. *Oxford, Pergamon Press*, p. 171.

Castoria, R., De Curtis F., Lima G., Caputo L. Pacifico S., De Cicoov (2001). *Aureobasidium pallulans* (LS-30) and antagonist of postharvest pathogens of fruits: Study on its mode of action. *Postharvest Biol. Technol.*, **22**: 7-17.

Chahal, S. S. (2010). Plant pathology in India- A roadmap to meet future needs. *Ind. Phytopath.* **63**(3): 241-249.

Chakraborty, B. N., R. Das Biswas and Sharma, M. (2007). Induction of resistance in Tea plants against *Alternaria alternata* by foliar application of leaf extracts. *J.Mycol Pl Pathol.*, **37**(1): 60-64.

Chander, M. and Thind, T. S. (1995). Development of carbendazim resistance in *Gloeosporium ampleophagum* and strategies for its management. *Ind. J. Mycol. Pl. Pathol.*, **24**(1): 25-33.

Chandra, H., N. K.Dubey, A. Asthana, R. D. Tripathi and Dixit, S.N. (1981). Effect of some homoeopathic drugs against spore germination of some fungi. *Natural Acad. Sci. Letter.*, **4**(4): 161-164.

Chaudhary, K. C. and Putto, B. L. (1984). Fungicide resistant strain of *Venturia inaequalis* in Kashmir, a prediction. *British Crop protection Confer.* (*Pest and Dis.*)., **6A**-12: 509-514.

Chet, I.Inbar J. and Hadar (1997): Fungal antagonists and mycoparasites. In the mycota, Enviornmental and Microbial relation. Wicklow D.T. and B. Soderstrom (Eds).Vol.4 Springer-vertag, Berlin, Germany, pp. 165-184.

Cole, R. J., A. J. Gilchrist and Soper D. (1971). The control of diseases of apple and pears with thiophanate methyl. Proc.7th Br. *Insectic. Fungic. Conf.* **2**: 623.

Cong, Jan Yang, Xin-Gang Zhang, Guan-Ying Shi (2011). Isolation and identification of endophytic bacterium W4 against tomato *B.cinerea* and antagonistic activity stability. *Afric. J Micro Res.* **5**(2): 131-136.

Cook, R. J. and Baker K.F. (2010). The nature and practice of biological control of plant pathogens. *Amer.Phytopath.*, p.539.

D'Ercole, N., Nipoti P., Finessi L.E. and Manzali D. (1988): Review of several years of research in Italy on the biological control of soil fungi with *Trichoderma* spp. *Organization Europeenne et Mediterraneenne pour Ia protection des plants Bulletin.*, **18**(1): 95-102.

Dahiwale, M. A., Baviskar R. N. and Suryawanshi, N. S. (2009): Integrated management of carbendazim resistant *Altrenaria alternata* causing fruit rot of Pomeranate. *Life Sci. Journal Bioinfolet.*, **6**(1): 44-45.

Dahiwale, M. A. and Suryawanshi, N.S. (2010): Integrated management of carbendazim resistant *Alternaria alternata* using homoeopathic medicine. *Bionano Frontier.*, **3**(2): 330-331.

Dahiwale, M. A. and Suryawanshi, N.S. (2010): In vitro evaluation of essential oils in the management of grey mould of grapes caused by *Botrytis cinerea. Bionano Frontier.*, **5(2)**: 290-292.

Dahiwale, M. A. and Suryawanshi, N.S. (2012): Biological control of postharvest diseases of fruit. *Bionano Frontier.*, **5**(1): 161-163.

Dahiwale, M. A. and Suryawanshi, N.S. (2013). Development of Thiophanate methyl resistance in *Botrytis cinerea* and strategy for its management. *Fungi and Human Welfare.* 62-67.

Dahiwale, M. A. and Suryawanshi, N.S. (2014). Grey mould of grape caused by *Botrytis cinerea-* control using homeopathic medicine. *Fungi and Agriculture,* pp. 3-5.

Dahiwale, M. and Suryawanshi, N.S. (2011). Integrated management of carbendazim resistant *Alternaria alternata.Bioinfolet.*, **8**(1): 67-68.

Davidse, L.C. (1981). Resistance to acylalanine fungicides in phytophthora megasperma sp. Medicaginis. *Neth.J.Pl.Pathol.*, **87** : 11-24.

De Kock, de P. J. and Holz G. (1994). Application of fungicides against post harvest *Botrytis cinerea* Bunch rot of table grapes in the Western Gape. *S. Africa. J. Enol. Vitic.*, **15**(2): 33-40.

De Kock, P. J. and Holz G. (1991). Colonization of table grape by *B. cinerea* in the western cape province. *Phytophylactica.*, **23**: 73-80.

De Meyer, G., Bigirimana, J., Elad, Y. and Hofte M. (1998). Induced systemic resistance in *Trichoderma harzianum* T39 biocontrol of *Botrytis cinerea. Eur. J. Plant Pathol.*, **104**: 279-286.

De Warrd, M. A. (1994). Resistance to fungicides which inhibit sterol 14a-demethylation an historical perspective in: Fungicide resistance, Heaney S., Slawson, D.D., Hollomon D.W., Smith M., Russell P.E., Parry D.W. eds. *British Crop Protection Council, Farnham, Survey*, pp. 3-10.

Dekker, J. (1977). Resistance in: Systemic fungicide (ed. R.W.Marsh Longman) London and New York.

Dekker, J. (1982). Introduction to course on fungicide resistance In: Fungicide resistance in crop protection (Dekker, J and Georgopoulos S.G. eds.), *CAPD'Wageningen, Netherlands*, pp. 128-238.

Delen N. and M. Yildiz (1981). Fungicide resistance in Turkey. Proc. resistance to fungicide in plant pathogens, Wageningen, Netherlands 28th July-5/6 August 1981.

Dennis, C., (1983a). Soft fruits. In: C.Dennis (ed.). Post harvest pathology of fruit and vegetables, *New York, Academic Press.*, 23-42.

Deokate, A. S., Khillare V.C. andGangawane, L.V. (2004). Management of carbendazim resistant *Gloeosporium ampleophagum* causing Anthracnose of grape using herbal extracts short communication. *J. Mycol. Pl. Pathol.*, pp. 80-82.

Deokate, A. S., Khilare, V. C. and Gangawane, L. V. (2002). Management of anthracnose of grape by carbendazim in combination with Agrochemicals, *Indian Phytopath.* **55**(1): 107-108.

Dixon, R.A. (2001). Natural products and Plant disease resistance. *Nature,* **411**: 843-847.

Dovas, C, Skylakakis G and Georgopoulous (1976). Adaptability of the benomyl resistant populations of *Cercospora beticola* in Northern Grcece.*Phytopath*., **66**: 1452-1456.

Droby, S. and Lichter, A. (2004). Post harvest *Botrytis* infection: etiology, development and management. In: Elad Y., Williamson B., Tudzynski P., Delan N. (Eds). *Botrytis*: Biology Pathology and Control. *Kluwer Academic Publishers, London, UK.* 349-367.

Eckert, J. W. (1982). Case study 5: *Penicillium* decay of citrus fruits.In: Fungicide resistance in crop protection, Dekker J and Georgopoulos, S.G. eds. Pudoc, Wagningen. 231-250.

Elad, Y., J. Chet and Katan, J. (1990). *Trichoderma harzianum* a biocontrol agent effective against *Sclerotium rolfsii* and *Rhizoctonia soloni. Phytopath.,* **70**: 119-121.

Elad, Y., Zimond G., Zaqs. Y., Zuriel S and Chet, I. (1993). Use of *Trichoderma harzianum* in combination or alternation with fungicides to control cucumber grey mould (*B.cinerea*) under commercial greenhouse conditions. *Pl.Pathol.,* **42**: 324-332.

Elad, Yigal, Ezra Shabi and Talma Katan (1988). Negative cross resistance between benzimidazole and N-phenylcarbamate fungicides and control of *B.cinerea* on grapes. *Digital object Identifier* (*DOB*). 10.1111/j.1365-3059 tb 02206.

Eleni, K. Kulakiotu, Constantine C. Thanassualopoulos and Evangelos, Sfakiotakis, M. (2004). Post harvest biological control of *Botrytis cinerea* on kiwifruit by volatile of Isabella grapes. *Amer. Phytopath.,* **94**: 1280- 1285.

Ernest, Eknow Abano and Livingston Kobina Sam - Amoh (2012). Application of antagonistic microorganism for the control of postharvest decays in fruits and vegetables. *Interna. J. Advanced B.iol. Research.,* **2**(1): 1-8.

Fisher, N. and Griffin, M.J. (1984). Benzimidazole (MBC) resistance in *Septoria tritici. ISPP Chemical Control Newletter.,* **5**: 8-9.

Formigini, A. G., Gastangna and Ciocca, P. (1971). Two years of trials in Italy on 1, 2-bis-(3-thioxycabomyl-2-thioureido) benzene a new systemic fungicide. *Meded Fac. Labb. Rij. Gent.* **36**: 96.

Foschi, S., Rapparini G. and Beneivelli, A. (1972). New results on scab and powdery mildew control on apples. *Informative Agrario.,* **28**: 8489.

Gangawane, L. V. and Deshpande, K. B. (1977). Agonomic treatment and changes in rhizosphere mycoflora II. Effect of chemical fertilizers. *Acta. Agronomic Acad. Sci. Hungery,* **26**: 68-78.

Gangawane, L. V. (1982).Fungicide resistance in plant crop protection. *Pesticides,* **15**: 12-16.

Gangawane, L. V. (1990). Fungicide resistance in plant pathogens in India. In: Proc. International Crop Plant Protection in Tropics. MAPPS, Malaysia Publ. 117-121. *Indian Phytopath.,* **40**: 551-553.

Gangawane, L. V. (1997). Management of fungicide resistance in plant pathogens *Ind. Phytopath.,* **50**(3): 305-315.

Gangawane, L. V. and Reddy, B. R. C. (1985). Resistance of *Aspergillus flavus* to certain fungicides. *ISPP Chemical Control Newsletter No.***6**: 23.

Gangawane, L. V. and Reddy, B. R. C. (1987). Development and Management carbendazim resistance in *Aspergillus flavus. Ind. Phytopath.* **40**(20): 297.

Gangawane L. V. and R. S. Saler (1981). Resistance to fungicides in *Aspergillus flavus. Neth. J.* t of *Pl. Pathol.* **87**: 254.

Gangawane, L.V. and Saler, R.S. (1988). Some physiological aspects of *Aspergillus flavus* isolates resistant and sensitive to fungicides. *Indian Bot. reptr.* **7**(1-2): 61-64.

Geeson, J. D. (1978). Mutational tolerance to carbendazim in *Botrytis cinerea. Ann Appli. Bio.* **90**: 59-64.

Georgopoulos, S. G. (1981). Genetical and biochemical background of fungicide resistance. Proc. Resistnce to fungicide in in plant pathogens. *Wageningen, Netherland,* 28th July 5/6 Aug. 1981.

Giannopolities, C. N. (1978). Occurrence of strains of *Cercospora beticola* resistant to triphenyltin fungicides in Greece. *Plant Disease Reporter.,* **62**: 205-208.

Gibbs, J. N. and Bresier, C. M. (1980). Futher studies on carbendazim tolerance in *Ceratocystis ulmi Proc. Ann. Appl. Bio.,* **940**: 38-310.

Gilchrist, A. J. and Cole, R. J. (1971). Control of disease of soft fruit and other crops in the United Kingdom with thiophanate methyl. *Proc. 6th Br. Insectic. Fungic. Conf.*, **2**: 332.

Gilpatrick, J. D. (1982). Case study 2: *Venturia* of pome fruits and *Monilinia* of stone fruits. In: Fungicide resistance in crop protection, Dekker, J and Georgopoulos S.G. eds. Pudoc, Wageningen, pp. 195-206.

Giuseppe, Lima, Filippo De Curtis, Raffaello Castoria and Vincenzo De Cicco. (2003). Integrated control of apple post harvest pathogens and survival of biocontrol yeasts in semi- commercial conditions. *European J. of Plant Pathology.*, **109**: 341-349.

Harman, G. E., Howel C. R., Viferbo A., Chat Y. and Lorito, N. (2004). *Trichoderma* species opportunistic, avirulent plant symbionts. *Nature Rev.*, **4**: 41-43.

Harman, G. E., Latorre B., Agosin E., R. San Martin, Riegel D. G., Nielson P. A., Tronsmo A. and Pearson, R. C. (1996). Biological and integrated control of *Botrytis* Bunch rots of Grape using *Trichoderma* spp. *Biological control.*, **7**(3): 259-266.

Hayes, W. J. and E. R. Laws (1991). Handbooks of pesticide toxicology. *Academic Press, New York.*

Hilber, U. W. and M. Hilber - Bodmer (1998). Genetic basis and monitoring of resistance of *Botryotimia fuckeliana* to anilinopyrimidine. *Plant Dis.*, **82**(5): 496-500.

Hilber, U. W.and H. Schuepp (1996 a). A reliable method for testing the sensitivity of *Botryotinia fuckeliana* to anilinopyrimidines *in vitro. Pestc. Sci.*, **47**: 241-247.

Hilber, U.W.and H. Schuepp (1996 b). Mycelial growth test for assessment of anilinopyrimidine sensitivity of grey mold. (*Botryotinia fuckeliana*) *EPPO Bull.*, **26**: 184-186.

Hiwale, D.K.(2004). Studies on management of *Sclerotium rolfsii* sacc.fruit rot of *Cucumis sativus* L in maharashtra.Ph.D. Thesis Dr. Babasaheb Ambedkar Marathwada University, Aurangabad.

Horsten, J. and Fehrmann (1980). Fungicide resistance of *Septoria nodorum* and *Cercosperella herpotrichoides* in cereals. *Dissertation, Agri. Univ. Wageningen, Netherlands*, p.107.

Hunter, T., Brent K. J., Carter G. A. and Hutcheson, J. A. (1987). Effect of fungicide spray regimes on incidence of dicarboximide resistance in grey mold (*Botrytis cinerea*) on strawberry plant. *Annual of Applied Bio.*, **110**(3): 515-525.

Ines Hammami, Nesrine Kamaun and Ahmed Rebai (2011). Biocontrol of *Botrytis cinerea* with essential oil and methanol extract of *Viola odorata* L. Flower. *Scholors Research Library Archieves of Applied Science Research.*, **3**(5): 44-51.

Ippolito, A., El-Ghaouth A., Wilson C. L. and Wisniewski, M. (2000). Control of post harvest decay of apple fruit by *Aureobasidium pallulans* and induction of defence responses. *Post harvest Bio.Technol.*, **19**: 265-272.

Ishii, K. (1970). 7th International congress of plant protection, Paris, ed.congress services. *Ampelot Brionne*, p. 200.

Jadeja, K. B., Nandolia, I. U., Dhruj, I. U. and Khandar, R. R. (1999). Efficacy of four triazole fungicides in the control of leaf spots and rust of groundnut. *Indian Phytopath.,* **52**(4): 421-442.

Janisiewicz, Wojciech J. and Lise Korsten (2002). Biological control of postharvest diseases of fruit. *Annual Rev.Phytopath.,* **40**: 411-441.

Jobling, Jenny (2000—). Essential Oils: A new idea for postharvest disease control. *Good Fruit and Vegetables magazine.,* **11**(3): 50.

Jones, A.L. and G.R.Ehert. (1981). Resistance of *Coccomyces hiemalis* to benzimidazole fungicides. *Plant Dis.,* **64**: 767-769.

Kadam, K. S. (1997). Integrated management of *Alternaria* fruit rot of grapes in Maharashtra. *Ph.D. Thesis Dr. BAMU, Aurangabad.*

Kaku, K., Takagaki M., Shimizu T. and Nagayama, K. (2003). Diagnosis of dehydratase inhibitors in melanin biosynthesis inhibitor (MBI-D) resistance by primer-introduced restriction enzyme analysis in scylatone dehydratase gene of *Magnaporthe grisea. Pest Management Science.,* **59**: 843-846.

Kalamarakis, A. E., N. Petsikos. Pantagiotarau, B. Mavroidis and Ziogas, B. N. (2000). Activity of fluazinum against strains of *Botrytis cinerea* resistant to benzimidazoles and/or dicarboximides and to a benzimidazole-N-phenyl-carbamate mixture. *J. Phytopath.,* **148**: 449-455.

Kamble, S. S. (1991). Studies on fungicide resistance in late blight and charcoal rot of potato. *Ph.D. Thesis. Dr. BAMU, Aurangabad.*

Kamble, S. S.and Gangawane, L.V. (1994). Effect of passage on the development of benomyl resistance in *Fusarium udum in vitro. Indian Phytopath.,* **47**(3): 354.

Kapoor, A. S., Paul, Y. S. and Singh, A. (2006). Integrated management of white rot root rot-wilt disease complex of pea. *Indian Phytopath.,* **59**(4): 467-474.

Kareppa, S.H. and Sangai, S.K. (1996). Compatibility between extract of *Glosscardia bosvalle* and carbendazim in the management of some seed borne pathogens of Jawar. *Nature and Biosphere.,* **1**(1): 31-34.

Katan, J. (1980). In proc. On foundation for post graduate couse. CAPD wageningen. Netherlands.28 July-6 August 1981.

Kato, T. (1988). Resistance experiences in Japan. In: Fungicide resistance in North America Delp C.J.ed. *Amer. Phytopath. Soc. St. Paul, Minnesota,* pp. 16-18.

Khan, Mohib, Vemuri Sastry, Murlidhar Shingare and Panuganti Amareshwar (2007). Antibacterial activity of essential oils from fruits of *Piper* species. *Bioinflolet.,* 253-255.

Khanna, H. R. and Chandra, S. (1992). Effect of homoeopathic on respiration of germinating fungal spores. *Ind. Phytopath.,* **45**(3): 348-353.

Khanna, K. K. and Chandra, S. (1989). Further investigations of the control of storage rot of mango, guava and tomato fruits with homeopathic drugs. *Ind. Phytopath.,* **13**: 436-440.

Khilare, V. C. and Gangawane, L. V. (1997). Application of medicinal plant extracts in the management of thiophanate methyl resistant *Penicillium digitatum* causing green mold of Mosambi. *J. Myco. Pl. Pathol.,* **27**(2): 134-137.

Khilare, V. C., Ade, A. B., Deokate, A. S. and L. V. Gangawane (2004). Biochemical changes in host by carbendazim resistant *Gloeosorium ampelophagum* causing anthracnose of grape. *Indian Phytopath.,* **57**(4): 497-498.

Khoury, W. M. and Saad, A. T. (1986). Resistance to *Botrytis cinerea* to fungicide in Leebanon. *Phytopath. Mediterana.,* **25**(1-3): 145-148.

Kuck, K. H., Kruger B. W., Roblenbroiach H. J.and Brendes, W. (1997). KBR2738-nouvean fungicide de la famille des hydroxyanilides, active sur *Botrytis cinerea* et *Monilinia* spp. ANPP-Cinquieme conference international sur les maladies des plantes, Tours 3-4-5 Decembre 1997: 1055-1062.

Kumar, Ashok, Ravindra Shukla, Priyanka Singh and Dubey, N. K. (2010). Chemical composition antifungal and antiaflatoxigenic activities of *Ocimum sanctum* L. essential oil and its safety assessment as plant based antimicrobial. *Food and Chemical Toxicology.,* **48**(2): 539-543.

Leroux, P. and Clerjeau, M. (1985). Resistance to *Botrytis cinerea* and *Plasmopara viticola* (Berk and Curt) Berl and de Toni to fungicides in French vineyards. *Crop Protection.,* **4**(2): 137-160.

Leroux, P., Chapeland F., Arnold A. and Gredt, S. (1998): Resistance de *Botrytis cinerea* aux fongicides. Du Laboratoire au vignoble et vice versa. *Phytoma.,* **504**: 62-67.

Leroux, P., R. Fritz, D. Debien, C. Albertini, C. Lanen, J. Bach, M.Gredt and Chapeland, F. (2002). Mechanisms of resistance to fungicides in field strains of *Botrytis cinerea. Pest Manage. Sci,* . **58**: 876-888.

Locke, T. (1986). Current incidence in the UK of fungicide resistance in pathogens of cereals. *Proceedings 1986 British Crop Protection Conference,* pp.781-786.

Long, P. G. (1971). Evaluation of benomyl Thiobendazole, benzene thiophanate and thiophanate for control of banana stem end rot disease (*Gloeosporium musarum*). *Aust. Jr. Expt. Agr. Anim.Husb.,* **11**: 559.

Lorenz, G. (1988). Dicarboximide fungicides: History of resistance development and monitoring methods. In: fungicide resistance in North America, Delp. C.J. ed. *American Phytopathol.* 45-51.

Marandi, Rasul Jalili, Abbas Hassani, Youbert Ghosta, Ali Abdollahi, Alireza Pirzad and Fatemeh Sefidkon (2011). Control of *Penicillium expansum* and *Botrytis cinerea* on pear with *Thymus kotschyanus, Ocimum basilicum* and *Rosmarinus officinalis* essential oils. *J. Medicinal Plants Research.,* **5**(4): 626-634.

Millard, W. A. and Taylor, C. B. (1927). Antagonism of microrganims as the controlling factor in the inhibition of scab by green manuring Ann. *Appl. Biol.,* **14**: 202-215.

Mishra, A. K. and Dubey, N. K. (1994). Evaluation of some essential oils for their toxicity against fungi causing deterioration of stored food commodities, *Applied and Environmental Microbiol.,* **60**: 1101-1105.

Mlikota, Franka, P. Male and S. B. Cvjetkovi (1996). Effectiveness of five fungicide on grapevine grey mold and their effects on must fermentation. *Journal of Wine Research.* **7**(2): 103-110.

Mohanty, A. K., Panda R.N., Kar A. K. and Sethi, P. N. (1995). Effect of certain homoeopathic drugs on the growth of *Phomopsis vexans* causing fruit rot of brinjal. *Crop. Res. (Hissar).*, **9**(1): 165-167.

More, D. R. (2006). *In vitro* antifungal activity of plant extracts on molds of sweet orange (*Citrus sinensis*). *Bioinfolet.*, **3**(2): 121-122.

Mughal, M.A., T.Z. Khan and Nasir, M.A. (1996). Antifungal activity of some plant extracts. *Pakistan J. Phytopathol.*, **8**: 46.

Mukhopadhyay, A.N. (1995). Exploitation of *Gliocladium virens* and *Trichoderma harzianum* for biological seed treatment against soil borne diseases. *Ind. J. Myco. Pl. Pathol.*, **25**(1&2): 124.

Mukhopadhyay, A.N. (2011). *Trichoderma* for plant disease management a reality or myth. *64th IPS meeting and National Symposium2-4 Dec.* 2.

Nagmani, A., C.Manoharachari, Agrawal D. K. and Chawdhary, P. N. (2002). Monographic contribution on *Trichoderma*. *Associated Publishing Company, New Delhi.*, pp.1-47.

Nakkeeran, S., Krishnamoorthy A.S., Ramamoorthy V. and Renukadevi, P. (2002). Microbial inoculants in plant disease control. *J. Ecobiology.*, **14**: 83-94.

Nasreen, S. (1997). Studies on resistance of fungal pathogens to certain fungicides-I. *Ph.D. Thesis, Marathwada University, Aurangabad.*

Niaz, I., Kazami A. R. and Jilani, G. (1994). Effect of turmeric derivatives on radial growth of different fungi. *Sarhad J. Agric.*, **10**(5): 571-573.

Noble, M., Maggarvie Q. D., Hams A. F. and Leaf, L. L. (1966). Resistance to mercury of *Pyrenophora avenae* in Scottish seed oats. *Pl.Pathol.*, **16**: 23-28.

O' Brien, R. G. and Glass, R. J. (1986). The appearance of dicarboximide resistance in *Botrytis cinerea* in Queensland. *Aust. Pl. Pathol.*, **15**: 24- 25.

O' Neill, T. M., Niv A., Elad Y., and Shtienberg, D. (1996). Biological control of *Botrytis cinerea* on tomato stems wounds with *Trichoderma harzianum*. *Euro J. Pl. Pathol.*, **102**: 635-643.

Ogawa, J. M., W. E.Yates, Manji B.T. and Cowden, R. E. (1972). Ground and air craft application of thiophanate methyl on control of stone fruit brown rot blossom blight. *Phytopath.*, **62**: 781.

Ogawa, J. Manji B.T. and El Behadli (1976). Tolerance of plant pathogens to fungicides and bactericides. *Fungicide and Nematicide Test.*, **31**: 3-8.

Pan, S, and Sen (1980). Relative virulence of parental and fungicide tolerant isolate of *Macrophomina phaseolina* toward soyben. *Indian Phytopath.*, **33**: 642-642.

Pande, S., G. Singh, J. Narayana Rao, M. A. Bakr, P. C. P. Chaurasia, S. Joshi, C. Johansen, S. D. Singh. J. Kumar, M. M. Rahman and Gawda, C. L. L. (2001).

Integrated management of *Botrytis cinerea* mold of chickpea. *Information Bulletin: 61, International crops Research Institute for the Semi Arid Tropics, Patancheru. Andhra Pradesh India*. 9.

Pandey, K. K., Pandey, P. K. and Mishra, K. K. (2006). Bio-efficacy of fungicides against different fungal bioagents for tolerance level and fungistatic behavior. *Indian Phytopath.,* **59**(1): 68-71.

Paola, Díaz Dellavalle, Andrea Cabrera, Diego Alem, Patricia Larrañaga, Fernando Ferreira and Marco Dalla Rizza. (2010). Antifungal activity of medicinal plant extracts against phytopathogenic fungus *Alternaria* spp. *Chilean Journal Agricultural Research.,* **71**(2): 231-239.

Patel, R.B. (1988). Post harvest diseases of orange fruits and their control.*Assoc. Microbiol. (Abstr.) Ind. Ann. Confr*. 153.

Patel, D. S. (2009). Chemical management of fruit spot of pomegranate caused by *Colletotrichum gloeosporioides* Penz. and Sacc. *Indian Phytopath, .,* **61**(2): 252-253.

Patil, Vandana B. and Shivaji Kamble (2011). Effect of passage on the development of carbendazim resistance in *Macrophomina phaseolina* causing charcoal rot of sweet potato. *Internat. J. Acad. Research.,* **3**(1): 130-132.

Patni, C, N, Kolte S.T. and Awasthi, R.P. (2005). Efficacy of totanicals Alternaria blight(Alternaria brassicae) of mustarad. *Indian Phytopath.,* **58**(4): 426-430.

Pattnaik, M.M., Kar M. and Sahu, R.K. (2012). Bioefficacy of some plant extract on growth parameters and control of disease in *Lycopersicum esculntum. Asian J.Plant Sci. and Res.,* **2**(2): 129-142.

Paul, P. K. and Sharma, P. D. (2002). *Azadirachta indica* leaf extract induces resistance in barley against leaf stripe disease. *Physiological and Molecular Plant Pathol.* **61**(1): 3-13.

Paulitz, T. C. and Belager, R. R. (2001). Biological control in green house system. *Ann. Rev. Phytopath.,* **39**: 103-133.

Paulus, A.O. Nelson, J and Besemer, S. (1976). Resistance of *Septoria nodorum* to benzimidazole fungicides. *Pl.Dis. Repotr.,* **60**: 695-697.

Prasad, H. C., Mazumdar, R. and R. Chakraborty (1993): Research on two medicinal plants from Ayurvedic system of medicine *Azadirachta indica* A.Juss (syn. *Melia azadirachta*) and *Melia azedarach* Linn-past, present and future. *Deerghayu International,* pp. 24-28.

Prusky, D. and Gullino, M. L. (2010). Plant Pathology in the 21st Century. *Postharvest Pathology,* Vol. 2, DOI 10.1007/978-1-4020-8930-5_10, © Springer Science + Business Media B.V.

Pusey, P. L. and C. L. Wilson (1984). Post harvest biological control of stone fruit brown rot by *Bacillus subtilis. Plant Dis.,* **68**: 753-756.

Qais, K. Zewain, Bahadur, P. and Sharma, Pratibha (2005). Integrated disease management of stalk rot of cauliflower. *Indian Phytopath.,* **58**(2): 167-173.

Raju, K and Naik, M.K. (2006). Effect of preharvest spray of fungicides and botanicals on storage disease of onion. *Indian Phytopathology*., **59**(2): 133-141.

Reddy, B. M.V., Angers P., Gosselin A. and Arul, J. (1997). Rapid characterization and use of essential oils from *Thymus vulgare* against *Botrytis cinerea* and *Rhizopus stolonifer* in strawberry fruits. *Phytochemistry*., **47**(8): 1515-1520.

Redmond, J. C., Marous J. J. and J. D. MacDonald (1987). Biological control of *Botrytis cinerea* on Roses with epiphytic microorganism. *Plant Dis.*, **71**: 799-802.

Rosslenbroich, H. J. and Stuebler, D. (2000). *Botrytis cinerea* - history of Chemical control and novel fungicides for its management. *Crop Protection.*, **19**: 557-561.

Rosslenbroich, H. J., Bronde W., Kruger B.W., Kuck K.H., Pontzen R. Stenzel K. and Suty, A. (1998). Fenhexamid (KBR 2738) a novel fungicide for control of *Botrytis cinerea* and related pathogens. *The 1998 Brighton Conference Pests and Diseases.*, pp. 327-334.

Rueegg, J., Lauber H. P., Siegfried W., Viret, O. and Hilber, U. (1997). Experience with anilinopyrimidines in Switzerland. *Pestic. Outlook*, **8**(3): 28-33.

Saini, N. K., Singhal M, Srivastava B., Sachdeva K and Singh, C. (2011). Antimicrobial activity of *Tecomaria capensis* leaves extract *International J. Pharmaceutical Sci. Review and Res.*, **7**(1): 120-124.

Salome, Zaldua and Eugenio Sanfuentes (2010). Control of *Botrytis cinerea* in *Eucalyptus globulus* mini cutting using *Clonostachys* and *Trichoderma* strains: *Chilean J. Agri. Research.*, **70**(4): 576-582.

Samane, Mohmmadi and Mohammad Hossien Aminifard (2011). In *vitro* and in *vivo* antifungal activities of three essential oils against grey mould disease in cucumber (*Cucumis sativus*)., **10**(5): 287-293.

Schena, L., Ippolito, A., Zahavi, T., Cohen, L., Nigro, F. and S. Droby (1999): Genetic diversity and biocontrol activity of Aureobasidium pullulans isolates against postharvest rots. *Postharvest Biol. Technol.*, **17**: 189-199.

Seiya, Saito and Shunji Suzuki (2009). Fungicide resistant profile of *B.cinerea* in vineyard. *J. Asev Jpn.*, **20**(1-2): 8-14.

Serey, Ricardo A., Rene Torres and Bernardo Latorre, A. (2007). Pre and post infection activity of new fungicides against *Botrytis cinerea* and other fungi causing decay of table grapes. *Cien Inv. Agr.*, **34**(3): 215-224.

Shabi E. and Yephet Y. Ben (1976). Tolerance of venturia pirina to benzimidazole fungicide. Plant Dis. Reptr., **60**: 451-454.

Sharma, Nidhi and Trivedi, P.C. (2005). Evaluation of different fungal antagonists, plant extracts and oil cakes against *Fusarium oxysporum f. sp. cumini* causing wilt of cumin (*Cuminum cyminum* L.). *J. Indian Bot. Soc.*, **84**: 30-32.

Sharma, Pankaj and Gour, H.N. (2009). Effect of seed dressing fungicide on microflora incidence during storage of pea (*Pisum sativum*). *J. Mycol. Pl. Pathol.*, **39**(1): 169-171.

Sharma, R.N., R.P. Maharshi and Gour, R. B. (2010). Management of post harvest spoilage of kinnow fruits by preharvest spraying of fungicide. *Indian Phytopath.*, **63**(3): 278-281.

Shinde, Saheb (2011). The antifungal activity of five *Terminalia* species checked by paper disc method. *Intern. J. Pharma. Reser. Development.*, **3**(2): 36-40.

Shinde, Saheb and S.S. Wadje (2011). Efficacy of *Terminalia* bark extracts against seed borne pathogens checked by paper disc method. *RJPBCS*, **2**(2): 602-607.

Singh, A.K., Dixit A., Sharma M. L. and Dixit, S.N. (1980). Fungitoxic activity of some essential oils. *Econ. Bot.*, **34**: 186-190.

Singh, J. and Majumdar, V.C. (2001). Efficacy of plant extracts against *Alternaria alternata* the incident of fruit rot of pomegranate (*Punica granatum* L.) *J. Mycol. Pl. Pathol.*, **31**: 346-349.

Singh, K.P. and Kumar, J. (1999): Efficacy of different fungicidal spray schedules in combating apple scab severity in Uttar Pradesh Himalayas. *Indian Phytopath.*, **52** (2) : 142-147.

Singh, Priyanka, Bhavana Srivastava, Ashok Kumar and N.K. Dubey (2008). Fungal contamination of raw materials of some herbal drugs and recommendation of *Cinnamomom comphora* oil as herbal fungitoxicant. *Microbial Ecology.*, **63**(3): 555-560.

Sirirat, Siripornvisal, Wimdpun Rungprom and Sanit Sawatidikarn (2009). Antifungal activity of essential oils derived from some medicinal plants against grey mold (*Botrytis cinerea*). *Asian J. Food Agro Industry* S229-S233.

Soares, D. J. (2008). Grey mould caused by *Botrytis cinerea* on *Oxalis latifolia* from Brazil *Australasian Plant Dis.*, **3**: 85-86.

Sofia, Vio-Michaelis1, Gastón Apablaza-Hidalgo, Miguel Gómez, Raúl Peña-Vera and Gloria Montenegro (2012). Antifungal activity of three Chilean plant extracts on *Botrytis cinerea. Botanical Sciences.*, **90**(2): 179-183.

Solunke, B. S., Kareppa, B. M. and Gangawane, L. V. (2001). Integrated management of Sclerotium rot potato using carbendazim and plant extracts. *Ind. J. Plant Protection.*, **29**(1-2): 142-143.

Soylu, E.M., Soylu S. and S.Kurt (2006). Antimicrobial activities of the essential oils of various plants against tomato late blight disease agent *Phytophthora infestans. Mycopathologia.*, **161**: 119 -128.

Srinivas, C., Niranjana, S. R., Praveen Kumar, L., Chandra Nayaka, S. and Shetty, H. S. (2006). Effect of fungicides and bioagents *Colletotrichum capsici* on seed quality of chilli. *Indian Phytopath.*, **59**(1): 62-67.

Staub, T. (1994). Early experience with phenylamine resistance and lesions for continued successful use. In: Fungicide resistance, Heaney S., Slawson, D.D., Hollomon D.W., Smith M., Russell P.E., Parry D.W. eds. British *Crop Protection Council, Farnham, Surrey*, pp. 131- 138.

Suryawanshi, A.P. Harde, A.L. Gholve U.M. Deshmukh R.W. and Kuldhar, D.P. (2011). Bioefficacy of fungicides, Botanicals and Bioagents against *Alternaria brassicola. Phyto.Pathol* : 155.

Suryawanshi, B.B. (2005). Studies on major diseases of vanilla and their management. M.Sc. (Agri) Thesis submitted to B.S.K.K.V. Dapoli, Maharashtra.

Suryawanshi, N. S. (1998). Studies on antibiotic resistance in *Xanthomonas campestries pv. citri.* causing citrus canker and its management. *Ph.D.Thesis. Dr.BAMU, Aurangabad.*

Suryawanshi, N.S. and Gangawane, L.V. (2005). Antibiotics resistance in bacterial pathogen and its integrated management. *Bioinfolet.*, **2**(2): 229-239.

Suty, A., Pontzen R. and Stenzel, K. (1997). KBR 2738. Mode d'action et sensibilite de *B.cinerea*. ANPP. Cinquieme conference international sur les maladies des plants tours 3, 4, 5 Decembre, pp. 561-568.

Tegos, G., Stermitz F.R., Lomovskaya O. and K. Lewis (2002). Multidrug pump inhibitors remarkable activity of plant antimicrobial. *Antimicrobial Chemotherapy.*, **46**: 3133-3141.

Telmore, K.M., Suryawanshi, N.S., N.F. Shaik and Kamble, S.S. (2013).Integrated management of leaf spot of Betelvine caused by Benomyl resistant *Fusarium semitectum.Bionano frontier.*, **6**(2): 297-299.

Thakur, V.S., G.K Gupta and R. Malhotra (1995). Antisporulant activity of ergosterol biosynthesis inhibiting fungicides in the management of apple scab. *Ind. Phytopath.*, **48**(1): 35-39.

Thind, T. S. (2009).Crop of Disease: Identification and Management L. V. Gangawane, V. Khilare and Jayashri Deshpande). *Saraswati Book Publisher, Aurangabad.*44-46.

Thind, T.S. (2007). Changing cover of fungicides umbrella in crop protection. *Indian Phytopath.***60**: 421-433.

Tiwari, R.K.S. Ojha B.M. and Chandravanshi, S.S. (2004). Efficacy of fungicide in controlling leaf spots (*Cercospora arachidicola*) and (*C.Personata*) and Rust in ground nut. *Mycol. Plant.Pathol.*, **34**(2): 520-521.

Tripathi, H.S. and Rathi, Y.P.S. (2000). Resistance to Botrytis grey mould in chickpea: screening techniques and identification of resistance sources. *J. Mycol. Pl. Pathol.* **30**(2): 231-232.

Tripathi, Pramila and Dubey, N.K. (2004). Exploitation of natural products as an alternative strategy to control postharvest fungal rotting of fruits and vegetables. *Postharvest Biology and Technology.*, **32**: 235-245.

Tripathi, Pramila, Dubey N.K. and Shukla, A.K. (2008). Use of some essential oils as post harvest botanical fungicides in the management of grey mould of grapes caused by *Botrytis cinerea. World J Mirobiol Biotechnol.*, **24**: 39-46.

Tripathi, R. K. and E. Schloesser (1982). The mechanism of resistance of *Botrytis cinerea* to methyl benzimidazole-2-yl-carbamate. Z.Pflanznker. Pflanznker. (Hugessen), **89**: 51-56.

Upadhyyay, M.P.Manudhar K.L. and Shreshtha, R.B. (1980). Antifungal activity of garlic agsinst fungi isolated from human eyes. *J.Gen.Appl. Microbiol.,* **26**: 421-424.

Vagelas, I., A.Papachatzis, H. Kalorizou and Wogiatzi, E. (2009). Biological control of *Botrytis* fruit rot (Grey mould) on strawberry and red pepper fruits by olive oil mill wastewater. *Biotechnol. Eq.,* **23**(4): 1489-1491.

Vargas, J. M. Jr. (1973). A benzimidazole resistant strain of *Erysiphe graminis. Phytopathology.* **63**: 1366-1368.

Vihol, J.B. Patel, K.D. Jaiman R.K. and Patel, N.R. (2009). Efficacy of Plant extracts, biological agents and fungicides against Alternaria blight of cumin. *Mycol. Pl. Pathol.,* **39**(3): 516-519.

Wada, T., Kuzuma S. and Takenaka, M. (1970). Sensitivity of *Fusarium moniliforme* isolates to perfurazoate. *Ann. Phytopathol. Soc. Jpn.,* **56**: 449-456.

Wang, Chunmei, Jie Zhang, Hao Chen, Yongjian fan and Zhiqi Shi (2010). Antifungal activity of eugenol against *Botrytis cinerea. Tropical Plant Pathol.,* **35**(3): 137-143.

Washington, W. S., N. Shanmuganathan and Forbes, C. (1992). Fungicide control of Strawberry fruit rots and the field occurrence of resistance of *Botrytis cinerea* to iprodione, benomyl and dichlorofluanid. *Crop Protection,* **11**(4): 355-360.

Wilson, C. L., Solar J. M., El. Ghaouth A. and Wisniewski, M. E. (1997). Rapid evaluation of plant extracts and essential oils for antifungal activity against *Botrytis cinerea. Plant Dis.,* **81**: 204 -210.

Wilson, Charles L., Michael E. Wisniewski, Charles L. Biles, Randy McLaughlin, Edo Chalutz and Samir Droby (1991). Biological control of post harvest diseases of fruits and vegetables: alternatives to synthetic fungicides. *Crop Protection.,* **10**: 172-177.

Wilson, M. and Lindow, S.E. (1993). Interactions between the biological control agent *Pseudomonas fluorescens* strain A506 and *Erwinia amylovora* in pear blossoms. *Phytopath.,* **83**: 117-123.

Wilson, C.L., M. Wisniewski, S. Droby, and Chalutz, E. (2011). Historical perspective on biological control of postharvest diseases-past, present, and future. *Acta Hort.* **90**: 23-28.

Wisniewski, M.E. and Wilson, C. L. (1992). Biological control of post harvest diseases of fruits and vegetables: recent advances. *Hort. Science.,* **27**: 94-98.

Yamada, S., T. Kuramato and Tanaka, H. (1972). Studies on control of citrus fruit decay in storage VII. Effect of the fungicide grow spray to Sastuma mandarin (*Citrus unshiu marcovitch*) *Fruit Hortic. Res. Sta. Ser. Bull.* (*Okitsu*)., **12**: 207-228.

Yang, Cong-Jun, Xin- Gang Zhang, Guan- Ying Shi, Hao- Yu-Zhao, Long Chen, Ke Tao and Tai Ping Hon (2011). Isolation and Identification of endophytic bacterium W4 against tomato *Botrytis cinerea* and antagonistic activity stability. *African J.Microbiol Res.,* **5**(2): 131-136.

Yourman, L.F., S.N. Jeffers and Dean, R.A. (2000). Genetic analysis of isolates of *B.cinerea* sensitive and resistant to benzimidazole to dicarboximide. *Ecol. Population Bio.*, **90**(8): 851-859.

Yusuf, Yanar, Izzet Kadioglu, Ayhan Gokce, Ibrahim Demirtas, Nezhum Goren, Halit Cam and Mark Whalan (2011). *In vitro* antifungal activities of 26 plant extracts of mycelial growth of *Phytophthora infestans* (Mout.) de bery. *African J. Biotech.*, **10**(14): 2625-2629.

Zaker, M. and Mosallanejad, M. (2010). Antifungal activity of some plant extracts on*Alternaria alternata*, the causal agent of *Alternaria* leaf spot of potato. *Pakistan Journal of Biological Sciences.*, **13**: 1023-1029.

2015, Major Constrains and Verdict of Crop Productivity Pages 51–63
Editor: Dr. U.N. Bhale
Published by: ASTRAL INTERNATIONAL (P) LIMITED, NEW DELHI

Chapter 3

Sesame: Identifying an Orphan Crop in International Arena through a Sustainable Compromise between Quality and Productivity

Yogranjan[1]*, Gyanesh K. Satpute[2], Rakesh S. Marabi[3] and Sudhakar P. Mishra[4]

[1]Plant Biotechnology,
JNKVV, College of Agriculture, Tikamgarh – 472 001, M.P., India
[2]Plant Breeding and Genetics,
Directorate of Soybean Research (ICAR), Indore – 452 001, India
[3]Entomology, JNKVV, College of Agriculture, Jabalpur – 485 004, M.P., India
[4]Agricultural Biochemistry, Department of Crop Sciences,
MGCGVV, Chitrakoot, Satna – 485 780, M.P., India

ABSTRACT

Sesame (Sesamum indicum L.) is one of the world's oldest spice and oilseed crop grown mainly for its seeds that contain approximately 50 per cent oil and 25 per cent protein. The presence of some antioxidants (sesamum, sesamolin and sesamol) makes the oil to be one of the most stable vegetable oils in the world. However, despite its nutritive as well as pharmaceutical significance, it is still at an early stage in its breeding history. Introduction, selection, hybridization and characterization of existing genetic diversity of sesame for crop improvement were included in the breeding efforts in recent decades only. Periodically, application of innovative breeding methods including mutational techniques

* *Corresponding Author:* E-mail: yogranjan@gmail.com

has proved to reduce our dependence on existence of genetic variability within a species and overcome the limitations of conventional breeding. More recently, biotechnological techniques have been introduced to sesame breeding programs. Several marker assisted selection (MAS) techniques such as isozymes, random amplified polymorphic DNA (RAPD) and inter-simple sequence repeats (ISSR), etc. have also been used in sesame breeding to study genetic variability of sesame to increase selection efficiency. Protocols for sesame in vitro culturing and genetic transformation are also being optimized by using appropriate concentration of hormones and nutrients. This chapter looks at a broaden aspect of sesame production, its industrial and commercial significance, as well as its potential demand in the food security scenario. The benchmark technological interventions and further research thrust for promoting the crop in an international area are also highlighted.

Introduction

Sesame (*Sesamum indicum* L.) a species of the family Pedaliaceae, cherishes the longest history of cultivation. Sesame oil, otherwise also referred to as gingelly oil, is one of the major sources of edible oil in India and is culturally associated from the Vedic period. The Sanskrit word for oil, taila is derived from the Sanskrit word for sesame tila. Sesame seeds and its oil are being utilized as important food ingradients since about 6000 years. The Thebes Medicinal Papyrus (1552 B.C.), an old text found in Egypt, describes the medicinal effect of sesame seed as a source of energy. In traditional Indian medicine, ayurveda, sesame oil has been used as the basal oil for human body massage since 1100 B.C. (Weiss, 2000; Joshi, 1961). Japanese traditionally consider sesame as good health supplement (Namiki, 1995). Recently, scientific researchers have investigated many physiological and pharmaceutical aspects of sesame seeds. Studies on antioxidants, anti-aging, the synergistic effect with tocopherols, serum lipid lowering, blood pressure lowering, and other functions have also been reported (Namiki, 2007). Despite such high value being placed on sesame seed, there have been very limited scientific studies validating these claims.

Origin, Geographic Distribution and Global Production

The exact origin of sesame is a contentious issue; however, mounting evidence has shown that sesame originated in India (Bedigian, 2004) or Africa (Kobayashi, 1986). Genetic, morphological, and phytochemical evidence support the hypothesis that domesticated sesame originated on the Indian subcontinent (Bedigian *et al.*, 1986). On the other hand, the assumption of Africa being the primary centre gets a valid support through the presence of a wide genetic diversity of sesame in African continent (Hiltebrandt, 1932).The sesame crop is now cultivated throughout the world in about 70 countries of Africa, Asia and Australia (Figure 3.1). It is grown worldwide over an area of 79 million hectares producing 63,000 tones seeds (FAOSTAT, 2012). India, China Myanmar, and Sudan are the leading sesame-producing countries of the world. The crop is generally adapted to tropical regions of world, where it is mainly grown for edible seeds and oil (Weiss, 2000). Despite its ideal adaptation to dry climates, sesame can also be grown in humid, tropical and subtropical regions.

Current world production is estimated at about 610 thousand metric tons annually, placing sesame behind soybean, peanut, cottonseed, sunflower, linseed and rapeseed, in the quantity of world oilseed production (FAOSTAT, 2012). India

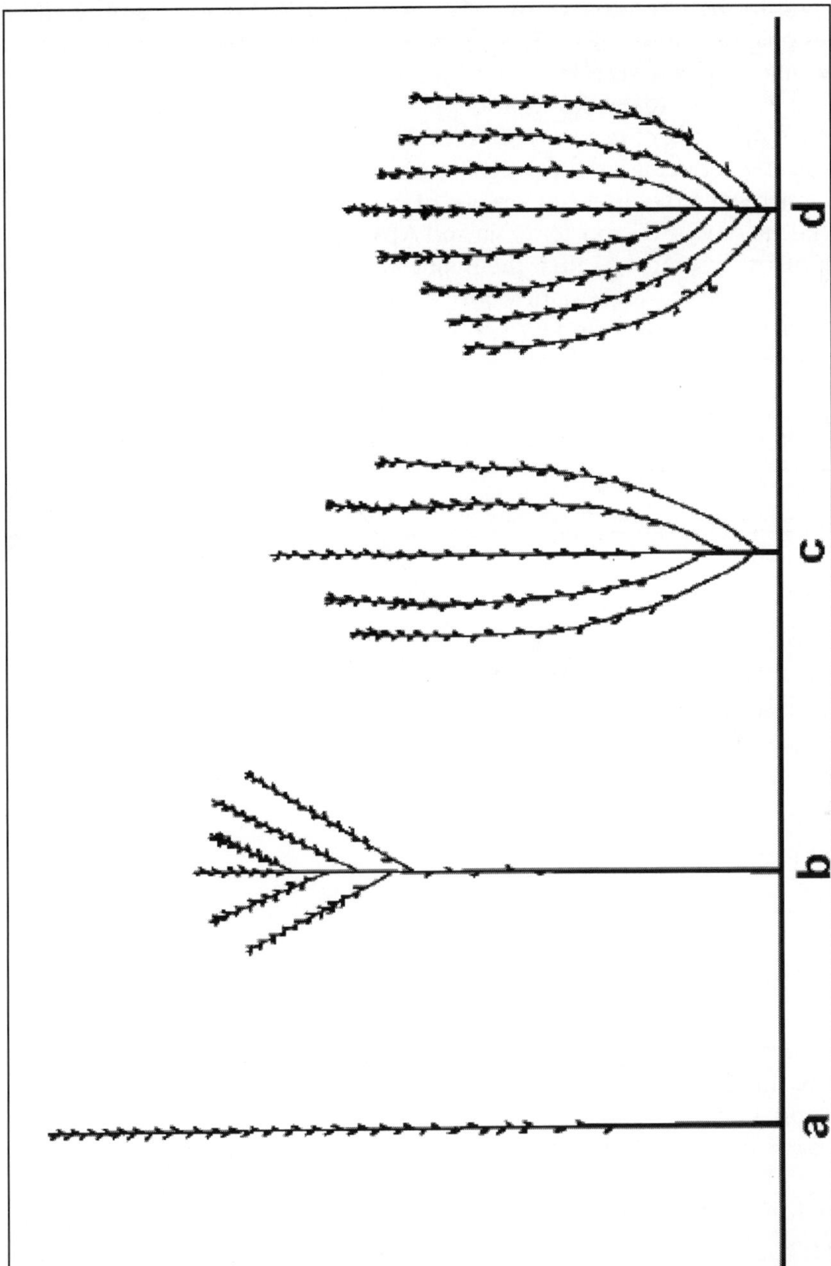

Figure 3.1: Schematic Diagram of Sesame Growth Habits.

ranks second in the world only after Myanmar with 18.20 Lakh ha area and 6.10 Lakh tones production with Rs. 1510.4 crore export (FAOSTAT 2012). The Average yield of sesame on global scale is 5.11 Qt/hectare, while in India, it is 3.30 Qt/hectare (FAOSTAT, 2012). Acreage, yield and production on a world basis has largely remained the same over the past five years. The relative standing of the world's 10 leading sesame-producing countries in 2012 is summarized in Table 3.1. India devotes the greatest acreage to sesame, but has one of the lower records for yield per hectare. Sudan is the major world exporter. Asia and Africa produce nearly 90 per cent of the world supply of sesame. In India, the major sesame producing states are Gujarat, West Bengal, Karnataka, Rajasthan, Madhya Pradesh, Tamil Nadu, Andhra Pradesh, Maharashtra, Uttar Pradesh and Orissa (DAC, 2007). Most of the seed is consumed in the countries where it is produced; less than 5 per cent of world production enters export trade.

Table 3.1: The World's 10 Leading Sesame-Producing Countries (FAOSTAT 2012)

Country	*Total Production (Lakh Tonnes)*	*Acreage (Lakh Ha)*	*Productivity (Qt/Ha)*
Myanmar	6.20	15.70	3.9
India	6.10	18.20	3.3
China	6.00	4.90	12.2
United Rep. of Tanzania	4.56	6.51	6.9
Uganda	1.88	3.15	5.9
Sudan	1.87	8.20	2.2
Ethiopia	1.81	2.39	7.5
Nigeria	1.58	3.30	4.7
Mozambique	1.17	2.20	5.3
Burkina Faso	1.00	1.65	6.0

Taxonomy, Cytogenetics and Growth Habit

The genus *Sesamum*, is a member of order Lamiales, of family Pedaliaceae. The genus *Sesamum* consists of many species and the most cultivated is *Sesamum indicum* L. There are three cytogenetic groups of which 2n = 26 consist of the cultivated *S. indicum* along with *S. alatum, S. capense, S. schenckii, S.malabaricum*; 2n = 32 consist of *S. prostratum, S. laciniatum, S. angolense, S.angustifolium;* while *S. radiatum, S. occidentale, S. schinzianum* belong to 2n =64. Mainly due to the difference in chromosomal numbers across the three cytotaxonomic groups, there is limited cross compatibility among the species.

Sesame is a short-day plant and is normally self-pollinated, although cross-pollination ranging from 5 to over 50 per cent occurs (Van Rheenen, 1980 and 1980a; Pathirana, 1994). It is an erect to semi erect herbaceous annual plant that has both indeterminate and determinate growth habit with the plants reaching heights of up to two meters (Figures 3.1a-d). Most varieties show an indeterminate growth habit, which is shown as a continuous production of new leaves, flowers and capsules as long as the environment remains suitable for growth (Carlsson *et al.*, 2008). The leaves on round or squaral stem are pointed with entire to serrate margin. Flowers range in size containing small sized tubular calyx and five-lobed corollas and color, *e.g.*, white, violet, red or maroon. Corolla is campanulate having lower corolla lobe longer than the upper one with one sterile and four functional epipetalous stamens. The growth period range from 70 to 150 days depending on the variety and the conditions of cultivation (Ashri, 1998).

Agro-ecology

Sesame can grow well in many ecological regions of tropical and subtropical climate, and also grows in most kind of soils. However, it thrives best on well-drained soil with a moderate fertility and a pH between 5.5 and 7.0. It can adapt and produce seed well under fairly high temperatures (Ashri, 1998).The crop is highly drought tolerant, however, moisture levels before planting and flowerings have great impact on the seed yield. Sesame requires a minimum of 300–400 mm of rainfall per season, but is sensitive to wet condition and has a very low salt tolerance (Carlsson *et al.*, 2008).

Nutritional Composition

Sesame seeds are rich in oil and protein with high dietary energy value of 6,355 kcal kg^{-1}. Chemical composition of seed shows that it contains 57–63 per cent oil, 23–25 per cent protein, 13.5 per cent carbohydrate and 5 per cent ash (Tunde-Akintunde and Akintunde, 2004; Elleuch *et al.*, 2007). It is rich source of various nutritional elements including iron, magnesium, manganese, copper and calcium, and important vitamins B_1 (thiamine) and E (tocopherol). Due to its flavor and stability as well as high-quality cooking value, sesame fat is of great significance in the food industry. The presence of sesamin and sesaminol lignans in its nonglycerol fraction contributed to oxidative stability and antioxidative activity of sesame oil (Wu 2007). These antioxidative agents terminate the process of oxidative damage in cells by capturing oxidative radicals. Antioxidants are also used as dietary supplements to cure cancer and heart diseases. The most abundant fatty acids present in sesame oil include: oleic acid (43 per cent), linoleic acid (35 per cent), palmitic acid (11 per cent) and stearic acid (7 per cent) contributing toward 96 per cent of total fatty acids (Elleuch *et al.*, 2007).

Oil Content and Fatty Acid Composition

Among oil crops, sesame is one of the highest in oil content. Generally, the oil content in sesame ranges from 34 to 63 per cent (Yermanos *et al.*, 1972; Ashri, 1998; Baydar *et al.*, 1999; Uzun *et al.*, 2007; Were *et al.*, 2006). Genetic and environmental factors greatly influence the oil content and fatty acid compositions in sesame (Carlsson *et al.*, 2008). Late maturing cultivars are reported to have higher oil content than early cultivars (Yermanos *et al.*, 1972) and the indeterminate cultivars accumulated more oil than determinate ones (Uzun *et al.*, 2002). Sesame contains high levels of antioxidants such as sesamol, sesamin, sesamolin and sesaminol and therefore also termed as the queen of the vegetable oils. (ASGA. 2011). The composition of sesame oil consist of mainly four fatty acids (palmitic–C16:0, stearic–C18:0, oleic–

C18:1 and linoleic–C18:2), while other fatty acids appear in very small amounts (Ashri, 1998). Even though different fatty acid compositions of sesame oil have been reported the major fatty acids are oleic and linoleic acids (Kamaleldin *et al.*, 1992). Yermanos *et al.* (1972) reported oleic acid level ranging from 32.7 per cent to 58.2 per cent and linoleic acid from 27.3 per cent to 59.0 per cent. Were *et al.* (2006), reported variation in the quantities of palmitic, stearic, oleic and linoleic acids, with palmitic and stearic acids ranging from 7.2 to 9.6 per cent and 3.7 to 5.6 per cent, respectively, while high levels of oleic acid and linoleic acid ranged from 31.6 to 41.9 per cent and 42.9 to 53.9 per cent. Linolenic acid was found but in very small amount (0.5 per cent).

Economic Importance of Sesame

Although sesame seeds are used as an ingredient in many different food supplies, a major part of the sesame seed production is processed into oil and meal (Morris, 2002). Sesame oil is, as mentioned before, an excellent vegetable oil because of its high contents of antioxidants such as sesamin, sesamol and sesamolin and its fatty acid composition (Suja *et al.*, 2004). The antioxidants make the oil very stable and it has therefore a long shelf life (Chung *et al.*, 2004; Suja *et al.*, 2004). The high levels of unsaturated (UFA) and polyunsaturated fatty acids (PUFAs) increase the quality of the oil for human consumption (Nupur *et al.*, 2010). Moreover, high level of PUFAs in sesame oil is claimed to reduce blood cholesterol, high blood pressure and play an important role in preventing atherosclerosis, heart diseases and cancers (Hibasami *et al.*, 2000; Miyahara *et al.*, 2001; Hemlatha and Ghafoorunissa, 2004;). People in Vietnam use sesame oil for cooking fish, meat, and frying vegetable. Sesame seed flour has a high protein content, with high levels of the essential amino acids *viz;* methionine and tryptophan, and has three times more calcium than milk (Morris, 2002). The meal that remains after oil extraction from sesame seeds is an excellent feed for poultry and livestock. A potential problem, though, with sesame when used for human consumption is that it contains one of the top food allergens. Its allergenicity emanates from protein sources, such as the 14 kDa 2S albumin precursor, which belongs to one of four protein families known to be allergenic and several oleosins (15 and 17 kDa), also known to cause hypersensitivity (Wolff *et al.*, 2003; Leduc *et al.*, 2006). Therefore, food products that contain sesame as an ingredient are now required to be labeled as potential allergenic in Europe and Canada (Carlsson *et al.*, 2008).

Non-Food Application

Sesame oil has found use in several non-food applications such as in paints, cosmetics, solvents and soap (Bedigian, 2004). In Africa, sesame flowers have been used to prepare perfumes (Morris, 2002). The antioxidant sesamin is used as a synergist for pyrethrum or rotenone insecticides and increases the toxicity of insecticides when spayed against flies (Haller *et al.*, 1942). Hasan *et al.* (2000) reported that chlorosesamone extracted from roots of sesame has antifungal properties and was used as a fungicide. Sesame oil has been investigated as a source for biodiesel and found to give a product with fuel properties in parity with mineral diesel but with superior environmental performance (Ahmad *et al.*, 2010).

Production Constraints in Sesame

Several biotic and abiotic stresses represent the important factors causing low production of sesame. Diseases that are known to be important in sesame are *Fusarium oxysporum* f.sp. sesami, *Phytophthora parasitica* var. sesami, *Macrophomina phaseolina* ssp. Sesamica (Charcoal rot), *Peilicularia rolfsii* (Southern blight),*Cercospora sesami, Alternaria sesami, Pseudomonas syringae* pv. *sesami, Xanthomonas campestris* pv. *sesami,* phyllody (Mycoplasma Like Organism = MLO), powdery mildew (Oidium sp. and others) and Sesame Mosaic Virus (Tripathi *et al.*, 1998; Ojiambo *et al.*, 1999; Langham and Wiemers, 2002; John *et al.*, 2005; Choudhary *et al.*, 2006). Green peach aphid (*Myzus persica*), white fly (*Bemisia argentifolium, Bemisia tabacci*), cabbage looper (*Pieris rapae*) and army worm (*Cupis unipuncta*), *Antigastra catalaunalis* (webworm, leaf webber, capsule borer), sphingid moth (*Acherontia styx*), and gall-midge are some major examples of pests that attack sesame (Ssekabembe, 2007).

The majority of the world's sesame is shattering, besides this, the prominant abiotic constraints in sesame production include non-synchronous maturity, lack of wider adapting cultivars, poor stand establishment, lack of fertilizer responses, profuse branching, and low harvest index (Ashri, 1998; Zerihun, 2013).

Prospects for Improvement of International Consumers' Preferred Traits in Sesame

The high quality premium edible as well as industrially important oil of sesame has fetched world-wide interest in recent years. Presence of lignans *viz*; sesamin (0.4-1.1 per cent), sesamolin (0.3-0.6 per cent) and traces of sesamol and tocopherol in the oil are accountable for placing the sesame at uppermost place of global demand. On account of these components, natural products of sesame have been in practice since ancient times as health food additives for preventing diseases and promoting well beings. However, despite the multilateral utilization and implications of sesame, the indigestible ingredient *viz*; Oxalic acid and Phytic acid present in seed coat of its seed make it unfit for direct consumption. These phytoconstituents, indeed have captured, in the current era also, the interest of consumers and scientists of the medical and pharmaceutical industries due to their antitumor, antimutagenic as well as anticarcinogenic properties (Shahidi *et al.*, 2006). Generally sesame varieties with high oil content of around 50 per cent in their seed along with higher lignans being there and less than 2.0 per cent FFA in oil are considered of good quality to fetch awesome prices and demand in international market.

Technological Interventions to Meet Global Command with Respect to Quality and Productivity in Sesame

Despite sesame, being an ancient oil crop, it is still at an early stage in its breeding history. More focused breeding efforts were undertaken only in recent decades, and even these only in very few research stations. Characterization of available or existing genetic diversity of crops played a significant role in bringing out a variety with desirable market oriented physio-chemical traits. Information on genetic diversity is noteworthy when working to improve qualitative profile of the crop varieties. Available genetic diversity is either directly used for evaluation leading to selection or desired traits are combined into a single plant *via* hybridization and backcrossing. Sesame germplasm evaluation for a variety with high oil content and least/zero anti-nutritional factors are based on genetic heritability estimates. Mutational techniques are routinely employed for broadening genetic diversity of sesame breeding material. Application of innovative breeding methods helps to reduce our dependence on existence of genetic variability within a species and overcome the limitations of conventional breeding.

Molecular Technologies

Various marker assisted selection (MAS) techniques such as isozymes, random amplified polymorphic DNA (RAPD), AFLP, inter simple sequence repeats (ISSR) and simple sequence repeats (SSR) etc. are also used in sesame breeding to study genetic variability of sesame to increase selection efficiency (Laurentin and Karlovsky, 2007; Ali *et al.*, 2007). The possibility of converting putative variety specific RAPD and ISSR markers to co-dominant Sequence Characterized Am-plified Region/ Sequence Tagged Site (SCAR/STS) markers has been explored to develop robust variety specific markers (Kumar and Sharma, 2011).

Identification of novel genes involved in the biosynthesis of sesame-specific flavor or lignans, and a better understanding of the metabol-ic pathways from photosynthates towards oil which can be stored and used, are desirable as aids to improve the quality and quantity of oil in sesame cultivars. Expressed Sequence Tags (ESTs) generated by large-scale single-pass cDNA se-quencing have proven valuable for the identification of novel genes in specific metabolic pathways. In order to elucidate the metabolic pathways for lignans in developing sesame seeds and to identify genes involved in the accumulation of storage products and in the biosynthesis of antioxidant lig-nans, Suh *et al.* (2003) obtained 3,328 ESTs from a cDNA library of 5-25 days old immature sesame seeds. ESTs were clustered and analyzed by the BLASTX or FASTAX pro-gram against the GenBank NR and *Arabidopsis* proteome databases. They carried out a comparative analysis between developing sesame and *Arabidopsis* seed ESTs for gene ex-pression profiles during development of green and non-green seeds. Analyses of these two seed EST sets helped to identify similar and different gene expression profiles during seed development, and to identify a large number of sesame seed-specific genes. Seed-specific expression of several can-didate genes was confirmed by northern blot analysis. Suh *et al.* (2003) identified EST candidates for genes possibly involved in biosynthesis of sesame lignans, sesamin and sesa-molin, and suggested a possible metabolic pathway for the generation of cofactors required for synthesis of storage lipid in non-green oilseeds. Molecular characterization in congruence with methodical screening of available germplasm would certainly be helpful in the identification and/or development of varieties with low or zero anti-nutritional factors.

In vitro Techniques and Genetic Transformation

Studies on sesame *in vitro* culture are limited, but some initial success has been reported, (van Zanten, 2001). Some protocols have been developed to produce somatic embryos from hypocotyl derived calli, plants have been regenerated following multiple shoot induction in seed cultures, and callus has been induced from anthers at the uni-nucleate stage of microspores. Attempts have been made to overcome disease problems through interspecific hybridization between the cultivated species and the disease resistant wild relatives *Sesamum alatum* and *S. radiatum*, where by using embryo rescue, a few progenies were obtained (van Zanten, 2001). A major barrier to successful utilization of *in vitro* techniques for enhancing sesame breeding research is overproduction of secondary metabolites (Baskaran and Jayabalan, 2006).

There also exists considerable potential for sesame crop transformation for desired genes from other sources through genetic trans-formation techniques. Sesame has been shown to be susceptible to *Agro-bacterium tumefaciens*, which opens up the avenues for *Agro-bacterium* mediated genenic transformation in the crop (Taskin *et al.*, 1999). There are reports (Ogasawara *et al.*, 1993; Jin *et al.*, 2005) showing successful establishment of hairy root cul-tures by *Agrobacterium rhizogenes* on sesame. Chun *et al.* (2009) established hairy root cultures, and a peroxidase gene was cloned from the hairy roots. Although there are several studies ongoing to have successful transgenic crops, the technologies used for transformation have been relatively crude (Tiwari, 2011). There is a need to improve the efficiency of transformation, to limit the presence of unnecessary genes in the products, to direct insertion to specific sites, and to give more control over when, where, and how much expression of the transgene occurs.

Conclusion, Opportunity and Future Prospects

Sesame shows great potential in the global market with its important contribution of edible-cum- therapeutic oil as well as its valuable usage in industries. In recent years, physio-biochemical studies with sesame oil and its constituents, especially lignan and fatty acids have been promising. Owing to its lignin antioxidants, enriched level of poly unsaturated fatty acids (PUFA), presence of essential amino acid rich proteins and vital minerals, sesame oil has put forward the best nutraceutical option. The qualitative physio-chemical features of sesame oil lead to form versatile products that find applications in livestock nutrition, and technical industries.

The results of the research and development in the recent past years on sesame have been encouraging. However, when viewed against the objective of meeting international preference in terms of quality and productivity, we believe that a more focused research is still needed to optimize the production and processing through the development of improved high yielding sesame varieties with tolerance to biotic and abiotic stresses. From the health and consumption perspectives, technologies for reduction of anti-nutritional factors in the oil in one hand, while enhancing the level of PUFA and essential amino acids on the other need to be developed in the international research system. In addition, there is an earnest need to continue further research mediated through classical as well as molecular breeding for the improvement of consumer preferred traits. An adequate harvesting techniques for both shattering and non-shattering varieties; breeding of cultivars for good characteristics such as high oil and protein composition, neutral day length response and thermo sensitivity, high stable yield and good response to fertilizers, identification and evaluation of chemicals used for insect pest and disease control for toxicity to the crop and products consumed by humans and livestock would certainly be the core of future research. It is expected that in the future such activities will broaden gradually the share of global sesame production among other oilseeds, thus justifying the significance of an orphan crop like sesame in alleviating human food security in the world.

References

Ahmad, M., Khan, M.A., Zafar, M. and Sultana, S. (2010). Environment friendly Renewable Energy from Sesame Biodiesel. *Energy Sources Part a-Recovery Utilization and Environmental Effects*, **32**(2): 189-196.

Ali, G. M., Yasumoto, S., and Seki-Katsuta, M. (2007). Assessment of genetic diversity in sesame (*Sesamum indicum* L.) detected by Amplified Fragment Length Polymorphism markers. *Electronic Journal of Biotechnology*, **10**(1): 12-23.

ASGA Sesame Markets. [online] (2011) Available from: http: // www.sesamegrowers.org/usesofsesame.htm. [Accessed 27/08/2013].

Ashri A. (2010). Sesame breeding.In: J Janick, ed, Plant *Breeding Reviews*, Volume 16. John Wiley and Sons Inc.

Ashri, A. (1998). Sesame breeding. *Plant breeding reviews*, **16**: 179-228.

Baskaran, P., and Jayabalan, N. (2006). *In vitro* mass propagation and diverse callus orientation on *Sesamum indicum* L.-an important oil plant. *J Agric Tech*, **2**: 259-269.

Baydar, H., Marquard, R., Turgut, I. (1999). Pure line selection for improved yield, oil content and different fatty acid composition of sesame, *Sesamum indicum*. *Plant Breeding*, **118**: 462–464.

Bedigian D (2003) Sesame in Africa: origin and dispersals. In: Neumann K, Butler A, Kahlheber S (eds) *Food, Fuel and Fields - Progress in African Archaeobotany, Africa Praehistorica.* Heinrich-Barth-Institute, Cologne, pp. 17–36.

Bedigian, D. (2004). History and lore of sesame in Southwest Asia. *Economic Botany*, **58(3)**: 329-353.

Bedigian, D., and Harlan, J. R. (1986). Evidence for cultivation of sesame in the ancient world. *Economic Botany*, **40**(2): 137-154.

Carlsson, A. S., Chanana, N.P., Gudu, S., Suh, M.C., and Were, B.A.I. (2008) Sesame. *In*: Kole C, Hall TC (ed) Compendium of transgenic crop plant - Transgenic oilseed crops. Wiley Blackwell, Texas-USA. 2. pp. 227-246.

Choudhary, C.S., Prasad, S.M. and Kudada, N. (2006). Evaluation of sesame genotypes for resistance against *Corynespora* blight. *Journal of Research, Birsa Agricultural University*, **18**(1): 149-151.

Chun, J.A., Lee, J.W., Yi, Y.B., Park, G.Y. and Chung, C.H. (2009). Induction of hairy roots and characterization of peroxidase expression as a potential root growth marker in sesame. *Preparative Biochemistry and Biotechnology*, **39**: 345-359.

Chung, J., Lee, J. and Choe, E. (2004). Oxidative stability of soybean and sesame oil mixture during frying of flour dough. *J. Food Sci.*, **69**(7): 574-578.

Directorate of Economics and Statistics, Department of Agriculture and Cooperation (2007) http: //www.dacnet.nic.in/eands/APY_96_To_06.htm

Elleuch M, Besbes S, Roiseux O, Blecker C, Attia H (2007) Quality characteristics of sesame seeds and by-products. *Food Chemistry,* **103**: 641–650.

FAOSTAT, Food and agriculture organization of the united nations. [online] (2012) Available from: http: //faostat.fao.org/site/567/default.aspx#ancor. [Accessed 27/04/2014].

Haller, H.L., McGovran, E.R., Goodhue, L.D. and Sullivan, W.N. (1942). The synergistic action of sesamin with pyrethrum insecticides. *The Journal of Organic Chemistry,* **07**(2): 183-184.

Hasan, A., Begum, S., Furumoto, T. and Fukui, H. (2000). A new chlorinated red naphthoquinone from roots of *Sesamum indicum*. *Bioscience Biotechnology and Biochemistry,* **64**(4): 873-874.

Hemalatha, S.and Ghafoorunissa (2004). Lignans and tocopherols in Indian sesame cultivars. *J Am Oil Chem Soc.,* **81:** 467-470.

Hibasami, H., Fujikawa, T., Takeda, H., Nishibe, S., Satoh, T., Fujisawa, T. and Nakashima, K. (2000). Induction of apoptosis by *Acanthopanax senticosus* HARMS and its component, sesamin in human stomach cancer KATO III cells. *Oncol. Rep.,* **7**(6): 1213-1216.

Hiltebrandt V.M. (1932). Sesame (*Sesamum indicum*) in (Russian) Bulletin of applied botany, *Genetics, and Plant Breeding*, **9** (2): 1-114.

Jin, U.H., Chun, J.A., Han, M.O., Lee, J.W., Yi, Y.B., Lee, S.W. and Chung, C.H. (2005). Sesame hairy root cul-tures for extra-cellular production of a recombinant fungal phytase. *Progress in Biochemistry and Biophysics,* **40**: 3754-3762.

John, P., Tripathi, N.N. and Naveen, K. (2005). Evaluation of sesame germplasm/ cultivars for resistance against charcoal rot. *Research on Crops,* **6**(1): 152-153.

Joshi, A. B. (1961). *Sesamum. A Monograph*. Indian Central Oil Seeds Committee, Hyderabad, India.

Kamaleldin, A., Yousif, G., Iskander, G.M. and Appelqvist, L.A. (1992). Seed lipids of *Sesamum indicum* L. and related wild-species in Sudan.1. Fatty-acids and Triacylglycerols. *Fett Wissenschaft Technologie-Fat Science Technology* **94**(7): 254-259.

Kobayashi T (1986) Goma no kita michi. (The path of sesame). Iwanami Shoten, Japan (in Japanese)

Kumar, V., and Sharma, S. N. (2011). Comparative potential of phenotypic, ISSR and SSR markers for characterization of sesame (*Sesamum indicum* L.) varieties from India. *Journal of Crop Science and Biotechnology,* **14**(3): 163-171.

Langham, D.R. and Wiemers, T. (2002). Progress in mechanizing sesame in the US through breeding. In: Janick, J., *et al.* (Eds.) *Trends in new crops and new uses*. pp. 157-173 ASHS Press, Alexandria, VA.

Laurentin, H., and Karlovsky, P. (2007). AFLP fingerprinting of sesame (*Sesamum indicum* L.) cultivars: identification, genetic relationship and comparison of AFLP informativeness parameters. *Genetic resources and crop evolution*, **54**(7): 1437-1446.

Leduc, V., Moneret-Vautrin, D.A., Tzen, J.T.C., Morisset, M., Guerin, L. and Kanny, G. (2006). Identification of oleosins as major allergens in sesame seed allergic patients. *Allergy*, **61**(3): 349-356.

Miyahara, Y., Hibasami, H., Katsuzaki, H., Imai, K. and Komiya, T. (2001). Sesamolin from sesame seed inhibits proliferation by inducing apoptosis in human lymphoid leukemia Molt 4B cells. *Int. J. Mol. Med.*, **7**(4): 369-371.

Morris, J. B. (2002). Food, industrial, nutraceutical, and pharmaceutical uses of sesame genetic resources. *Trends in new crops and new uses. ASHS press, Alexandria VA*, 153-156.

Namiki M (1995) The chemistry and physiological functions of sesame. *Food Rev Int.*, **11**: 281–329

Namiki, M. (2007). Nutraceutical functions of sesame: a review. *Critical reviews in food science and nutrition*, **47**(7): 651-673.

Nupur, M., Bhat, K.V. and Srivastava, P.S. (2010). Variation in fatty acid composition in indian germplasm of sesame. *Journal of the American Oil Chemists' Society*, **87** (11): 1263-1269.

Ogasawara, T., Chiba, K. and Tada, M. (1993). Production and high yield of napthoquinone by a hairy root culture of *Sesamum indicum*. *Photochemistry*, **33**: 1095-1098.

Ojiambo, P.S., Ayiecho, P.O. and Nyabundi, J.O. (1999). Effect of plant age on sesame infection by *Alternaria* leaf spot. *African Crop Science Journal*, **7**(1): 91-96.

Pathirana, R. (1994). Natural CrossPollination in Sesame (*Sesamum indicum* L.). *Plant breeding*, **112**(2): 167-170.

Purseglove, J.W. (1969). *Tropical crops dicotyledons*. Volume 1, Longman, London, pp. 430-435.

Shahidi, F., Liyana-Pathirana, C. M., and Wall, D. S. (2006). Antioxidant activity of white and black sesame seeds and their hull fractions. *Food Chemistry*, **99**(3): 478-483.

Ssekabembe, C. K. (2007). Comparison of research on sesame (*Sesamum indicum*) and nakati (*Solanum aethiopicum*) at Makerere University. In: *African Crop Science Conference Proceedings*, **8**: 2063-2069.

Suh, M.C., Kim, M.J., Hur, C.G., Bae, J.M., Park, Y.I., Chung, C.H., Kang, C.W. and Ohlrogge, J.B. 2003. Comparative analysis of expressed sequence tags from *Sesamum indicum* and *Arabidopsis thaliana* developing seeds. *Plant Molecular Biology*, **52**: 1107-1123.

Suja, K.P., Abraham, J.T., Thamizh, S.N., Jayalekshmy, A. and Arumughan, C. (2004). Antioxidant efficacy of sesame cake extract in vegetable oil protection. *Food Chem.* **84**(3): 393-400.

Taskin, K.M. and Turgut, K. (1997). *In vitro* regeneration of sesame (*Sesamum indicum* L.). *Turkish Journal of Botany*, **21**: 15-18.

Tiwari, S., Sunil Kumar, Iti Gontia, (2011). Biotechnological approaches for sesame (*Sesamum indicum* L.) and niger (*Guizotia abyssinica* Lf Cass.)." *Asia-Pac J Mol Biol Biotechnol*, **19**: 2-9.

Tripathi, S.K., Tamrakar, P. and Chatterji, D. (1998). Evaluation of sesamum cultivars for resistance to *Cercospora* leaf spot. *Crop Research (Hisar)*, **15**(2-3): 265-269.

Tunde-Akintunde, T.Y., Akintunde, B.O. (2004). Some physical properties of sesame seed. *Biosystems Engineering*, **88** (1): 127–129.

Uzun B., S. Ulger, M. I. Cagirgan (2002). Comparison of Determinate and Indeterminate Types of Sesame for Oil Content and Fatty Acid Composition, *Turk J Agric For*, **26**: 269-274.

Uzun, B., Arslan, C., Karhan, M., Toker, C. (2007). Fat and fatty acids of white lupin (*Lupinus albus* L.) in comparison to sesame (*Sesamum indicum* L.). *Food Chemistry*, **102**: 45–49.

Van Rheenen, H. A. (1980). Aspects of natural cross-fertilization in sesame (*Sesamum indicum* L.). *Tropical Agriculture, Trinidad and Tobago*, **57**(1) : 53-59.

Van Rheenen, H.A. (1980 a). Relation between capsule age, capsule length and viability of seed in sesame (*S. indicum*) consultation on sesame improvement. Rome, Italy, Dec., 8-12, 1980.

Van Zanten, L. (2001). Sesame improvement by induced mutations: Results of the co-ordinated research project and recommendation for future studies. *Sesame improvement by induced mutations. Final report of an FAO/IAEA coodinated research project organized by the joint FAO/IAEA division of nuclear techniques in Food and Agriculture (1993-1998)* pp. 1-12.

Weiss EA (2000) Oilseed crops, 2nd edn. Oxford, Blackwell Science, pp. 131–164.

Were, B.A., Onkware, A.O., Gudu, S., Welander, M. and Carlsson, A.S. (2006). Seed oil content and fatty acid composition in East African sesame (*Sesamum indicum* L.) accessions evaluated over 3 years. *Field Crop. Res.*, **97**(2-3): 254-260.

Wolff, N., Cogan, U., Admon, A., Dalal, I., Katz, Y., Hodos, N., Karin, N. and Yannai, S. (2003). Allergy to sesame in humans is associated primarily with IgE antibody to a 14 kDa 2S albumin precursor. *Food and Chemical Toxicology*, **41**(8): 1165-1174.

Wu, W. H. (2007). The contents of lignans in commercial sesame oils of Taiwan and their changes during heating. *Food chemistry*, **104**(1): 341-344.

Yermanos D.M., Hemstreet S., Saleeb W., Huszar C.K. (1972). Oil content and composition of the seed in the world collection of sesame introductions. *J Am Oil Chem Soc.*, **49**(1): 20–23.

Zerihun, J. (2013). Sesame (*Sesamum indicum* L.) Crop Production in Ethiopia: Trends, Challenges and Future Prospects. *Science, Technology and Arts Research Journal*, **1**(3): 01-07.

2015, Major Constrains and Verdict of Crop Productivity *Pages 64–74*
Editor: Dr. U.N. Bhale
Published by: ASTRAL INTERNATIONAL (P) LIMITED, NEW DELHI

Chapter 4

Role of Organic Amendments in Plant Disease Management: An Overview

Rajkumar H. Garampalli*, Heminder Singh and M.C. Ravikumar

Department of Studies in Botany, University of Mysore, Manasagangotri, Mysore – 570 006, Karnataka, India

ABSTRACT

Plants evolved their own defence against plant pathogens through their biomolecules. When organic amendments are applied in the crop fields, these biomolecules release in to the soil and stimulates the activity of microbes which are antagonistic to plant pathogens therby reducing the activity of phytopathogens in the soil. In the begining it was believed that organic amendments are commonly used to improve the nutrient quality of the soil, but off late it was proved that organic amendments can be used in plant disease management, particularly soilborne diseases. Soil amendment with organics controls the disease by killing the pathogens in the soil thus protecting the present crop and reducing the inoculum for the future crops. It has been demonstrated that organic amendments are used to control many soil borne plant diseases caused by viruses, bacteria, fungi and nematodes. Organic amendments are found effective in suppressing soil borne diseases such as damping-off and root rots caused by Pythium ultimum, Rhizoctonia solani, Phytopthora spp. and wilts caused by Fusarium wilt and Verticillium dahliae.

Keywords: *Organic amandments, Disease management, Botanicals, Compost.*

* *Corresponding Author.* E-mail: rajkumarhg@gmail.com

Introduction

Composts and manures derived from the processing of plant and animal parts have been in use as crop nutrient source since the beginning of agriculture. Bordeaux mixture was developed as the first successful chemical fungicide in 1885. Early fertilizers and fungicides were cheap, powerful and easy to transport in bulk. Better, quick and consistent results given by these chemicals, made them widely popular and increased their demand. Artificial, synthetic fertilizers had been created during 18th century, initially with superphosphates and then the ammonia derived fertilizers were mass produced during World War I by Haber Bosch process (Horne and Anthony, 2008). So 1940's has been called as 'pesticide era' (Horne and Anthony, 2008). Widespread use of such chemicals initially gave encouraging results by boosting agricultural production world over. But within a few decades their drawbacks became apparent by appearance of environmental and human health problems.

Organic movement began a few decades back as a reaction to agriculture's growing reliance on synthetic fertilizers and chemicals. Organic biochemicals are environment friendly, easily biodegradable and safe for humans. Organics thus are capable of restoring ecological balance if become popular. Therefore, there is a need to go back to traditional methods of farming. During the past two decades, considerable progress has been made in restoring cultural practices in agriculture. Though, composts are in use as plant nutrient source since centuries, their use in plant disease control is recent. It is generally known to agriculturists since a long time that the use of organic compost in field produces healthier plants with reduced disease incidence but this area has recently received focus of research and has been on fast-track since then.

Methodological Approach

The role of organic amendments in plant disease control has long been suspected but has recently been established by scientific research. The nursery industry first observed that composted tree bark seemed to suppress *Phytophthora* root rots (Ceuster and Hoitink, 1999). Wilhelm (1951) showed that the use of blood meal and fish meal totally reduced the incidence of *Verticillium* wilt of tomato. Such observations opened a new area of possibilities of formulating organic composts of special composition, tailored for control of particular diseases. Since then there has been much documentation of data regarding the control of certain diseases with specific composts. In 1970s, suppression of a plant disease in field conditions by tailored compost was reported from Nagano valley in Japan, where *Fusarium* crown rot of Chinese yam was suppressed in sandy soil, amended with larch bark (Ceuster and Hoitink, 1999).

The killing of microsclerotia of *Verticillium dahliae* was achieved in test tube microcosms by meat and bone meal amendments of soil (Tenuta and Lazarovits, 2002). The control of cucumber *Fusarium* wilt was achieved by the application of pig manure and rape seed bio-organic fertilizer to field plots (Zhang *et al.*, 2008). The control of snap bean common root rots (*Aphanomyces euteiches*) was achieved in field conditions by the use of fresh paper and composted paper amendments (Leon *et al.*, 2006). The control of *Pythium* spp. on cucumber was achieved in pot experiments by the use of different composts (Fuchs and Larbi, 2004). The suppression of white pumpkin *Rhizoctonia* damping off was achieved in plastic trays by the use of vermicompost (Rivera *et al.*, 2004). The suppression of lettuce drop (*Sclerotinia* sp.,) was achieved in field plots by the use of composted sewage sludge (Lumsden *et al.*, 1986). The reduction of potato scab, *Verticillium* wilt and nematodes was achieved by the use of soymeal, meat and bone meal amendments in field plots (Lazarovits *et al.*, 1999).

The traditional Indian medicinal plant neem (*Azadirachta indica*) has now been established by research to possess broad spectrum antimicrobial properties. Neem based products used as soil amendments, have been found to be effective against nematodes by numerous reports. Beneficial effects of application of oil seed cakes of neem (*Azadirachta indica*), castor (*Ricinus communis*), mustard (*Brassica compestris*) and duan (*Eruca sativa*) on management of parasitic nematode like *Meloidogyne incognita*, *Rotylanchulus reniformis* and *Tylenchorhynchus brassicae* and pathogenic fungi like *Fusarium oxysporum* f. *ciceri*, *Macrophomina phaseolina* and *Rhizoctonia solani* has been demonstrated in field conditions (Tiyagi and Alam, 1996). Management of *M. incognita* infesting *Vigna radiata* and *Cicer arietinum* in preliminary trials (in ice cream cups), green house trials (earthen pots) and field trials (micro plots of size 0.1 sq. mt.) was reported by the use of crop-seed treatment with dried powdered neem seed and soaking of crop-seeds in aqueous extracts of neem cake, neem seed kernel and seed coat (Majumdar and Mishra, 1996). Suppression of tobacco damping off (*Pythium aphanidermatum*) was reported in green house micro plots (sized 0.5 sq. m.) and nursery conditions by neem based products used for the pre-sowing soil treatment in form of cakes and soil drench (Shenoi *et al.*, 1996). Amendments have also been used for yield enhancement, control of weed, pests and air borne diseases. Beneficial effects of *Parthenium hysterophorus* compost on soil nutrients yield of soya bean and cow pea, control of weeds and pests (stem borer and leaf roller) in rice fields was reported (Ramaswami, 1997). Suppression of *Phytophthora* spp., *Fusarium* spp. and *Septoria* spp. in tomato fields was achieved by use of nine compost derived from wastes and by products of olive oil, wine and *Agaricus* mushroom agro-industry (Ntougias *et al.*, 2008).

Neem based extract formulations have also been found to be effective against pests and other foliar pathogens like fungi, bacteria and viruses. Management of pest and diseases of cotton (*Aphis gossypii, Bemisia tabaci*) and soybean (*Aproaerema modicelli*, Yellow mosaic virus) was reported by spraying of neem based products (Vidya *et al.*, 1996). Management of rice pests (*Cnaphalocrocis medinalis, Nilaparvata lugens*) and diseases (sheath blight by *Rhizoctonia solani*) was achieved by neem cake application to nursery soil and neem seed kernel extract, neem oil spraying on field plants (Ouerdraogo *et al.*, 2001). Reduction of rice bacterial blight (*Xanthomonas compestris* pv. *oryzae*) was demonstrated in field conditions by spray of plant extracts like neem cake extract 5 per cent, neem seed kernel extract 2 per cent, *Prosopis* spp. leaf extract 10 per cent, *Acacia* spp. leaf extract 10 per cent, Plantomycin 500 ppm, Plantomycin 1000 ppm and *Bacillus subtilis* suspension (Eswaramurthy *et al.*, 1996). Efficacy of neem based extracts and other products for management of root rots of *Elettaria cardamomum* caused by *Rhizoctonia solani* and *Phytophthora meadii* was reported under *in vitro* conditions (culture plates), green house conditions (pot experiments) and field conditions (Dhanpal *et al.*, 1996). Soil application of Melia azadirach controlled the meloidogyne incognita on cucumber (Cavoski *et al.*, 2012). Nematicedal effect of four neem formulations such as neem seed powder (NSM), organic, watery and essential oil extracts. The NSP killed all nematodes (100 per cent) at lowest concentration, 10µg/ml after, 48 h. (Kosma *et al.*, 2011).

Since resistance in plants is highly versatile and elastic, even the susceptible plant can be protected from the disease by the developing systems to induce and enhance the plants own defence mechanisms (Garcia-Brugger, *et al.*, 2006). The induction of systemic acquired resistance results in the development of a long lasting, broad spectrum immunity in non- infected tissues against a broad range of plant pathogens including viruses, bacteria and fungi (Vallard and Goodman, 2004; Walters *et al.*, 2005; Kuc, 2006). There are many reports on various types of induced resistance in a variety of host-pathogen systems using biotic or abiotic agents (Oostendorp *et al.*, 2001; Agostini, *et al.*, 2003). There are several reports indicating the resistance inducing activity of plant extract as well as components of plant extracts (Doke and Chai, 1985; Doubrava *et al.*, 1988; Yamada, *et al.*, 1990; Devaiah, *et al.*, 2009; Guleria and Kumar, 2006; Govindappa, *et al.*, 2011; Kagale *et al.*, 2011). Natural products like plant extracts are preferred over chemicals as an inducer of resistance.

There are a few reports about evaluation of multiple composts and their formulations for management of particular diseases. The suppression of potato *Verticillium* wilt in field plots was achieved by application of 3 green composts made from *Pisum sativum; Brassica oleracea; Sorghum vulgare* (Ochiai *et al.*, 2008). The *in situ* and *in vitro* suppression of *Pythium debaryanum*, *Fusarium oxysporum* and *Sclerotium bataticola* was achieved by application of leafy fruit compost, garden compost, crops compost and their water extracts (EL-Masry *et al.*, 2002). The control of *Pythium* spp. was achieved in pot experiments by application of composts of *Iberis amara, Rapistrum rugusum, Cleome hassleriana* (Luca and Lusia, 1996). Tomato bacterial canker (*Clavibacter michigansis*) was successfully suppressed by the use of tomato and pepper residues combined with cattle and chicken manure (Yogev *et al.*, 2009). Nico *et al.* (2004) reported the management of root rot nematodes of tomato by use of composted dry cork and dry grape mark at 25 per cent, 50 per cent 75 per cent and 100 per cent (v/v) amendment rates in greenhouse conditions. Suppressive effects of four waste residues from *Ricinus communis, Erica sativa, Linum usitatissimum and Triticum aestivum* was studied at the amendment rates of 1.25 per cent, 2.5 per cent, 3.75 per cent against nematode infestation of brinjal (Ismail *et al.*, 2012). Suppressive effects of six composted amendments and compost teas from farm yard manure, poultry manure, vermicompost, spent mushroom compost, *Lantana camara* and *Urtica* spp. was reported by Joshi *et al.* (2009), against root rot and angular leaf spot of french bean in field condition. An experiment conducted to evaluate the effect of the 18 composts (from source materials like green and yard waste, straw, bark, bio waste, municipal sludge) on seven patho-systems (*Verticillium dahliae* on egg plant, *Rhizoctonia solani* on cauliflower, *Phytophthora nicotianae* on tomato, *Phytophthora cinnamomi* on lupin, *Cylindrocladium spathiphyllion* on *Spathiphyllium* sp.) at amendment rates of 20 per cent (v/v), found the compost to be generally suppressive in most of the treatments (54 per cent) and rarely disease stimulating (3 per cent) (Termorshuizen *et al.*, 2006). Formulations of plant extracts have been reported to have more activity against pathogens than individual extracts by Opraeke (2007a), Opraeke (2007b), Opraeke *et al.* (2006), Shanker and Uthamasamy (2010), De Britto *et al.* (2012). Microcosm experiment of four organic amendments namely, sugar cane bagasse, Sugar cane sludge, Plant residue and sewage sludge against soil nematodes and banana parasitic nematode. Among, these four amendments sugar cane bagasse effectively controlled (96 per cent) the nematodes (Tabarant *et al.*, 2011). Oka *et al.* (2007) showed that in pot and green jouse experiments organic amendments with soil solarization, effectively controlled the root rot nematodes *Meloidogyne incognita* and *Meloidogyne javanica.* Soil solarisation, Poultry manure and Methyl bromide tested, against *Fusarium wilt* in green house conditions in three experiments. Poultry manure alone controlled the disease incidence 93 to 100 per cent and 89 to100 per cent in experiment 1, 2 and 3 (Melero-vara *et al.*, 2011). Water extract of Tithonia diversiflora inhibited the Meloidogyne incognita egg hatch by 100 per cent after nine days after inoculation in-vitro, tithonia residue at a rate of 30 tonnes/ha effectively suppressed the reproduction, number of eggs and juveniles and galling (Odeyemi and Adewale, 2011)

The Mechanisms of Action of Organics

Various modes of action of organic amendments on phytopathogenic fungi reported in literature are:

1. Direct antimicrobial action of toxins released by compost biomass or produced in soil by microbial action (Singh, 1996).
2. Supply of nutrients suitable for antagonists, favouring them in competition (www. epa. gov; Singh, 1996).
3. Hyperparasitism or antimicrobial action of antagonists (Singh, 1996).
4. Soils physico-chemical manipulation in favour of antagonists (Singh, 1996)
5. Activation of disease resistance genes in plants (Dutta *et al.*, 2008; www. epa. gov)
6. Absorption and transport of synthetic biochemicals by plant roots to aerial parts to provide resistance against air borne pathogens (Fuch and Larbi, 2004; Ceuster and Hoitink, 1999).
7. Release of fumigant biochemicals during microbial activity in soil (Lazarovits, 2001; Pavlica *et al.*, 1978).
8. Stimulation of germination of fungal resting structures by aldehydes, etc. in absence of host (Singh, 1996; Singh *et al.*, 2013).

Background Attitudes and Beliefs

Drawbacks often held against organic composts, despite the demonstrated benefits, are that, the results are inconsistent, transportation costs of bulky composts are high and their low shelf-life. Inconsistency of results has been due to inconsistency in composition and preparation practices practiced by farmers. Standardization of composition and preparation methods for local soil and climate conditions will solve this problem. To reduce transportation cost, the bulky portion of compost, *i.e.*, cattle dung manure may be prepared separately from the minor additive *i.e.*, plant biomass amendment. Cattle dung manure may be prepared by farmers in their own backyard in compost pit. It would remove the need of transportation of bulky manure. Plant biomass amendments with fine tuned composition may be marketed. The cattle dung manure should be taken out of the pit just before maturation and mixed with plant additive to produce the final tailored compost. It is done, so that it can be applied to the field 6 weeks before sowing, to allow leaching of some biochemicals and colonization of soil by compost microbes. The maturation will take place in the field itself. To increase the shelf-life of marketed compost, addition of dried and powdered plant biomass to the cattle dung manure was tested for efficacy. There are some reports of successful dry biomass addition to the soil, without composting, for control of diseases. Fritz (2007) reported that, rape seed powder application to field furrows controlled pea root rot (*Aphanomyces euteiches*). Similarly, soymeal, meat and bone meal application controlled potato scab, potato *Verticillium* wilt and nematodes as reported by Lazarovits *et al.* (1999). Suppressive effects of dried plant residues from *Brassica rapa*, *Arachis hypogaea*, *Trifolium pratense* against sugar beat damping off (*R. solani*) at 1 per cent (w/w) rate was reported by Kasuya *et al.* (2006), by the mechanism of promotion of soil antagonists.

Conclusion

Critics, including father of 'Green Revolution' Norman Borlaug claim that organic farming cannot feed the world population (Trewavas, 2001; Trewavas, 2004; Quinones *et al.*, 1997). Further, some studies also claim that organic farms yield is lower than that of conventional farming (Maeder *et al.*, 2002; Badgley *et al.*, 2007). So, the answer lies in making least possible use of inorganic chemicals and methods. The central idea behind writing this review is to enthuse the reader to develop cultural methods which are cheap, environment friendly, safe for human and animal health, which do not require much technical expertise to apply and can even be developed further by farmers themselves in the working conditions, away from hi-tech lab facilities.

References

Agostini, J.P., Bushong, P.M., Timmer, L.W.(2003). Glasshouse evaluation of products that induce host resistance for control of scab, melanose, and Alternaria brown spot of citrus. *Plant Disease*, **87**: 69 - 75.

Badgley, C., Moghtader, J., Quintero, E., Zakem, E., Chappell, M.J., Avilés-Vásquez, K., Samulon, A., Perfecto, I. (2007). Organic agriculture and the global food supply. *Renew. Agr. food Syst.*, **22**: 86-108.

Cavoski I., Chami Z. Al, Bouzebboudja F., Sasanelli N., Simeone V., Mondelli D., Miano T., Sarais G., Ntalli N.G. and Caboni P.(2012). *Melia Azederach* controls *Meloidogyne incognita* and triggers plant defence mechanisms on cucumber. *Crop Protection*, **35:** 85-90.

Ceuster, T.J.J.D., Hoitink, H.A.J.(1999). Using compost to control plant diseases. Biocycle, (June), p. 61.

De Britto, A.J., Gracelin, D.H.S., Benjamin, P., Kumar, J.R.(2012). Antibacterial potency and synergistic effects of a few south Indian spices against antibiotic resistant bacteria. *Indian Journal of Natural Products and Resources*, **3**: 557-562.

Devaiah, S.P., Mahadevappa, G.H., Shetty, H.S.(2009). Induction of systemic resistance in pearl millet against downy mildew pathogen by *Datura metel* extract. *Crop Protection*, **28**: 783 – 791.

Dhanpal, K., Joseph, T., Naidu, R.(1996). Use of neem in integrated management of rot diseases of small cardamom (*Elletaria cardamomumn* Maton.), in: Singh, R.P., Chari, M.S., Raheja, A.K., Raus, W.K. (Eds.), Neem and Environment Vol. 2, Oxford and IBH Publishing Co. Pvt. Ltd., New Delhi, pp. 793-798.

Doke, N., Chai, H.B.(1985). Activation of superoxide generation and enhancement of resistance against compliable races of *Phytophthora infestans* in potato plants treated with digitonin. *Physiological Plant Pathology*, **27**: 323 - 334.

Doubrava, N.S., Dean R.A., Kuc, J.(1988). Induction of systemic resistance to anthracnose caused by *Colletotrichum lagenarium* in cucumber by oxalate and extracts from spinach and rhubarb leaves. *Physiological and Molecular Plant Pathology*, **133**: 69-79.

Dutta, S., Mishra, A.K., Dilip kumar, S.(2008). Induction of systemic resistance against fusarial wilt in pigeonpea through interaction of plant growth promoting rhizobacteria and rhizobia. *Soil Biol. Biochem.*, **40**: 452-461.

EL-Masry, M.H.S., Khalil, A.I., Hassouna, M.S., Ibrahim, H.A.H.(2002). In situ and in vitro suppressive effect of agricultural composts and their water extracts on some phytopathogenic fungi.*World J. Microb. Biot.*, **18**: 551-558.

Eswaramurthy, S., Mariappan, V., Muthuswamy, M., Alagianagalingam, M.N., Subramanyam, K.S.(1996). Efficacy of neem products in control of rice bacterial blight. in: Singh, R.P., Chari, M.S., Raheja, A.K., Raus, W.K. (Eds.), Neem and Environment Vol. 2, Oxford and IBH Publishing Co. Pvt. Ltd., New Delhi, pp. 783-785.

Fritz, V.(2007). Use of Green Manure Cover Crops Improves Disease Management. M.N.Fruits and Vegetables, IPM News, 4, 5, June 8.

Fuchs, J.G., Larbi, M., (2004). Disease control with quality compost in pot and field trials, in: *Proceeding of first International Conference, Soil and Compost Eco-biology*, Leon, Spain.

Garcia-Brugger, A., Lamotte, O., Vandelle, E., Bourque, S., Lecourieux, D., Poinssot, B., Wendehenne, D., Pugin, A.(2006). Early signaling events induced by elicitors of plant defences. *Molecular Plant-Microbe Interactions.*, **19**: 711-724.

Govindappa, M., Umesha, S., Lokesh, S., (2011). *Adathoda vasica* leaf extract induces resistance in rice against bacterial leaf blight disease (*Xanthomonas oryzae* pv. *oryzae*). *International Journal of Plant Physiology and Biochemistr*, **3**: 6-14.

Guleria, S., Kumar, A., (2006). *Azadirachta indica* leaf extract induces resistance in sesame against Alternaria leaf spot disease. *Journal of Cell and Molecular Biology*, **5**: 81-86.

Horne., Anthony, P., (2008). Integrated pest management for crops and pastures. CSIRO Publishing, p. 2.

Ismail, A.E., Dawood, M.G., El-Nagdi, W.M.A., (2012). Role of organic soil amendments with some non-conventional plant additives on the growth of eggplant and their role against Meloidogyne incognita infection. *Arch. Phytopathol. Plant Prot.,* p.1-10.

Joshi, D., Hooda, K.S., Bhatt, J.C., Mina, B.L., Gupta, H.S.(2009). Suppressive effect of composts on soil-borne and foliar disease of French bean in the field in western Indian Himalayas. *Crop Prot.,* **28:** 608-615.

Kagale, S., Marimuthu, T., Kagale, J., Thayumanavan, B., R. Samiyappan, R.(2011). Induction of systemic resistance in rice by leaf extracts of *Zizyphus jujube* and *Ipomoea carnea* against *Rhizoctonia solani. Plant Signaling and Behaviour,* **6:** 919-923.

Kasuya, M., Oliver, A.R., Ota, Y., Tojo, M., Honjo, H., Fukai, R.(2006). Induction of soil suppressiveness against *Rhizoctonia solani* by incorporation of dried plant residues into soil. *Phytopathol.,* **96**: 1372-1379.

Kosma, P., Ambang, Z., Begoude, B.A.D., Ten Hoopen, G.M., Kuate, J., Akoa, A. (2011). Assessment of nematicedal property and phytochemical screening of neem seed formulations using *Radopholus similis* parasitic nematode of plantain in Cameroon. *Crop Prot.,* **30**: 733-738.

Kuc, J.(2006). What's old and what's new in concepts of induced systemic resistance in plants and its application, in T. Tuzun and E. Bent (Eds.), Multigenic and induced systemic resistance in plants (USA: Springer), p. 9-20.

Lazarovits, G.(2001). Management of soil-borne plant pathogens with organic soil amendments: A disease control strategy salvaged from the past. *Can. J. Plant Pathol.,* **23**: 1-7.

Lazarovits, G., Conn, K.L., Potter, J.(1999). Reduction of potato scab, *Verticillium* wilt, and nematodes by soymeal and meat and bone meal in two Ontario potato fields. *Can. J. Plant Pathol.,* **21**: 345 -353.

Lazarovits, G., Tenuta, M., Conn, K.L.(2001). Organic amendments as a disease control strategy for soil-borne diseases of high-value agricultural crop. *Australas. Plant Path.,* **30**: 111- 117.

Leon, M.C.C., Stone, A., Dick, R.P., (2006). Oraganic soil amendments: Impact on Snap bean common root rot (*Aphanomyces euteiches*) and soil quality. *Appl. Soil Ecol.* **31**: 199 -210.

Luca, L., Lusia, M.M.(1996). Allellopathic effect of glucosinolate containing plant green manure on *Pythium* spp. and total fungal population in soil. *Soil Biol. Biochem.* **28:** 1207-1214.

Lumsden, R.D., Millner, P.D., Lewis, J.A.(1986). Suppression of lettuce drop caused by *Sclerotinia minor* with composted sewage sludge. *Plant Dis.,* **70**: 197 -201.

Maeder, P., Fliessbach, A., Dubois, D., Gunst, L., Fried, P., Niggli, U.(2002). Soil fertility and biodiversity in organic farming. *Science*, **296**: 1694–1697.

Melero-vara, J.M., Lopez-Herrera, C.J., Prados-Lijero, A.M., Vela-Delgado, M.D., Navas-Becerra, J.A., Basallote-Ureba, M.J.(2011). Effects of soil amendment with poultry manure on cornation fusarium wilt in green houses in southwest spain. *Crop Prot.*, **30**: 970-976.

Mojumder, V., Mishra, S.D. (1996). Management of *Meloidogyne incognata* infesting pulse crop through seed treatments with neem products, in: Singh, R.P., Chari, M.S., Raheja, A.K., Raus, W.K. (Eds.), Neem and Environment Vol. 2, Oxford and IBH Publishing Co. Pvt. Ltd., New Delhi, pp. 699-703.

Nico, A.I., Jimenez-Diaz, R.M., Castillo, P. (2004). Control of root-rot nematodes by composted agro industrial wastes in potting mixtures. *Crop prot.*, **23:** 581-587.

Ntougias, S., Papadopoulou, K.K., Zervakis, G.I., Kavroulakis, N., Ehaliotis, C.(2008). Suppression of soil-borne pathogens of tomato by composts derived from agro-industrial wastes abundant in Mediterranean regions. Biol. *Fertil. Soils.*, **44:** 1081–1090.

Ochiai, N., Powelson, M.L., Crowe, F.J., Dick, R.P.(2008). Green manure effects on soil quality in relation to suppression of *Verticillium* wilt of potatoes. *Biol. Fertil. Soils.*, **44:** 1013-1023.

Odeyemi, I.S., Adewale, K.A.(2011). Phythonematoxic properties and nematicidal potential of Tithonia diversifolia extract and residue on Meloidogyne incognita infecting yam (*Discoria rotundata*). *Phytopathol. Plant Prot.*, **44**(18): 1745-1753.

Oka, Y., Shapira, N., Fine, P. (2007).Control of root rot nematodes in organic farming systems by organic amendments and soil solarisation. *Crop prot.* **26:** 1556-1565.

Oostendorp, M., Kunz, W., Dietrich B., Staub, T.(2001). Induced disease resistance in plants by chemicals. *European Journal of Plant Pathology*, **107**: 19-28.

Oparaeke, A.M.(2007a). Synergistic activity of aqueous extracts mixtures of some Nigerian plants against *Maruca vitrata* and *Clavigralla tomentosicollis* on field cowpea, *Vigna unguiculata* (L.) Walp, *Arch. Phytopathol. Plant Prot*, **40**: 257-263.

Oparaeke, A.M.(2007b). Effect of crude extracts of mixtures of different plant species in the control of *Megalurothrips sjostedti* Trybom. and pod setting of cowpea plants, *Arch. Phytopathol. Plant Prot*, **40**: 201-206.

Oparaeke, A.M., Dike, M.C., Amatobi, C.I.(2006). Field activity of three mixture levels of plant extract formulations for the management of post-flowering insect pests of cowpea, *Vigna unguiculata* (L.) Walp–The Flower Thrips, *Megalurothrips sjostedti* (Trybom). *Journal of Sustainable Agriculture*, **28:** 45-54.

Ouerdraogo, E., Mando, A., Zombre, N.P.(2001). Use of compost to improve soil properties and crop productivity under low input agricultural system in West Africa. *Agr. Ecosyst. Environ.*, **84**: 259-266.

Pavlica, D.A., Hora, T.S., Bradshaw, J.J., Skogerboe, R.K., Baker, R., 1978. Volatile compounds from soil influencing activities of soil fungi. *Phytopathology.* **68**: 758-765.

Quin'ones, M.A., Borlaug, N.E., Dowswell, C.R., 1997. A fertilizer-based green revolution for Africa, in: Buresh, R.J., Sanchez, P.A., Calhoun, F. (Eds.), Replenishing soil fertility in Africa. Soil Science Society of America and ICRAF, Special publication 51, Madison. pp. 81 – 96.

Ramaswami, P.P. (1997). Potential uses of *Parthenium,* in: Proceedings of the First International conference on *Parthenium* management, Dharwad, University of Agricultural Sciences, 1, pp. 77-80.

Rivera, M.C., Wright, E.R., Lopez, M.V., Fabrizo, M.C.(2004). Temperature and dosage dependent suppression of damping off caused by *Rhizoctonia solani* in vermicompost amended nurseries of white pumpkin. *Int. J. Exp. Bot*., **53**: 131- 139.

Shanker, C., Uthamasamy, S.(2010). Evaluation of some medicinal plants and their mixtures for their bio-efficacy against crop and stored product pests. *Arch. Phytopathol. Plant Prot,* **43**(2): 140-148.

Shenoi, M.M., Wazid, S.M.A., Sreenivas, S.S.(1996). Evaluation of *Azadirachta indica* and *Lawsonia inermis* for antifungal activity and studies on neem products for control of damping off in tobbaco nurseries, in: Singh, R.P., Chari, M.S., Raheja, A.K., Raus, W.K. (Eds.), Neem and Environment Vol. 2, Oxford and IBH Publishing Co. Pvt. Ltd., New Delhi, pp. 745-753.

Singh, H., Ravikumar, M.C., Garampalli, R.H., 2013. Enhancement of downy mildew of sorghum by some organic soil amendments: a caution to organic farmers. *Arch. Phytopathol. Plant Prot*. **46**, 1030 -1035.

Singh, R.S.(1996). Introduction to principles of plant disease, third ed. Oxford and IBH Publishing Co. Pvt. Ltd., New Delhi.

Tabarant, P., Villenave, C., Risede, J.M., Rojer-estrade, J., Thuries, L., Dorel, M., 2011. Effect of four amendments on banana parasitic nematodes and soil nematode communities. *Applied soil ecol.,* **49**: 59-67.

Tenuta, M., Lazarovits, G., 2002. Ammonia and nitrous acid from nitrogenous amendments kill the microsclerotia of *Verticillium dahliae*. *Phytopatholgy,* **92:** 255 -264.

Termorshuizen, A.J., van Rijn, E., van der Gaag, D.J., Alabouvette, C., Chen, Y., Lagerlöf, J., Malandrakis, A.A., Paplomatas, E.J., Rämert, B., Ryckeboer, J., Steinberg, C., Zmora-Nahum, S.(2006). Suppressiveness of 18 composts against 7 pathosystems: Variability in pathogen response. *Soil biol. Biochem*., **38:** 2461-2477.

Tiyagi, S.A., Alam, M.M., 1996. Efficacy of neem cake against soil nematodes and fungi on chick pea, in: Singh, R.P., Chari, M.S., Raheja, A.K., Raus, W.K. (Eds.), Neem and Environment Vol. 2, Oxford and IBH Publishing Co. Pvt. Ltd., New Delhi, pp. 627-641.

Trewavas, A.(2001). Urban myths of organic farming. *Nature,* **410**: 409-410.

Trewavas, A.(2004). A critical assessment of organic farming and food assertions with particular respect to the UK and the potential environmental benefits of no-till agriculture. *Crop Prot.*, **23**: 757-781.

Vallard, G.E., Goodman, R.M., 2004. Systemic acquired resistance and induced systemic resistance in conventional agriculture. *Crop Science,* **44**: 1920 - 1934.

Vidya, F., Sethuraman, K., Mariappan, V., Karunakaran, K.R., Vidhyasekaran, P. (1996). Management of pest and diseases of cotton and soybean using neem products, in: Singh, R.P., Chari, M.S., Raheja, A.K., Raus, W.K. (Eds.), Neem and Environment Vol. 2, Oxford and IBH Publishing Co. Pvt. Ltd., New Delhi, pp. 729-733.

Walters, D., Walsh, D., Newton, A., Lyon, G.(2005). Induced resistance for plant disease control: maximizing the efficacy of resistance elicitors. *Phytopathology,* **95**: 1368-1373.

Wilhelm, S. (1951). Effect of various soil amendments on the inoculum potential of *Verticillium* wilt fungus. *Phytopathology,* **41**: 684-690.

Yamada, T., Hiramoto, T., Tobimatsu, R., Shiraishi, J., Oku, H.(1990). Elicitor like substances present in barley and wheat seeds. *Journal of Phytopathology,* **128**: 89-98.

Yogev, A., Raviv, M., Kritzman, G., Hadar, Y., cohen, R., Krishner, B., Kaatan, J. (2009). Suppression of bacterial canker by composts. *Crop Prot.*, **28**: 97-103.

Zhang, C., Li, B., Gaikwad, A.S.(2008). Avicin D selectively induces apoptosis and downregulates p-STAT-3, bcl-2, and survivin in cutaneous T-cell lymphoma cells. *J. Invest. Dermatol.*, **128**: 2728–2735.

Websites Referred

www.epa.gov. Available from: http://www.epa.gov/wastes/conserve/composting/pubs/disease.pdf (United States Environment Protection Agency, October, 1997). (Accessed on 22 Sept, 2013).

2015, Major Constrains and Verdict of Crop Productivity Pages 75–97
Editor: Dr. U.N. Bhale
Published by: ASTRAL INTERNATIONAL (P) LIMITED, NEW DELHI

Chapter 5

Role of Biotechnology in Crop Improvement

T. Ugandhar[1]*, M. Venkateshwarlu[2] and T. Srilatha[3]

[1]*Department of Botany,*
S.R.R. Govt Degree and P.G. College, Karimnagar – 505 001, India
[2]*Department of Botany,*
University College, Kakatiya University, Warangal – 506 009, India
[3]*Department of Botany,*
Govt Degree and P.G. for Women, Warangal – 506 009, India

Introduction

The last decade has witnessed remarkable change which has taken plant biotechnology from study of basic science to large scale commercial applications. This is true for almost every aspect of plant biotechnology such as development of molecular markers to speed up plant breeding practices and using knowledge of genes and their expression to generate and commercialize transgenic crops. In general, the role of biotechnology in crop improvement can be divided into two categories:

1. Those directed towards same goals as conventional plant breeding like improved yield, quality, resistance to pests and diseases, tolerance to abiotic stresses etc by molecular breeding or production of transgenic crops.
2. Novel applications such as use of plants as bioreactors to generate pharmaceuticals, vaccines or biodegradable plastics.

* *Corresponding Author.* E-mail: tugandharbiotech@gmail.com

A) Molecular Breeding

By 2025, the global population will exceed seven billion. In the interim per-capita availability of arable land and irrigation water will decrease from year to year as biotic and abiotic stresses increase. Food security, best defined as economic, physical and social access to a balanced diet and safe drinking water will be threatened, with a holistic approach to nutritional and nonnutritional factors needed to achieve success in the eradication of hunger. Science and technology can play a very important role in stimulating and sustaining an Evergreen Revolution leading to long-term increases in productivity without any associated ecological harm (Borlaug, 2001; Swaminathan, 2007). The objectives of the plant breeder can be realized through conventional breeding integrated with various biotechnology developments (Damude and Kinney, 2008; Xu *et al.*, 2009c).

Plant breeding can be defined as an evolving science and technology (Figure 5.1). It has gradually been evolving from art to science over the last 10,000 years, starting as an ancient art to the present molecular design-based science. Plant breeding is becoming quicker, easier, more effective and more efficient (Phillips, 2006). Plant breeders will be well equipped with innovative approaches to identify and/or create genetic variation, to define the genetic feature of the genes related to the variation (position, function and relationship with other genes and environments), to understand the structure of breeding populations, to recombine novel alleles or allele combinations into specific cultivars or hybrids, and to select the best individuals with desirable genetic features which enable them to adapt to a wide range of environments. Sequencing data for many plants is now readily available and the Gene Bank database is doubling every 15 months. Over 20 plant species including many important crops are in the process of being sequenced (Phillips, 2008). The next challenge is to determine the function of every gene and eventually how genes interact to form the basis of complex traits. Fortunately, DNA chips and other technologies are being developed to study the expression of multiple or even all genes simultaneously. High throughput robotics and bioinformatics tools will play an essential role in this endeavor.

New information about our crop species is expanding our capabilities to use molecular genetics. For example, we did not previously realize how similar broadly related species are in terms of their gene content and gene order. Since these species cannot usually be crossed, there was no means of assessing their relatedness. With the advent of DNA-based molecular markers, the extensive genetic mapping of chromosomes became readily possible for a variety of species. We learned that the genomes were highly similar and that this similarity allowed the prediction of gene locations among species. For example, rice has become the model or reference species for the cereals as many of the gene sequences on the rice chromosomes are shared with other cereals such as maize, sorghum, sugarcane, millet, oats, wheat and barley (Xu *et al.*, 2005). Knowing the complete DNA sequence of a model or reference genome allows genes/traits from this model to be tracked to other genomes. We have come to realize that the differences between species of plants are not due to novel genes, but to novel allelic specifications and interactions (Figures 5.1 and 5.2).

Art-based Plant Breeding

↓

Collection of wild plants for food

↓

Selection of wild plants for cultivation
(starting from 10,000 years ago)

↓

Large-scale breeding activities supported by commercial seed production enterprises

↓

Hybridization combined with selection

↓

Evolution through natural selection
(1700s–1800s)

↓

Mendelian genetics

↓

Quantitative genetics

↓

Mutation

↓

Polyploidy

↓

Tissue culture
(1900s)

↓

Gene cloning and direct transfer

↓

Genomics-assisted breeding

↓

Molecular Plant Breeding
(2000s and beyond)

Figure 5.1: The Steps of Evolution of 'Plant Breeding'. With the availability of more sophisticated tools, the art of plant breeding became science-based technology, molecular plant breeding.

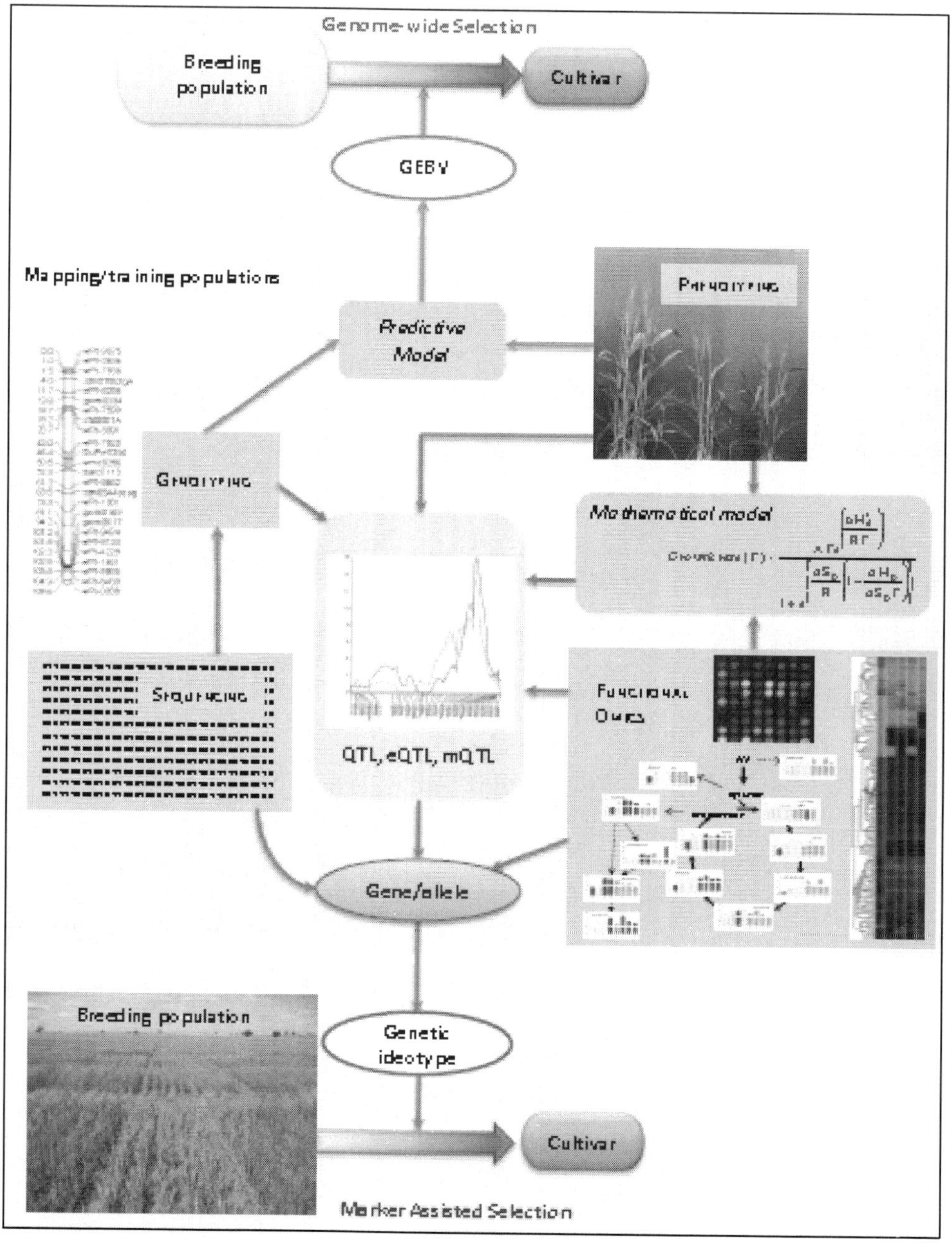

Figure 5.2: Future Molecular Breeding Strategies.

Molecular Breeding and Marker-Assisted Selection

The process of developing new crop varieties can take almost 25 years. Now, however, biotechnology has considerably shortened the time to 7-10 years for new crop varieties to be brought to the market. One of the tools which can make it easier and faster for scientists to select plant traits is marker-assisted selection (MAS).

Molecular Shortcut

The differences that distinguish one plant from another are encoded in the plant's genetic material, the DNA. DNA is packaged in chromosome pairs (strands of genetic material), one coming from each parent. The genes, which control a plant's characteristics, are located on specific segments of each chromosome. Together, all of a plant's genes make up its genome.

Some traits, like flower color may be controlled by only one gene. Other more complex characteristics, however, like crop yield or starch content, may be influenced by many genes. Traditionally, plant breeders have selected plants based on their visible or measurable traits, called the phenotype. This process can be difficult, slow, influenced by the environment, and costly – not only in the development itself, but also for the economy, as farmers suffer crop losses.

As a shortcut, plant breeders now use marker-assisted selection (MAS). To help identify specific genes, scientists use what are called molecular or genetic markers. The markers are a string or sequence of nucleic acid which makes up a segment of DNA. The markers are located near the DNA sequence of the desired gene and are transmitted by the standard laws of inheritance from one generation to the next (Figure 5.1). Since the markers and the genes are close together on the same chromosome, they tend to stay together as each generation of plants is produced. This is called genetic linkage. This linkage helps scientists to predict whether a plant will have a desired gene. If researchers can find the marker for the gene, it means the desired gene itself is present.

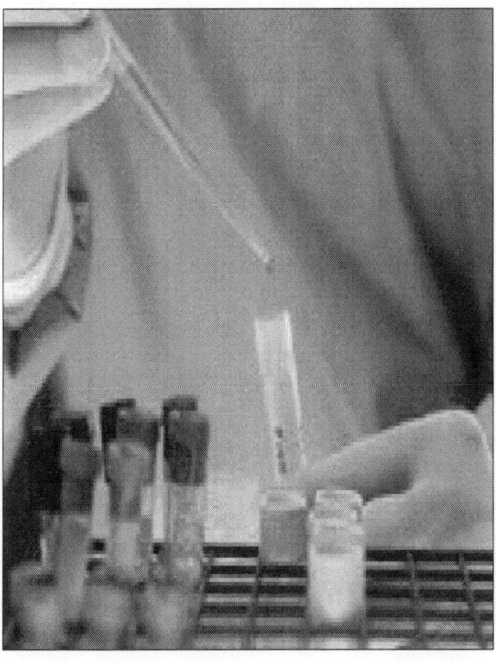

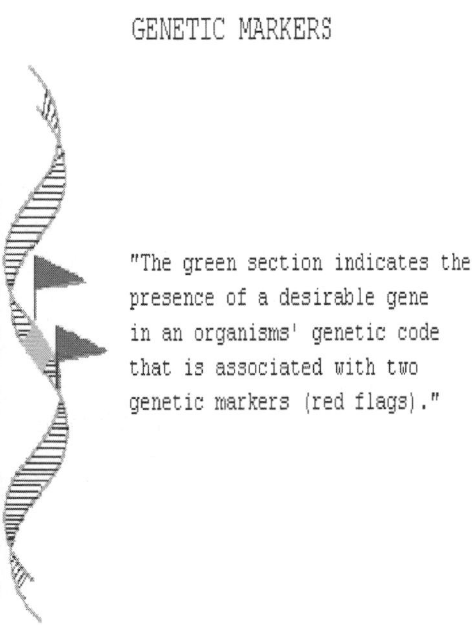

As scientists learn where markers occur on a chromosome, and how close they are to specific genes, they can create a genetic linkage map. Such a map would show the location of markers and genes, and their distance from other known genes. Scientists can produce detailed maps in only one generation of plant breeding.

Using very detailed genetic maps and better knowledge of the molecular structure of a plant's DNA, researchers can analyze only a tiny bit of plant tissue, even from a newly germinated seedling. Once the tissue is analyzed, scientists know whether that seedling contains the appropriate gene. If it doesn't, they can quickly move on and concentrate on analysis of another seedling, eventually working only with the plants which contain a specific trait.

It should be noted, however, that molecular breeding through MAS is somewhat limited in scope compared to genetic engineering or modification because:

1. It works only for traits already present in a crop;
2. It cannot be used effectively to breed crops which have long generation times (*e.g.* citrus);
3. It cannot be used effectively with crops which are clonally propagated because they are sterile or do not breed true (this includes many staples such as yams, bananas, plantain, sweet potato, and cassava).

Molecular Markers

Molecular marker is a DNA segment that is readily detected and whose inheritance can easily be monitored. The use of molecular markers is based on the naturally occurring DNA polymorphism. A DNA marker is a small region of DNA showing sequence polymorphism in different individuals with in a species (or) among different species The DNA markers are most widely used type of markers predominantly due to their abundance the first such DNA markers to be used for the fragments produced by the digestion with restriction endonucleases that means restriction fragment length polymorphism (RFLP) Based genetic markers. A wide range of molecular techniques is now available to detect the Polymorphism at DNA level.

These have been grouped into the following.

1. Non PCR based approaches *e.g*: - RFLP.
2. PCR based approaches *e.g*: - RAPD AFLP SSRS and SNPs etc (plants).

Restricted Fragment Length Polymorphism (RFLP)

This technique was first developed in 1980 and it was based on Southern hybridization technique. RFLP is used to describe the variation in the length of fragments obtained from the digestion of DNA from two or more organisms with the same endonuclease. The variation at DNA level is assayed by shearing the entire DNA with restriction enzymes. The restriction sites for particular enzymes are present at several places throughout the entire genome with the result that a large number of fragments of DNA are produced. The length of each segment depends on the distance between two adjacent restriction sites. Plants are able to replicate their DNA with high accuracy and rapidity, but many mechanisms causing changes in the DNA are operative. Simple base, pairs changes (or) large scale changes as a result of inversions, translocations, deletions (or) transposition may occur.

This will result in loss (or) gain of recognition site and inturn lead to restriction fragments of different lengths. These restriction fragments of different lengths between the genotypes can be defected on southern blots and by the use of suitable probe. An RFLP is detected as a differential movement of a band on the gel lanes from different species and strains. Each such bond is regarded as single RFLP locus. The pattern of RFLP generated will depend mainly on

1. The differentiation in DNA of selected strains (or) species
2. The restriction enzymes used
3. The DNA probe employed for southern hybridization

RFLP analysis comprises the following basic steps.

1. Isolation of DNA
2. Cutting of DNA into smaller fragments using restriction endonuclease enzymes
3. Separation of DNA fragments by gel electrophoresis
4. Transferring of DNA fragments on to a nylon and nitrocellulose membrane filter.
5. Visualization of special DNA fragments by using labeled probes
6. Analysis of results.

Application (or) Uses of RFLP

1. Identification and isolation of any gene known to be linked with an RFLP locus.
2. RFLP are co-dominant markers enabling heterozygotes to be distinguished from homozygotes.
3. The method is simple as no sequence specific information is required.
4. Used to measure genetic diversity between different populations or related species and also important in studies on evolution.
5. RFLP'S can be used to supplement regular plant breeding protocols through indirect selection.Used as a marker in chromosomal mapping.
6. Used as a diagnostic tool for diseases.

Limitations

1. Requires relatively large amount of highly pure DNA.
2. Needs a good supply of probes that can reliably detect variation.
3. Laborious and expensive to identify a suitable marker restriction enzymes combination for genomic or CDNA libraries.
4. Time consuming as they are not amenable to automation.

5. Required expertise in auto radiography because of using radio actively labeled probes

Random Amplified Polymorphic DNA (RAPD)

It is a PCR based molecule marker technique. Here, single short oligonucleotide primers are arbitrarily selected to amplify a set of DNA segments distributed randomly throughout the genome. RAPD amplification is performed in condition resembling those of PCR using genomic DNA from the species of interest and a single short oligonucleotide (usually 10 base sequences) primer. The DNA amplification product is generated from a region that is flanked by a part of 0 bp priming sites in the appropriate orientation. Genomic DNA from different individuals often produces different amplification patterns *i.e.* RAPDs. A Particular fragment generated for one individual but not for the other represents DNA polymorphism and can be used as a genetic marker. A DNA sequence different between individuals in a primer binding site may result in the failure of primer to bind, and hence in the absence of a particular band among the amplification product. The reaction products are conveniently analysed on agarose gels stained with Ethidium bromide and seen under UV light.

In inheritance studies, the amplification products are transmitted as dominant markers.

Thus presence of a RAPD band corresponding to a PAPD detminant allele against absence of a band that corresponds to a recessive allele. Thus heterozygous and homozugous dominant individuals cannot be differentiated with RAPD markers. In F2, the segregating population may be scored as follows. Therefore in F_2 segregating population may be scored a follows:

Band present AA (homozygote dominant) Aa (heterozygous) Band absent aa (recessive homozygote) therefore in F_2 these two classes would be expected to segregate in 3:1 ratio. This causes a loss of information relative to RFLP markers which show codominance.

Applications

1. Construction of genetic maps.
2. Mapping of traits.
3. Analysis of genetic structure of populations.
4. Finger printing of individuals.
5. Targeting markers to specific region of the genome.
6. Identification of somatic hybrids.
7. A useful system for evolution and characterization of genetic resources.

Advantages of RAPD

1. Simple quick to perform
2. Require relatively very small amounts of DNA
3. Involves no radio activity

Limitations

Unable to distinguish belong homozygous and heterozygous.

Several marker systems have been developed and are applied to a range of crop species. These are the Restriction Fragment Length Polymorphisms (RFLPs), Random Amplification of Polymorphic DNAs (RAPDs), Sequence Tagged Sites (STS), Amplified Fragment Length Polymorphisms (AFLPs), Simple Sequence Repeats (SSRs) or microsatellites, and Single Nucleotide Polymorphism (SNPs). The advantages and disadvantages of these marker systems are provided in (Table 5.1).

Table 5.1: Comparison of most Commonly Used Narker Systems (Adopted from Korzun, 2003)

Feature	*RFLPs*	*RAPDs*	*AFLPs*	*SSRs*	*SNPs*
DNA required (μg)	10	0.02	0.5-1.0	0.05	0.05
DNA quality	High	High	Moderate	Moderate	High
PCR-based	No	Yes	Yes	Yes	Yes
Number of polymorph loci analyzed	1.0-3.0	1.5-50	20-100	1.0-3.0	1.0
Ease of use	Not easy	Easy	Easy	Easy	Easy
Amenable to automation	Low	Moderate	Moderate	High	High
Reproducibility	High	Unreliable	High	High	High
Development cost	Low	Low	Moderate	High	High
Cost per analysis	High	Low	Moderate	Low	Low

These molecular techniques have been widely used to monitor differences in DNA sequence in and among species. They also allow the creation of new sources of genetic variation by introducing new and desirable traits from wild varieties into elite lines. While RFLP markers have been the basis for most genetic work in crop plants, AFLPs and SSRs are currently the most popular techniques used due to ease in detection and automation. The adoption of the new marker system, SNPs, is now highly preferred, with the increasing amount of sequence information, and the determination of gene function due to genomic research.

MAS for Pathogen Resistance in Tomato

One of the major constraints in tomato cultivation and production are severe harvest losses caused by a number of pathogens, including viruses, bacteria, fungi, and nematodes. Farmers have adopted control measures, such as applications of agrochemicals and use of resistant lines. Although conventional breeding has had a significant impact on improving resistance of tomato, the time-consuming process of making crosses and backcrosses, and the selection of the desired resistant progeny make it difficult to react adequately to the evolution of new virulent pathogens.

Molecular markers are now being widely used for breeding tomato. More than 40 genes that confer resistance to major classes of tomato pathogens have been mapped, cloned, and/or sequenced (Grube, *et al.*, 2000). These maps have allowed for

"pyramiding" resistance genes in tomato through MAS, where several resistance genes can be engineered into one genotype. Currently, tomato breeding through MAS has resulted in varieties with resistance or tolerance to one or more specific pathogens.

Genetic Markers

In conventional plant breeding, genetic variation is usually identified by visual selection. However, with the development of molecular biology, it can now be identified at the molecular level based on changes in the DNA and their effect on the phenotype. Molecular changes can be identified by the many techniques that have been used to label and amplify DNA and to highlight the DNA variation among individuals. Once the DNA has been extracted from plants or their seeds, variation in samples can be identified using a polymerase chain reaction (PCR) and/or hybridization process followed by Polyacrylamide gel electrophoresis (PAGE) or capillary electrophoresis (CE) to identify distinct molecules based on their sizes, chemical compositions and charges. Genetic markers are used to tag and track genetic variation in DNA samples.

Genetic markers are biological features that are determined by allelic forms and can be used as experimental probes or tags to keep track of an individual, a tissue, cell, nucleus, chromosome or gene. In classical genetics, genetic polymorphism represents allelic variation. In modern genetics, genetic polymorphism is the relative difference at any genetic locus across a genome. Genetic markers can be used to facilitate studies of inheritance and variation

Desirable genetic markers should meet the following criteria:

1. High level of genetic polymorphism;
2. Co-dominance (so that heterozygotes can be distinguished from homozygotes);

3. Clear distinct allele features (so that different alleles can be identified easily);
4. Even distribution on the entire genome;
5. Neutral selection (without pleiotropic effect);
6. Easy detection (so that the whole process can be automated);
7. Low cost of marker development and genotyping; and
8. High duplicability (so that the data can be accumulated and shared between laboratories).

Most molecular markers belong to the so-called anonymous DNA marker type and generally measure apparently neutral DNA variation. Suitable DNA markers should represent genetic polymorphism at the DNA level and should be expressed consistently across tissues, organs, developmental stages and environments; their number should be almost unlimited; there should be a high level of natural polymorphism; and they should be neutral with no effect on the expression of the target trait. Finally, most DNA markers are co- dominant or can be converted into co-dominant markers.

Table 5.2 lists the major molecular marker technologies that are currently available. Only a selection of widely-used representative the molecular mechanism of several major DNA markers and the genetic polymorphisms that can be generated by restriction site or PCR priming site mutation, insertion, deletion or by changing the number of repeat units between two restriction or PCR priming sites and nucleotide mutation resulting in a single nucleotide polymorphism (SNP). There are several comprehensive reviews that cover all the important DNA markers, *e.g.* Reiter (2001), Avise (2004), Mohler and Schwarz (2005) and Falque and Santoni (2007). Further information regarding the application of DNA markers in genetics and breeding can be found in Lörz and Wenzel (2005). After a brief review of the classical markers, DNA markers will be discussed in more detail in this section.

Molecular maps using markers RFLP, RAPD, SSR, ESTs for major crop species like rice, maize, tomato etc has been utilized very effectively in crop improvement programmes like marker assisted selection (MAS). Many markers have been identified which are closely linked to genes for agronomic traits of interest. These include markers for genes coding for:

1. **Pest and disease resistance** (against viruses fungi, bacteria, nematodes, insects) for *e.g.* RFLP has been used to map Tm-2a locus in tomato resistant to tobacco mosaic virus (TMV), gene Mi in tomato for resistance against root knot nematode, two genes in rice for blast resistance and in many other crops. RAPD assisted selection has been done for *pto* gene conferring resistance to Pseudomonas in tomato. IARI has developed new improved Basmati by marker assisted transfer of *Xa13 and Xa21* genes against bacterial leaf bight.
2. **Quality traits** For *e.g.* malting quality in barley, alkaloid levels etc.
3. **Abiotic stresses** *e.g.* tolerance to salinity or drought.
4. **Developmental traits** for *e.g.* flowering time, vegetative period.

Table 5.2: DNA Markers and Related Major Molecular Techniques

Southern blot-based markers

1. Restriction fragment length polymorphism (RFLP)
2. Single strand conformation polymorphic RFLP (SSCP-RFLP)
3. Denaturing gradient gel electrophoresis RFLP (DGGE-RFLP)

PCR-based markers

1. Randomly amplified polymorphic DNA (RAPD)
2. Sequence tagged site (STS)
3. Sequence characterized amplified region (SCAR)
4. Random primer-PCR (RP-PCR)
5. Arbitrary primer-PCR (AP-PCR)
6. Oligo primer-PCR (OP-PCR)
7. Single strand conformation polymorphism-PCR (SSCP-PCR)
8. Small oligo DNA analysis (SODA)
9. DNA amplification fingerprinting (DAF)
10. Amplified fragment length polymorphism (AFLP)
11. Sequence-related amplified polymorphism (SRAP)
12. Target region amplified polymorphism (TRAP)
13. Insertion/deletion polymorphism (Indel)

Repeat sequence-based markers

1. Satellite DNA (repeat unit containing several hundred to thousand base pairs (bp)
2. Microsatellite DNA (repeat unit containing 2–5 bp)
3. Minisatellite DNA (repeat unit containing more than 5 bp)
4. Simple sequence repeat (SSR) or simple sequence length polymorphism (SSLP)
5. Short repeat sequence (SRS)
6. Tandem repeat sequence (TRS)

mRNA-based markers

1. Differential display (DD)
2. Reverse transcription PCR (RT-PCR)
3. Differential display reverse transcription PCR (DDRT-PCR)
4. Representational difference analysis (RDA)
5. Expression sequence tags (EST)
6. Sequence target sites (STS)
7. Serial analysis of gene expression (SAGE)

Single nucleotide polymorphism-based markers

1. Single nucleotide polymorphism (SNP)

5. **Quantitative traits** *e.g.* high fibre strength in cotton
6. **Micronutrient uptake** *e.g.* high iron and zinc uptake in wheat

The role of molecular MAS for crop improvement is in increased speed and accuracy of selection, gene pyramiding, and reduced cost of field based selection. Thus rather than growing breeding lines in the field and testing for important traits over the growing season, it is possible to extract DNA from 50ng of seedling leaflet and test for presence or absence of a range of traits in that DNA sample in one day. Plants lacking the required traits can then be removed early in the breeding programme. With the availability of more validated molecular marker, MAS therefore, becomes a highly cost effective and efficient process. The same principle used in developing molecular markers can be applied for range of molecular diagnostics and DNA fingerprinting *viz.* identification of breeding lines and varieties, characterization of genetic resources and study of phylogenetic relationships.

Genetic Transformation – Gene Transfer Methods – Indirect Method of Gene Transfer – *Agro-bacterium* Mediated Gene Transfer Method Genetic Transformation of Plants or Transgenic Techniques

Genetic engineering over comes the limitations of traditional Breeding and allows scientists to use new traits from many kinds of plants and other living things such as fish, insects, bacterium and even humans. If a new gene is introduced to another organisms cell, the cell begin to produce that genes particular protein and will display the new character and trait directed by the gene.

Steps for developing new crop varieties:

Step-1: Finding a plant or other organism showing the desired characteristic.

Step-2: Identify the gene controlling that trait and its location on a chromosome (By using genome mapping).

Step-3: Marking the gene, so its presence can be detected quickly through "marker assisted selection". Specific genes or segments of DNA, called markers are used to mark the location of the genes to be transferred.

Step-4: Isolating the gene from the rest of the DNA so that only the desired information will be transferred. (By endonuclease enzymes).

Step-5: Producing large number of the gene and introducing the desired gene into cells of the organism to be enhanced (Agrobacterium gene transfer or by particle gun method).

Step-6: Identify which plant cells now contain the desired gene, through the use of markers as described in step-3.

Step-7: Growing the altered cells into full plants, using a process called tissue culture, to confirm that the desired trait is expressed properly.

Step-8: Using traditional plant breeding techniques to transfer the trait into a usable variety.

Transgenic Techniques

The common means of transferring genes into plants is by employing the natural ability of some bacteria to alter the DNA of organisms, which they infect. These bacteria have the ability to inject a segment of their DNA into that of an infected cell, there by affecting its behavior. For example, the common soil bacterium, *A. tumifaciens* causes a disease called crown gall is susceptible plants by inserting its genes into a plants DNA.

The bacterial genes disrupt the genetic code of the plants cells in the infected area causing them to grow into an abnormally shaped mass called a gall. The bacterial species produce small, circular pieces of DNA called plasmids, which function independently of other Bacterial DNA and can enter cells of other organisms, carrying genetic information with them that then merges with that of the host cell. To transfer genes from one cell to another, researchers use an enzyme to cut a gap in a bacterial plasmid then they insert a gene from the donor DNA strand in to the plasmid. Because the cut ends of both the plasmid and the donor gene segment are chemically 'sticky' they attach to each other or recombine to form a plasmid containing the new gene (hence the name, recombinant DNA (rDNA) Technology). The recombined plasmid now carries the gene that can result in the expression of the new trait when transferred into a plant cell by the plasmid. Once a gene is inserted into a plasmid, it can be introduced into a bacterium that is easily reproduced in the lab when the altered bacteria multiply; the plasmid containing the desired gene is also copied or cloned in every cell. Because bacteria grow quickly, this method makes it possible to make millions of copies of the desired gene in a short period.

Using the appropriate enzymes, the plant genes are extracted again from the bacteria and inserted into the plasmids of *A. tumefaciens*. Finally, the altered *A. tumefaciens* is mixed with the target plant cells. The foreign gene is then incorporated into DNA of some of the target plant cells.

Direct Methods of Transformation – Particle Bombardment/Gene Gun Method – Electroporation, Microinjection, Macroinjection, Liposome, Mediated etc.

Introduction of DNA into plant cells without the involvement of biological agents such as Agrobacterium and leading to stable transformation is called direct gene transfer. The various methods of direct gene transfers are

1. Chemical methods
2. Electroporation
3. Particle bombardment
4. Lypofection
5. Micro injection
6. Macro injection
7. Pollen transformation
8. Delivery via growing pollen tubes
9. Laser induced transformation
10. Fibre mediated transformation etc.

Some of these like involve lipofection, electroporation delivery of DNA to protoplast and thus require regeneration of plants from transformed protoplast.

1. Chemical Methods

It is based on ability of protoplast to uptake the foreign DNA from surrounding solution. An isolated plasmid DNA is mixed with protoplast in the presence of the poly ethylene glycol (PEG), PVA and Ca (PO4) polyvinyl alcohol which enhance the uptake of DNA by protoplast. After 15-20 min of incubation the protoplasts are cultured. On the presence of appropriate selective agents, the protoplasts are regenerated and the transgenic plants are further characterized for conformation.

2. Electroporation

Induction of DNA into cell by exposing them for a very brief period to high voltage electrical pulses to induce transiant pores in the plasma lemma is called Electroporation. Generally protoplasts are used since they have expand plasma membrane. A suspension of protoplast with a desired DNA is prepared. Then a high voltage current is applied through the protoplast DNA suspension. The electric current leads to the formation of small temporary holes in the membrane of the protoplasts through which the DNA can pass. After entry into the cell, the Foreign DNA gets incorporated with the host genome, resulting the genetic transformation.

The protoplasts are then cultured to regenerate in to whole plants. This method can be used in those crop species in which regeneration from protoplast is possible.

3. Particle Bambordment/Microprojectile/Biolistic/Gene gun/Particle Acceleration

The process of partical acceleration (or) biolistics acceleration of DNA into cells with sufficient force such that a part of it gets integrated in to DNA of target cells. The process of transformation employes foreign DNA coated with minute 0.2-0.7 μm gold (or) are tungsten particles to deliver into target plant cells. Two procedures have been used to accelerate the minute particles

1. By using pressurized helium gas.
2. By electro static energy released by a droplet of water exposed to a high voltage

This method is being widely used because of its ability to deliver foreign DNA into regenerable cells, tissue (or) organs irrespective of monocots (or) dicots Because of the physical nature of process there is no biological limitation to the active DNA delivery that makes it, genotype independent. This method allows the transport of genes into many cells of nearly any desired position in an experimental system without too much manual labour.

4. Lypofection

Introduction of DNA into cells via lyposomes is known as lipofection, lioposomes are small lipid artificial vesicles. The procedure of liposome encapsulation was developed to protect the foreign DNA during the transfer process The DNA enclosed in the lipid vesicles when mixed with protoplast under appropriate condition penetrates into the protoplast where lipase activity of the protoplast dissolves the lapid vesicles and DNA gets released for integration into the host genome. This method has not been commonly used as it is difficult to construct the lipid vesicles. The success depends upon the protoplast regeneration.

5. Microinjection

The DNA solution is injected directly inside the cell using capillary glass micropipettes with the help of micromanipulators of a microinjection assembly. It is easier to use protoplast than cells since cell wall interferes with the process of microinjection. The protoplasts are usually immobilized in agarose (or) on a glass slides coated with polylysine or by holding them under suction by a micropipette. The process of microinjection is technically demanding and time consuming a maximum of 40-50 protoplasts can be microinjected in one hour

6. Macroinjection

The injection of plasmid DNA into the lumen of developing inflorescence using hypodermic syringe is known as macro injection An aqueous solution of DNA was introduced into the developing floral tillers 14 days prior to meiosis. Transformed seeds were obtained from these injected tillers after cross pollination with other and injected tillers. However the mechanism by which the DNA entered the zygotic tissue yet unknown.

7. Pollen Transformation

Involves the gene transfer by soaking the pollen grains in DNA solution prior to their use for pollination. The method is highly attractive in view of its simplicity and general applicability but so far there is no definite evidence for a transgene being transferred by pollen soaked in DNA solution

8. DNA Delivery via Growing Pollen Tubes

The stigmas were cut after pollination exposing the pollen tubes, the DNA was introduced onto the cut surface that presumably diffused through the germinating pollen tube into the ovule. This method is simple easy and very promising provided consistent result and stable transformations are achieved The mechanism of DNA transfer into zygote through this method is not yet established.

9. Laser Induced Transformation

It is method of introducing DNA into plant cells with a laser microbeam. Small pores in the membrane are created by laser microbeam. The DNA from the surrounding solution may than enter into the cell cytoplasm through the small pores Lasers have been used to deliver DNA into plant cells but there is no information on transient expression or stable integration.

10. Fiber Mediated Transformation

The DNA is delivered into the cell cytoplasm and nucleus by silicon carbide fibres of 0.6 µm diameter and 10-80 µm length. The fibres mediated delivery of DNA into the cytoplasm is similar to microinjection. The method was successful with maize and tobacco suspension cell culture It is the most rapid and expensive method of DNA delivery provided stable integrations are achieved.

DNA Imbition by Cells Tissue, Embryos and Seeds

When dry isolated embryos of wheat barly, rye, pea and bean are imbibed in a DNA solution, they take up the DNA and show the expression of marker gene. Dry seeds, whose seed coats have been removed also, take up DNA when imbibed in a DNA solution. The imbibed seeds or embryos are germinated on appropriate selective medium to isolate the transformed embryos. It was thought that the DNA is taken up by the embryos through the cells injured during their isolation. The DNA then moves through plasmadesmata to other cells of embryos. Recently two DNA delivary systems namely

1. Agrobacterium mediated gene transfer.
2. Particle bombardment are widely used for the development of transgenic embryos.

B) Transgenic Plants

Transfer of gene from an organism into a plant cell and its integration into the genetic material of the later usually employing recombinant-DNA technique is known as Genetic engineering of plants It is the most potent biological approach which constitute the transfer of Specifically constructed gene assembles through various transformation techniques. The plants obtained through genetic engineering contain a gene (or) genes usually from an unrelated organism. Such genes are called transgenes and plants containing transgenes are known as transgenic plants.

The first transgenic plant was produced in 1983 when a tobacco line expressing Kanamycin resistant was produced. Soon after transgenic crop varieties resistant to herbicides, insects (or) viruses (or) expressing male sterility, delayed ripening, and slow fruit

Softening was developed. Flavr savr tomato was first transgenic variety to reach market. Fruits of this variety remain fresh for a prolonged period.

Transgenic Dicotyledonous Plants

1. *Nicotiana tabacum*
2. *Helianthus annuus*
3. *Beta vulgaris*
4. *Solanum tuberosum*
5. *Glycine max*
6. *Gossypium hirsutum*

Transgenic Monocotyledonous Plants

1. *Asparagus* sp.
2. *Zea mays*
3. *Oryza sativa*
4. *Avena sativa*

Herbicide Resistance in Transgenic Plants

Under normal circumstances, herbicides affect photosynthesis or biosynthesis of essential amino acids. Under field conditions, application of herbicides not only kills the unwanted weeds but also greatly affects the field crops. In order to protect the crops against exposure to herbicides, scientists after intensive research isolated a gene from Streptomyces hygroscopicus which encodes an enzyme, capable of inactivating the herbicide 'Basta'. Transgenic plants with this gene have been developed, demonstrating effectiveness of this gene for protection against herbicide 'Basta'. Thus, herbicide-tolerant crop plants have now been developed by genetically manipulating plant genomes resistant to specific herbicides.

Improved Resistance to Insect Pests and Microbial Diseases

Genes from *Bacillus thuringiensis* (Bt2) have been introduced into several crops, including tomato and cotton, and field-testing has demonstrated impressive results against many pests. Spore preparation of this bacterium is used as a biological insecticide during the last 20 years. Insecticidal activity depends on a toxic protein called delta endotoxins. The toxin gene (Bt2) from *Bacillus thuringiensis* has been isolated and used for Agrobacterium. Ti plasmid mediated transformation of tobacco, cotton and tomato plants. The transgenic plants were resistant to the *Manducta sexta*, a pest of tobacco. India had acquired technology from U.S.A. for introducing Bt toxin gene in cotton for the development of resistance against pests in this major cash crop of India. Widespread use of insecticides, fungicides and pesticides for crop protection undoubtedly has damaging effects on the environment and hence it is important to improve the control of pests and diseases by genetic means. Genetic modification of plants is an attempt for ecofriendly measures against environmental degradation.

Through genetic modification, the oil-producing soya bean was tailored to produce a wide range of industrial lubricants, cosmetic compounds and detergents that are biodegradable. A whole new area of biotechnology has been opened up and plants are made to synthesize many novel substances including functional human antibody fragments.

1) Genes have been Successfully Transferred to Many Crops for Resistance to Various Biotic Stresses

Such as the biopesticidal gene of bacterium *Bacillus thuringiensis* (Bt) has been incorporated into cotton for resistance against bollworm, maize against European corn borer, tomato, tobacco against Bt cotton is already in markets and rest are in process. Also, plant derived insecticidal genes like protease inhibitors; lectins etc have also been integrated into many pulse crops which are under field trials.

Also, transgenic plants have been especially successful for control of viral diseases. The approach followed is to identify those viral genes or gene products which when present a wrong time or in improper amount, will intefere with the normal functions of the infection process and prevent disease development. Few examples of transgenic plants with viral resistance genes are in tobacco against TMV, rice against rice yellow mosaic virus, potato against potato virus Y and potato leaf roll virus etc.

2) Genes Resistant to Abiotic Stresses like Herbicide Resistance

Herbicides are the method of choice to control weeds and hence maximize crop yields by reducing by reducing competition from weeds. This has necessitated development of safer, biodegradable herbicides and development of crop plants resistant to those herbicides. Also, because herbicide resistance genes are also effective selectable marker genes in culture, herbicide tolerant crop varieties were the first major transgenic trait to be produced and commercialized. Based on either expression of a herbicide insensitive gene, degradation of herbicide or over expression of herbicide target gene product, engineered resistance is now available to a range of herbicides for *e.g.* transgenic petunia plants resistant to glyphosate of Roundup herbicide was developed by transfer of a gene for EPSPS (5-enol-pyruvyl-shikimate-3 phosphate synthase) that overproduces this enzyme. Transgenic tomato and maize plants using herbicide detoxifying gene GST (glutathione- S transferase) have been successfully used.

Also, extensive research is going on for producing transgenic crops against salt, drought, chilling stresses in rice, wheat, tomato etc. In most cases, advances to generate stress tolerant plants by traditional breeding are slow because of involvement of many genes and physiological processes. For cold and drought tolerance, recent research has shown that a series of functionally different cold and drought response genes show common promoter regulatory sequences. These results are significant because they show that introduction of a single regulatory gene confer tolerance to different stresses.

3) Gene Transfers to Improve Quality of Food Products

Bruise resistant tomatoes were developed which expressed antisense RNA against polygalatouronase which attacks pectin in the cell walls of ripening fruit and thus softens the skin. This transgenic tomato was commercialized under the name of Flavr Savr. Also, tomatoes with delayed ripening were developed by using gene for ACC deaminase to degrade ACC which is immediate precursor to ethylene. This increases shelf life of tomato. Starch content in potatoes could be increased by 20-40 per cent by using a bacterial ADP glucose pyrophosphorylase gene.

4) Male Sterlity and Fertility Restoration in Transgenic Plants

which is required for hybrid seed production. For many crops, it is difficult to generate commercial hybrid seed by conventional means. Hybrid seed is attractive to seed companies as farmers must purchase new seed from them each year, since hybrid varieties donot breed true. It is now possible to engineer male sterlity bt expression of a ribonuclease gene (*barnase*) specifically during development of the tapetal layer that nourishes developing pollen grains. Developmental regulation of the ribonuclease by the TA29 tapetum specific promoter kills the tapetal cells leading to male sterility. Male sterile plants can be used as the female parent to produce hybrid seed. Fertility can be restored by expression of the *barstar* gene, which inactivates *barnase*. This technology can be used to produce hybrids of crops such as maize, sugarbeet or canola. It is not possible to produce hybrid canola conventionally but hybrid canola can exhibit hybrid vigor and increased yields.

5) Molecular Farming

A major role of biotechnology is the use of transgenic plants as factories for manufacturing speciality chemicals and pharmaceuticals such as transgenic tobacco plants carrying mannitol dehydrogenase gene from *E. coli* is used for increasing production of mannitol. Similarly production of Polyhydroxy butyrate (PHB) in plants provide attractive source of biodegradable plastics at low cost. Two genes which catalyse two steps in production of PHB, acetoacetyl CoA reductase (*phb B*) and polyhydroxy butyrate synthase (*phb C*) have been successfully transformed and expressed in *Arabidopsis thaliana*. Efforts are on to create transgenic plants as source of edible vaccines, antibodies, pharmaceuticals.

6) To Study Regulated Gene Expression

Transgenic plants have proven to be particularly useful tools in studies on plant molecular biology, gene regulation, identification of regulatory sequences involved in differential expression of gene activity. T-DNA and transposable elements are used as molecular tags to produce mutations by becoming inserted within genes thus making it non functional. Using this approach regulatory sequences of many structural genes have been cloned like light inducible genes like ribulose biphosphate carboxylase (*rbcS*), chlorophyll a/b binding protein (Cab), heat shock protein genes etc. The genes for seed storage proteins provide an excellent example of a cell and tissue specific expression.

Role of Transgenic Plants in Crop Improvement

Transgenic plants are those which carry additional stability integrated and expressed foreign genes, usually transferred from unrelated organisms. The combined use of recombinant-DNA technology, gene transfer methods and tissue culture technique has lead to efficient productions of transgenics in a variety of crop plants. Transgenic plants or genetically modified plants (GMP) have both basic and applied uses which are briefly summerised below:

1. Transgenes will be important in increasing the efficiency of crop production systems. For instance tragenic plants resistant to herbicides, insects, viruses and other biotic and abiotic stress have already been produced.
2. Transgenic breeding is an effective means of inducing male sterility in crop plants. *e.g.*, Barnase and barstar systems in *Brassica napus*.
3. Transgenic plants which are stable for food processing have also been produced. *e.g.*, Bruise resistant and delay ripening in tomato.
4. Several gene transfers have been aimed in improving the produce quality. *e.g.*, Protein and lipid quality : improved quality may be achieved by either supression or over production of endogenous genes.
5. Transgenic plants are aimed to produce novel biochemicals like interferon, Insulin, Immunoglobin etc useful biopolymers like poly hydroxy butyrate which are not produced by normal plants. These compounds are extracted from plants can be used as pharmaceutical (or) industrial substrates.

6. The utilization of transgenic plants as bioreacters (or) factories for the production of special chemical and pharmaceuticals is known as Molecular forming.
7. Transgenic plants have been produced that express a gene encoding antigenic protein from a pathogen. Use of transgenic plants as vaccines for immunization against pathogens fast emerging as an important objective.
8. Transgenic plants have proved to be extremely valuable tools in studies on plant molecular biology. Regulation of gene action, identification of regulatory, promoters sequences etc.

Limitations

1. Transgenic plants sometimes exhibit instable performance for character under consideration.
2. The transformants had the undesirable side effects of trans genes.
3. The position and integration of foreign gene in host genome effects the expression of transgene in the transformant.
4. Inability to transfer polygenic traits.
5. Transgenic breeding is a very expensive method of crop improvement.
6. Requires high technical skills.
7. Transgenic cells are recovered at a very low frequency in cell culture, more over regene ration of transgenic cells into whole plants is also difficult and time consuming task.
8. Transgenic breeding acts against natural evolution.
9. There are chances of developing new weed species through transgenic breeding.
10. Sometimes the foreign genes have adverse effects on genome of the recipient parent, in such cases it may give rise useless gene combinants which may become a problem.

Practical Application of Genetic Transformation

By genetic manipulation, it is possible to obtain plants with insecticidal property. Thus, application of chemical pesticides to the crop plants is reduced.

Genetic manipulation is also carried out in crops for making desirable storage proteins, vitamins, amino acids and for the suppression of antinutritional protein synthesis. Plants are made to produce large amount of secondary metabolites having high commercial value.

Losses during storage and transport of some crops can be as high as 80 per cent. This is mainly due to biological activities - bruising, heat and cold damage in soft fruits and vegetables. In tomato the enzyme polygalactronase breaks down cell wall constituents, thus leading to softening of the fruit during ripening. By inhibiting the polygalactronase by antisense genes the tomato can remain dormant fresh until mature and be transported in a firm solid state. Antisense RNA is a RNA molecule capable of controlling and expression of particular enzymes which are involved in ripening processes.

Genetic manipulation of flower and leaf colour, abundance of flowers, perfume and shape are now the major targets for decorative plant industries.

Although, biotechnology alone will not be able to generate sustainable food production, however, the "gene revolution' will have immense role in crop improvement and agriculture.

References

Avise, J.C. (2004). *Molecular Markers, Natural History and Evolution*, 2nd edn. Sinauer Associates, Inc., Sunderland, Massachusetts.

Borlaug, N.E. (2001). Feeding the world in the 21st century: the role of agricultural science and technology.Speech given at Tuskegee University, April 2001. Available at: http://www.agbioworld.org/biotechinfo/topics/borlaug/borlaugspeech.html(accessed17 November 2009).

Damude, H.G. and Kinney, A.J. (2008). Enhancing plant seed oils for human nutrition. *Plant Physiology* **147**, 962–968.

Falque, M. and Santoni, S. (2007). Molecular markers and high-throughput genotyping analysis. In: Morot- Gaudry, J.F., Lea, P. and Briat, J.F. (eds) *Functional Plant Genomics*. Science Publishers, Enfield, New Hampshire, pp. 503–527.

Grube, R.C., Radwanski, E.r., Jahn, M. 2000. Comparative genetics of disease resistance within the Solanaceae. *Genetics* **155**: 873-887.

Korzun, V. 2003. Molecular markers and their applications in cereals breeding. A paper presented during the FAO international workshop on "Marker assisted selection: A fast track to increase genetic gain in plant and animal breeding?". 17-18 October 2003, Turin, Italy.

Lörz, H. and Wenzel, G. (eds). (2005). *Molecular Marker Systems in Plant Breeding and Crop Improvement. Biotechnology in Agriculture and Forestry*, Vol. 55. Springer-Verlag, Berlin.

Mohler, V. and Singrün, C. (2004). General considerations: marker-assisted selection. In: Lörz, H. and Wenzl, G. (eds). *Biotechnology in Agricultural and Forestry*, Vol. 55. *Molecular Marker Systems in Plant Breeding and Crop Improvement*. Springer-Verlag, Berlin, pp. 305–317.

Phillips, R.L. (2006). Genetic tools from nature and the nature of genetic tools. *Crop Science* **46**: 2245–2252.

Phillips, R.L. (2008). Can genome sequencing of model plants be helpful for crop improvement? *Proceedings of 5th International Crop Science Congress*, 13–18 April 2008, Jeju, Korea. International Crop Science Society, Madison, Wisconsin.

Reiter, R. (2001). PCR-based marker systems. In: Phillip, R.L. and Vasil, I.K. (eds). *DNA- Based Markers in Plants*. Kluwer Academic Publishers, Dordrecht, Netherlands, pp. 9–29.

Swaminathan, M.S. (2007). Can science and technology feed the world in 2025? *Field Crops Research* **104**, 3–9.

Xu, Y., McCouch, S.R. and Zhang, Q. (2005). How can we use genomics to improve cereals with rice as a reference genome? *Plant Molecular Biology* **59**, 7–26.

Xu, Y., Skinner, D.J., Wu, H., Palacios-Rojas, N., Araus, J.L., Yan, J., Gao, S., Warburton, M.L. and Crouch, J.H. (2009c). Advances in maize genomics and their value for enhancing genetic gains from breeding. *International Journal of Plant Genomics* Volume 2009, ArticleID957602, 30pages.

Suggested Readings

Adams, R.P. and Adams, J. E. (eds.) 1992, Conservation of Plant Genes, DNA Banking and *In Vitro* Technology. Academic Press, New York.

Bhojwani, S.S. (eds.) 1990, Plant Tissue Culture: Applications and Limitations. Elsevier, Amsterdam.

Lewin, B. 2004 Genes VIII, Oxford University Press, New York.

Old, R.W. and Primrose, S.B. 1990. Principles of Gene Manipulation, An Introduction to Genetic Engineering. Blackwell Science Publishers, London.

Singer, M. and Berg, P. 1991. Genes and Genomes, University Science Books, Mill Valley, California and Blackwell Science Publishers.

Singh, B.D. 1998, Biotechnology, Kalyani publishers, New Delhi

Slater, A., Scott, N. and Fowler, M. 2003, Plant Biotechnology: The Genetic Manipulations of Plants, Oxford University Press.

Street, H. E. 1977, Plant Cell and Tissue Culture, Blackwell, London.

Vasil, I.K. and Thorpe, T. A. (eds.) 1994. Plant Cell and Tissue Culture. Kluwer Academic Publishers, London.

2015, Major Constrains and Verdict of Crop Productivity Pages 98–116
Editor: Dr. U.N. Bhale
Published by: ASTRAL INTERNATIONAL (P) LIMITED, NEW DELHI

Chapter 6

Downy Mildew of Maize: A Review

A.P. Suryawanshi*, Utpal Dey and D.P. Kuldhar

Department of Plant Pathology
Vasantrao Naik Marathwada Krishi Vidyapeeth,
Parbhani – 431 402, M.S., India

Introduction

True downy mildews are caused by a group of oomycetes that belong to the family peronosporace. All species of this family are obligate parasites of higher plants and cause downy mildew diseases on most cultivated grain crops, Vegetables, field crops, ornamentals, shrubs and Vines. The downy mildews have caused spectacular and catastrophic epidemics on several crops in the past and some of them continue to cause several losses. In recent years, the downy mildews of sorghum has appeared spread in the United States and has raised fears of future introduction of other, even more serious, downy mildews of grain crops now present in Asia and Africa.

Downy mildew caused by *Peronosclerospora sorghi* (Weston and Uppal.) C.G. Shaw, is one of the most destructive diseases of maize (*Zea mays* L.) in the humid sub-tropical and tropical regions of Asian countries and widely distributed (Williams, 1990; Rashid *et al.*, 2013). *Peronosclerospora sorghi* is a soil borne oomycetous fungus that infects sorghum and maize crops. The fungus invades the plants through the root system, resulting in plant leaves to turn chlorotic, usually within 2 weeks of sowing. Heavily infected plants can be dwarfed and a white downy growth may appear on both surfaces of infected leaves.

* *Corresponding Author.* E-mail: apsmkv@rediffmail.com

The pathogen is known to cause both local and systemic symptoms (Kling *et al.*, 1994; Hau *et al.*, 1995; Adenle and Cardwell, 2000; Odvody, 2002; Shaarawy *et al.*, 2002) which cause heavy economic losses. The disease was first reported in maize during 1970 at Samaru, Zaria, regions of Nigeria (Kings and Webster, 1970).

Occurrence, Distribution and Yield Losses

Geographic distribution, alternate hosts and yield losses of various downy mildews of maize are summarized in the Table 6.1.

Table 6.1: Geographic Distribution, Host Range, and Yield Losses of Various Downy Mildews of Maize.

Pathogen (Disease Name)	*Geographic Distribution*	*Host Range*	*Yield Loss*
Peronosclerospora sorghi (Sorghum D.M.)	Americas (North, Central, South), Asia, Africa, Europe, Australia	Cultivated and wild sorghum, Johnson grass, teosinthe, wild grasses (*Panicum, Penniselum, Andropogon* spp.)	Severe outbreaks in India, Israel, Mexico, Nigeria, Texas, Thailand, and Venezuela, caused as high as 90 per cent yield losses (Nigeria).
P. maydis (Java D.M.)	Indonesia, Australia	Teosinthe, wild grasses (*P, Tripsacum*s pecies).	Very serious in Indonesia, upto 40 per cent crop loss.
P. philippinensis (Philippine D.M.)	Philippines, China, India, Indonesia, Nepal, Pakistan, Thailand	Oats, teosinthe, sugarcane, sorghum	Very serious in Philippines, yield losses 15 - 40 per cent. Or more upto 70 per cent.
P. sacchari (Sugarcane D.M.)	Australia, Fiji, Taiwan, Japan, Nepal, New Guinea, India, the Philippines, Thailand	Sugarcane, teosinthe, sorghum and wild grasses	Serious in Australia and Asia, yield losses 30 to 60 per cent.
Sclerospora graminicola (Graminicola D.M.)	USA and Israel	Wild grasses, millet	Infect maize only in USA and Israel, minor one.
Sclerophthora macrospora (Crazy top D.M.)	Americas, eastern and southern Europe, parts of Africa and Asia	Oats, wheat, sorghum, rice, finger millet, various grasses	Rare in tropical areas, causes extensive loss only in localized area
Scleropthora rayssiae var. *zeae* (Brown stripe D.M.)	India, Nepal, Pakistan and Thailand	*Digitaria* species	Very serious in India, losses 60 per cent.

Asia: Bangladesh, China, India, Iran, Israel, Japan, Nepal, Pakistan, Philippines, Thailand and Yemen (CMI, 1988; Jeger *et al.*, 1998).

Africa: Benin, Botswana, Burundi, Egypt, Ethiopia, Ghana, Kenya, Malawi, Mauritania, Mozambique, Nigeria, Rwanda, Somalia, South Africa, Sudan, Swaziland, Tanzania, Uganda, Zambia and Zimbabwe (CMI, 1988; Jeger *et al.*, 1998; Kutama *et al.*, 2010).

North America: Mexico and USA (Alabama, Alaska, Arkansas, Georgia, Illinois, Indiana, Kansas, Kentucky, Louisiana, Michigan, Minnesota, Mississippi, Missouri, Nebraska, Nevada, New Mexico, Oklahoma, Tennessee and Texas (Bonde, 1982; CMI, 1988; Jeger *et al.*, 1998).

Central America and Caribbean: El Salvador, Guatemala, Honduras, Nicaragua, Panama and Puerto Rico (CMI, 1988; Jeger *et al.*, 1998).

South America: Argentina, Bolivia, Brazil, Colombia, Uruguay and Venezuela (CMI, 1988; Jeger *et al.*, 1998).

The major downy mildew diseases occurring in Asia (Table 6.1) include maize DM [*Peronosclerospora sorghi* (Weston and Uppal) C.G. Shaw], Philippine DM [*P. philippinensis* (Weston) C.G. Shaw], sugarcane DM [*P. sacchari* (T. Miyabe) Shirai and Hara], Java DM [*P. maydis* (Raciborski)] and brown stripe DM [*Sclеropthora rayssiae* var. *zeae* (Payak and Renfro)]. *P. sorghi* causes DM in both maize and sorghum crops in India and elsewhere, however, the *P. sorghi* strain in Thailand does not infect sorghum, and therefore the strain was re-classified as *P. zeae* (Yao *et al.*, 1991, 1992). The *P. sorghi* found in Rajasthan, a state in India, again does not infect sorghum but forms oospores on *Heteropogon contortus* (a wild grass, commonly known as spear grass) as a collateral host and was consequently re-classified as *P. heteropogoni* (Siradhana *et al.*, 1980). When maize is infected by the conidial stage of the fungus it is referred to as Rajasthan downy mildew (RDM) (White, 1999; Spencer and Dick, 2002). The disease has been observed in several districts of Rajasthan (Rathore *et al.*, 2002) and has also been found to be spreading to adjacent states. Sorghum downy mildew (SDM) is particularly prevalent in the peninsular of India, in the states of Karnataka, Tamil Nadu and Andhra Pradesh (Krishnappa *et al.*, 1995).

Downy mildews often cause rapid and severe losses of young crop plants still in the seedbed or in the field. They often destroy from 40 to 90 per cent of young plants in the field, causing heavy or total losses of crop yields. The severity of loss is directly correlated to the prolonged presence of wet, cool weather during which the downy,mildews sporulate profusely, cause numerous new infections, spread and rapidly kill young succulent tissues.

Globally, DM affected areas with significant economic losses are reported to be as high as 30 per cent, including lowland tropical, subtropical, mid-altitude, transition zone and highland environments (Jeffers *et al.*, 2000). Heavy losses (as high as 100 per cent) have been reported in maize from the DM pathogens in several countries in Asia and from other parts of the world (Bonde, 1982; Rifin, 1983; Rathore and Siradhana, 1987). In the Asian tropics where yield losses of 50 per cent or more are common, DM is considered the highest priority as a biotic constraint limiting maize productivity (Pingali, 2001). Despite the introduction of downy mildew resistant (DMR) cultivars and the use of metalaxyl fungicide, severe incidence of the disease still occurs in localized areas (Dalmacio, 2000). Seed treatments with fungicide, undertaken by commercial companies, makes the seeds expensive and generally beyond the financial reach of resource-poor farmers. Also, complete reliance on chemical treatment has previously proved to be an unsound practice as cases have been reported where crops grown with treated seed still sustained losses due to D.M.

(Dalmacio, 2000), indicating that resistance in pathogens against the fungicide is developing (Raymundo, 2000). These cost concerns, and the emerging problem of a build up of chemical resistance in the pathogen, point to the use of resistant varieties as a more cost-effective and environmentally safe alternative for managing the problem of D.M. in maize.

Symptomatology

Symptoms of downy mildew on maize (Plate 6.1) caused by the various pathogenic species are similar, although symptoms can vary depending on plant age, prevailing climatic conditions, and host germplasm. Infection of maize plants at the seedling stage (less than 4 weeks old) results in stunted and chlorotic plants and premature plant death. Leaves on older plants display characteristic symptoms of downy mildews which include mottling, chlorotic streaking and lesions, and white striped

Initial Symptoms

Green Ear Symptoms

Plate 6.1

leaves that eventually shred. 'Downy' growth is often observed on both leaf surfaces, but is more common on the lower leaf surface. On infected plants, leaves are narrower and more erect compared to healthy leaves. On infected plants are often stunted, tiller excessively and have malformed reproductive organs (tassels and ears), such plants may not seed, while tassels may exhibit 'bushy' growth.

There are two types of symptoms produced by *P. sorghi* in both sorghum and maize: systemic infection which occurs by oospores or conidia during the first four weeks of germination of the seed; and local lesions resulting from conidial infection of older plants, which may also give rise to systemic infection (Schuh *et al.*, 1986).

Systemic infection in maize seedlings is characterised by chlorosis that normally appears after seedling emergence. The first leaf is invariably free from infection; this may be because the first leaf outgrowing the pathogen, which requires time to invade the root and stem tissue, or the existence of a passive defence mechanism in the first leaf which prevents entry of the pathogen. If no infection is observed within 2 weeks, the second leaf escapes infection while the younger leaves may subsequently develop symptoms.

Infected maize plants are sometimes stunted and occasionally have white-striped leaves and abnormal seed set. Leaf chlorosis always includes the base of the blade, with the transverse margin, usually sharply defined between the diseased and healthy tissue. This symptom appears further up on the blades in successively formed leaves (half-leaf symptom). The leaves of infected plants tend to be narrower and more erect than those on healthy plants. In maize, most systemically infected plants are sterile, but occasionally some will set seed.

Systemic infection of young seedlings leads to stunted growth, chlorosis and premature death, limiting yields. Under severe infection, tassels are malformed and ears are barren, leading to extensive yield loss.

The affected plant parts exhibits following kinds of the symptoms:

1. Fruits/pods: abnormal shape
2. Inflorescence: abnormal leaves and abnormal forms,
3. Leaves: abnormal colours, abnormal forms, fungal growth and shredding,
4. Stems: internal discolouration
5. Whole plant: dwarfing

The Pathogen

Oogonia of *P. sorghi* are spherical, 40–55 µm in diameter and are embedded among mesophyll cells between fibrovascular bundles. Oospores measure 25–42.9 µm (av. 36 µm), are hyaline, spherical, with light yellow walls, reddish brown coloured and germinate by germ tube. Conidiophores are massive, wedge shaped, short, erect, fragile, hyaline, 180–300 µm long, usually dichotomously branched and emerge through stomata singly or in groups of 2, rarely 3on the lower sides of leaves. Conidia are thin walled hyaline, oval to spherical, 15–26.9 × 15–28.9 µm, and borne on elongated, tapered sterigmata. Conidia and oospores germinate by germ tube under high humidity (APS, 1973). Morphological characteristics of various downy mildew pathogens of maize are given in Table 6.2.

Table 6.2: Morphological Characteristics of Various Downy Mildew Pathogens of Maize

Pathogen (Disease Name)	*Morphological Characteristics*		
	Conidiophores/ Sporangiophores	*Conidia/Sporangia*	*Oospores*
Peronosclerospora sorghi (Maize D.M.)	Erect, dichotomously branched, 180 - 300µm in length, emerge singly or in groups from stomata.	Oval (14.4-27.3 × 15-28.9 µm), borne on sterigmata 13µm long.	Spherical (36µm dia), light yellow or brown coloured.
P. maydis (Java D.M.)	Clustered conidiophores (150 - 550µm in length), emerge from stomata, dichotomously branched (2-4 times).	Spherical to subspherical shaped (17-23µm x 27-39µm).	Not reported.
P. philippinensis (Philippine D.M.)	Erect, dichotomously branched (2-4 times), 150 to 400µm in length, emerge from stomata.	Ovoid to cyclindrincal (17-21µm x 27-38µm), slightly rounded at apex.	Rare, spherical (25 to 27µm dia.) smooth walled.
P. sacchari (Sugarcane D.M.)	160 to 170µm in length, erect, arise singly or in pairs from stomata.	Elliptical, oblong (15-23µm x 25-41µm) with round apex.	40–50 µm dia., globular, yellow.
Sclerospora graminicola (Graminicola D.M.)	Average length of 268 µm.	Borne on short sterigmata, elliptical (12-21 x 14-31µm) with distinctive papillate, operculum at apex.	Pale brown and 22 to 35 µm dia.
Sclerophthora macrospora (crazy top)	Very short (14µm).	Lemon shaped (30-65 x 60-100µm), operculate.	Pale yellow, circular (45-75µm).
Scleropthora rayssiae var. *zeae* (Brown stripe D.M.)	–	Oval to cyclindrical (18-26 x 29-67µm).	Spherical (29-37µm dia.), brown coloured.

The taxonomy of *P. sorghi* is difficult (Perumal *et al.*, 2008). There have been two incorrect identifications of this fungus in Australia, which subsequent examination showed to be *P. maydis* (Wang *et al.*, 2000).

The causal agent of downy mildew can be determined by detailed examination of spore-bearing structures (conidiophores or sporangiophores) and spores (conidia or sporangia). Geographic location can be used as a preliminary diagnostic. White stripe symptoms caused by downy mildew are not limited by veins and can therefore be distinguished from signs of iron deficiency.

Taxonomy of the Pathogen

The taxonomy of *P. sorghi* is confused and therefore identification of the downy mildew species in different parts of the world is in doubt (Perumal *et al.*, 2008).

P. sorghi from sorghum infects sorghum and maize, producing oospores in sorghum but not readily producing oospores in maize. However, *P. sorghi*, which is prevalent in the Nigerian Guinea Savana, does not readily form oospores in maize or in sorghum, although oospores have been detected in maize seeds from some areas (Adenle and Cardwell, 2000). It is, therefore, unlikely that this Nigerian pathogen over seasons as soil-borne oospores. It has been reported that *P. sorghi* may perpetuate itself by conidial infection in maize plants in the Guinea savannah area of Nigeria (Anaso, 1989) but seed borne infection emanating from farmer-saved seeds is also an important consideration (Adenle and Cardwell, 2000). Nonetheless, the species identification of this Nigerian *P. sorghi* remains in doubt (Perumal *et al.*, 2008), so the significance of its failure to produce oospores remains unknown.

Pathotypes of *P. sorghi* have been identified. A total of 16 worldwide isolates of *P. sorghi* were tested for virulence on 75 sorghum cultivars previously identified as resistant to three pathotypes from Texas, USA, and to some regional populations of *P. sorghi*, in an ICRISAT multi-location screening programme. All isolates were identified as distinct pathotypes on the basis of differential sporulation on host cultivars. Host-pathogen interactions were consistent with the gene-for-gene hypothesis (Pawar *et al.*, 1985). The relationship between the genes for resistance to *P. sorghi* pathotypes 1, 2 and 3 in the sorghum lines QL3-India and SC414-12 were investigated by making reciprocal crosses between them and susceptible lines. A pathotype confined to maize also occurs in some areas such as the southern humid zone of Nigeria where it causes substantial damage through systemic infection (Payak, 1973; Olanya and Fajemisin, 1993).

Life Cycle

Depending on the pathogen species, the initial source of disease inoculum can be oospores that overwinter in the soil or conidia produced in infected, overwintering crop debris and infected neighbouring plants. Some species that cause downy mildew can also be seed borne, although this is largely restricted to seed that is fresh and has high moisture content.

At the onset of the growing season, at soil temperatures above 20°C, oospores in the soil germinate in response to root exudates from susceptible maize seedlings. The germ tube infects the underground sections of maize plants leading to characteristic symptoms of systemic infection including extensive chlorosis and stunted growth. When oospores initiate infection, the first leaf generally remains disease free as it is able to outgrow the fungi. However, the whole plant will show disease symptoms if the pathogen was seed borne. Oospores are reported to survive in nature for upto 10 years.

Once the fungus has colonized the host tissues, sporangiophores (conidiophores) emerge from stomata and produce sporangia (conidia) which are wind and rain splash disseminated and initiate secondary infections. Depending on the species, sporangia germinate directly or release zoospores that initiate infection (Table 6.3). Sporangia are always produced during the night. They are fragile and cannot be disseminated more than a few hundred metres and do not remain viable for more than a few hours.

Table 6.3: Major Downy Mildew Pathogens of Maize and their Modes of Infection

Pathogen (Disease Name)	*Intial Source of Inoculum*	*Seed Borne*	*Sporangia Germinates to*	*Optimum Temp. for*	
				Sporangia Production	*Sporangia germination*
Peronosclerospora sorghi (Sorghum D.M.)	Oospores and sporangia	Yes	Germ tubes	17-29°C	21-25°C
P. maydis(Java D.M.)	Sporangia	Yes	Germ tubes	Below 24°C	Below 24°C
P. philippinensis (Philippine D.M.)	Sporangia	Yes	Germ tubes	21-26°C	19-20°C
P. sacchari (Sugarcane D.M.)	Sporangia	Yes	Germ tubes	20-25°C	20-25°C
Sclerospora graminicola (Graminicola D.M.)	Oospores and sporangia	No	Zoospores	17°C	17°C
Sclerophthora macrospora (crazy top D.M.)	Oospores and sporangia	No	Zoospores	24-28°C	12-16°C
Scleropthora rayssiae var. *zeae* (Brown stripe D.M.)	Oospores and sporangia	Yes	Zoospores	22-25°C	20-22°C

Germination of sporangia is dependent on the availability of free water on the leaf surface. If sufficient water is available, sporangia germinate and infect the plant through stomata on the leaf, sheaths, or stems in a couple of hours. Initial symptoms of disease (chlorotic specks and streaks that elongate parallel to veins) occur in 3 days. Conidia are produced profusely during the growing season. As the crop approaches senescence, oospores are produced in large numbers.

Disease Cycle

The pathogen (*P. sorghi*) can penetrate through oospores and alternate hosts such as *Saccharum spontaneum, Sorghum bicolar* and *S. halepense* (Singh and Chanbe, 1968). (*P. sorghi*) produces oospores in infected host tissues (Figure 6.1). Oospores are produced less frequently and abundantly on maize than on sorghum, and appear in both hosts only in systemically infected plants (Bigirwa *et al.*, 1998). Oospore populations of 8–95 nos/g soil oospores per gram have been reported as potential inoculums to cause infection (Pratt and Janke, 1978). Oospores are thought to survive for at least three years under a variety of conditions (Frederiksen, 1980), and are dispersed by wind. Oospores germinate in soil by germ tubes that infect underground parts of susceptible seedlings, which then become systemically infected. Infection by oospores does not occur if seedlings emerge in cool soils below 20°C. In arid areas, where soil temperature is higher, oospores initiate infection of seedlings, whereas in areas where a perennial host such as Johnson grass is present, infection can be initiated from conidia (Bigirwa *et al.*, 1998). Secondary infection occurs through the production of condia or sporangia.

Conidia are produced from systemically infected plants at temperatures between 17 and 29°C from midnight to about 8 am (Bock *et al.*, 1998), with an optimum temperature of 24 to 26°C, depending on the geographic isolate (Bonde *et al.*, 1985).

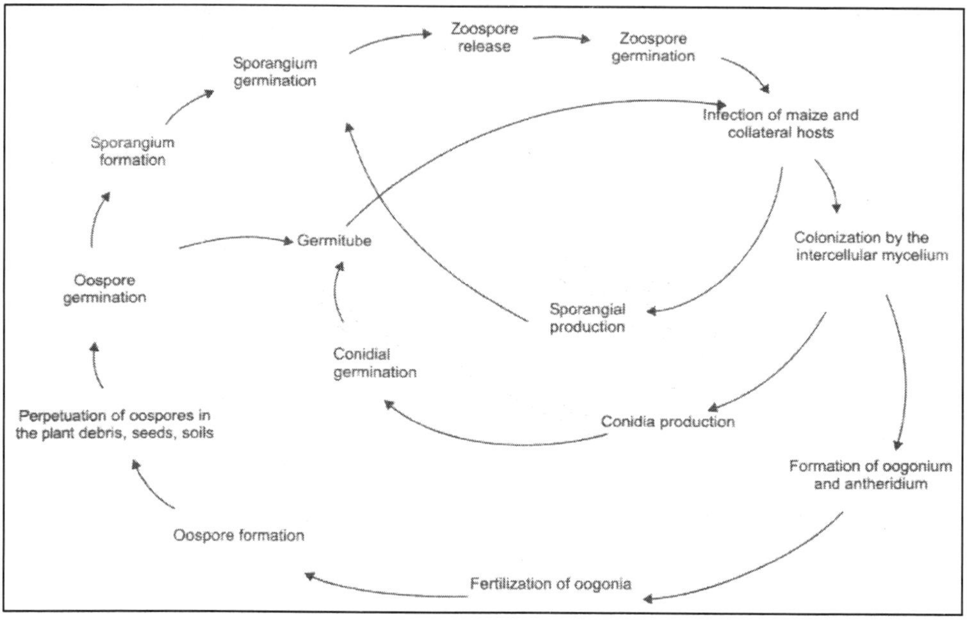

Figure 6.1: Disease Cycle of Maize Downy Mildew.
(*Courtesy*: Chaube and Pundhir, 2005).

Conidial germination requires a saturated atmosphere, and moderate temperatures of 15 to 25°C, depending on the isolate. High levels of systemic infection can occur between 11 and 32°C, with a wet period of 4 hours or longer. Conidia are well adapted to wind dispersal (Bock *et al.*, 1998), but lose their viability after 3 to 4 hours, so probably only play a role in short distance spread of the fungus.

The fungus can infect corn seed but is confined in mature seeds to the pericarp and pedicel. Jones *et al.* (1972) found that seed transmission of the disease occurs when infected seeds at the soft dough stage are planted in sterile soil. In contrast, seed transmission of the disease is prevented by reducing the moisture content to 9 per cent and by storage for 40 days prior to planting. When 400 seeds harvested from systemically infected maize plants were planted in sterile soil in a greenhouse, 256 plants showed systemic infection at and after the fifth leaf stage. These findings suggest that the presence of oospores within the seed resulted in the transmission of the fungus, regardless of seed moisture content (Rao *et al.*, 1984).

Epidemiology

Downy mildews are very significant maize diseases in tropical regions of Africa and Asia, where prolonged periods of leaf wetness and cultivation of alternate hosts are prevalent during the growing season. Cool, wet and humid conditions are optimal for disease development. In favourable conditions, disease cycles are rapid, leading to severe infection and spread of the disease. The disease seems to be confined to the areas of high relative humidity, high temperature, and high annual rainfall (Olanya *et al.*, 1993; Bock, 1995).

Perenating infected crop debris serves as a primary source of inoculum in subsequent seasons. Cultivation of alternate hosts in rotation or simultaneously aids in build up of the build pathogen pressure. Moist soils favor oospore germination and therefore damp soil as a result of irrigation or reduced tillage techniques will encourage disease development.

Dispersal of the Pathogen

Transmission occurs mainly through oospores which survive in the soil (Broyles, 1956), while it can also be transmitted from plant to plant by airborne conidia or through the seeds (Rao *et al.*, 1985) and also soil bone (Kutama *et al.*, 2008), except when stored under low moisture conditions (Sommartaya *et al.*, 1975). Soil temperature (Schuh *et al.*, 1987) and composition (Bigirwa *et al.*, 1998) play important role in determining disease spread, in sandy soils about 25°C being the optimum temperature for its spread.

P. sorghi is seed-borne, either as mycelium in immature seeds, or oospores in the pericarp and pedicel of the seed, so the entry potential is high. While there is evidence that seed transmission can be prevented by reducing the moisture content of the seed to 9 per cent and by storage for 40 days prior to planting. Other experiments suggest that the presence of oospores within the seed results in the transmission of the fungus, regardless of seed moisture content (Rao *et al.*, 1984). The fungus could also be introduced in bulk maize shipments as oospores in trash and soil.

The establishment potential of the pathogen is high in some parts of the maize and sorghum growing areas of Australia: Wang *et al.* (2000) found that areas of coastal Queensland were suitable for infection, but the main inland growing areas of Queensland and NSW were much less suitable. Oospores in or on seed, trash or soil are the likely primary inoculum. These spores require soil temperatures above 20°C for germination, and these temperatures occur in Australia in summer where maize crops are grown. Suitable temperatures for conidial production and germination also occur during this period but the moisture requirements are only met in coastal areas of Queensland.

The close proximity of maize and sorghum crops to many of the feed lots where imported maize would be used in Australia increases the risk of establishment. The wide distribution of Johnson grass in northern Australia, especially along roadways, also provides a perennial source of susceptible host material.

The spread potential of the pathogen is high, as oospores can be spread long distances by wind, in seed and crop residues, and in soil on farm machinery. Conidia can also spread the disease locally within crops and between adjacent crops and weed hosts.

The economic impact is likely to vary depending on location. Conditions are more likely to be favourable for disease development in coastal Queensland, where losses would be potentially high and less favourable in inland growing areas, where the losses are likely to be low. The disease is more likely to affect breeding programs, which are in the more conducive coastal areas.

Sorghum downy mildew is a serious disease of both maize and sorghum in the tropics and subtropics, and is the most widely distributed of the major downy mildew diseases of these crops. Severe outbreaks have occurred in India, Israel, Mexico, Thailand, Texas and Venezuela. In Nigeria, grain yield losses in maize from this disease ranged from 10 to 90 per cent. In Texas, the disease incidence can be 30 per cent or higher, under favourable environmental conditions.

Disease Management Principles

Cultural Control

Use Resistant and or Tolerant Maize Varieties

Do not rotate or simultaneously cultivate maize with alternate hosts of downy mildew and follow long crop rotation.Diseased infected crop debris and alternate/ collateral hosts should be destroyed.Plant when soil temperature is below 20°C, which is unfavourable for oospore germination.Ensure seed has low moisture content (below 9 per cent) before planting.Reduce crop density and control weeds to increase aeration within the crop and reduce moisture levels in the soil.

Mechanical Control

Drying of the seeds to less than 14 per cent moisture content reduces or eliminates seed-borne transmission. Rotations with more than three years between sorghum or maize crops and control of alternative grass hosts within the paddock will reduce soil-borne infection.

Biological Control

Biological control of plant pathogens, based on the management of natural sources to develop the antagonistic activity against harmful organisms is currently accepted as a key practice in sustainable agriculture and as potential alternative to the intensive use of chemical pesticides which are harmful to the environment. Several investigators reported the use of biological control against soil-borne disease (El-Kazzaz *et al.*, 2000 and Chincholkal and Mukerji, 2007). The biocontrol agents *viz.*, *Trichoderma viride*, *Trichoderma harzianum* and *Bacillus subtilis* represent the most thoroughly studied antagonistic micro-organisms. The mixture of bioagents has been used as a good biocontrol agents for many soil-borne and air-borne diseases such as late-wilt of maize (El-Assiuty *et al.*, 1991), leaf blight of maize (Abd El-Naby *et al.*, 2007) and root rot of tomato plants (Morsy *et al.*, 2009). *T. viride*, *T. harizianum* and *B. subtilis* were reported as the most effective with dual bio-agent as the individual one (El-Fiki *et al.*, 2004; Pertot *et al.*, 2008 and El-Shehawy, 2009). The mixtures, (*T. viride* + *T. harzianum*) and (*T. viride* + *B. subtilis*) @ 1:1 (v/v) were efficient to control of downy mildew of maize (Sadoma *et al.*, 2011).

Chemical Management

Since the discovery of systemic fungicides, our ability to control downy mildew diseases has improved considerably, although these are very difficult to control. However, following are some of the chemical management options for maize downy mildews.

Seed treatment with Ridomil 25 WP (@ 49/kg seed) or Apron 35 WS (@ 49/kg seed) and single spray of Ridomil MZ (@ 6 g/Litre) were reported to gave better control of the disease (Lal *et al.*, 1980).

Metalaxyl compounds have been widely used for the control of downy mildew in a number of crops since its introduction in 1970s (Exconde and Molina, 1978; Frederiksen, 1980; Raid and Datnoff, 1991; Wicks *et al.*, 1993; Singh, 1995). However, pathogens were reported to produce resistant strains to metalaxyl (Bruck *et al.*, 1982; Crute, 1987; Wicks *et al.*, 1994). Once resistant strains appear, they can survive for several years. Therefore, the risk of enhancing resistant populations with continued application of ineffective fungicide is very high (Dekker, 1987). Cultivar resistance has also been negated (Singh and Shetty, 1990), which restricted its use. Moreover, resistance that is effective in one region may not be effective when grown in another, because of pathogen and/or environmental variability (Williams, 1990). All these factors have emphasized the need for an alternative chemical for downy mildew control. One potential method of alternative control is to use a chemical that differs in its mode of action from that of metalaxyl. One such chemical is phosphonic acid H3PO3 (Wicks *et al.*, 1991; Magarey *et al.*, 1989). Previous studies indicated that phosphonic acid induced resistance in plants against the pathogen (Guest, 1984), thus the risk of pathogens developing resistance to the fungicide is greatly reduced.

Foliar sprays with metalaxyl provide effective control of *P. sorghi* on maize (Odvody and Frederiksen, 1984). Several waxes and plastic polymers, sprayed onto seedlings before inoculation with the pathogen, reduced infection (Ziv and Frederiksen, 1983). Metalaxyl, applied as a seed treatment and sprayed after planting, completely controlled both systemic infection and local lesions of *P. sorghi* on sorghum. Infected plants sprayed with metalaxyl after planting recovered and produced normal heads. Seed treatment with metalaxyl alone did not provide protection against late systemic infection on main shoots or nodal tillers, or against local lesions on leaves (Anahosur and Patil, 1980).

Panicker and Gangadharan (1999) controlled downy mildew of maize by foliar sprays of phosphonic acid formulations including Aliette and Akom and at a lower cost than using metalaxyl. The disease was effectively controlled (84 per cent) by three or four sprays with mancozeb, the first after emergence and then at weekly intervals, in field experiments using a susceptible sorghum line (Balasubramanian, 1976). Metalaxyl can be effective as a seed treatment against *P. sorghi* in maize. Plants grown from seeds treated with metalaxyl, or any mixture containing metalaxyl, remained free from the disease and had significantly higher grain yield than plants grown from seeds treated with thiram-based fungicides or from untreated seeds. Low rates of metalaxyl completely controlled downy mildew (Anaso *et al.*, 1989). However, metalaxyl only slightly reduced seed transmission in maize in Nigeria (Adenle and Cardwell, 2000), an indication of deep seated internal infection.

Resistance to metalaxyl has developed in *P. sorghi* in Texas, where seed treatment with metalaxyl no longer controls the pathogen (Isakeit *et al.*, 2003).

Downy mildew in maize caused by *Peronosclerospora sorghi* (Weston and Uppal) C.G.Shaw, was significantly controlled by foliar spray of phosphonic acid (20 per cent) and its formulations like Aliette (0.25 per cent) and Akomin (0.25 per cent) 20 days after sowing (DAS), under greenhouse and field conditions. The reduction in disease incidence and severity under field conditions was reflected by an increased yield of 54-73 per cent (Panicker and Gangadharan, 1999).

Host Plant Resistance

Genetic resistance continues to be the most economic and effective means for the management of DM disease in maize. Breeding for DM resistance in Asian maize for lowland tropics has been achieved with good success (De Leon *et al.*, 1993; George *et al.*, 2004). However, the key challenge in breeding for host-plant resistance has been the spill-over of resistance across DM pathogens and even variants within species, and also its prolonged and widespread use in an environment favourable for DM disease development. Nair *et al.* (2004) reported two inbred lines from Kasetsart University, Thailand (KUI1411 and KUI1414A), resistant to highly prevalent DM in Thailand (*P. zeae*), were found to be susceptible to *P. sorghi* at Mandya, India. Similar observations on differential responses of maize inbred lines to different DM pathogens have been made by others (Singburaudom and Renfro, 1982; Yen *et al.*, 2001; Nair *et al.*, 2001). This might be related to the difference in virulence of different isolates (Rathore *et al.*, 2002). Susceptibility of the released CMLs to *P. sorghi* might also be related to some breakdown of the host-plant resistance during last few years, as in an earlier screening almost all DMR CMLs except five (CML-425, CML-426, CML-432, CML-473 and CML-474) showed moderate to good levels of resistance to *P. sorghi* at Mandya during Kharif, 2003. DM resistance breakdown has been common in breeding lines and hybrids, which is largely attributed to the evolution of new virulence in the pathogen populations (Thakur *et al.*, 2009). The pathogens that possess a mixed reproductive system, both sexual and asexual (such as *Perenosclerospora* species), pose the greatest risk to the breakdown of resistance genes (McDonald and Linde, 2002). The life span of DM resistance in most of the commercial pearl millet hybrids was only about five years because of the robust genetic variability in the pathogen and rapid selection of virulence against specific host cultivars (Christ *et al.*, 1987; Thakur *et al.*, 2004). Therefore, host-plant resistance breeding programs require close monitoring of virulence changes in the pathogen and identification of new resistance sources to new virulent strains. Recently indentified DM resistant CMLs are CML-433, 465, 466 and 472 (Rashid *et al.*, 2013).

References

Abd El-Naby, A.E., Zakaria, M.M, Klafalla, G. and A. Zein El-Abedeen. (2007). Control of leaf blight disease of maize using two bioagents and some natural plant oils. *Ann. Agric. Sci. Moshtohtohar*, **45**(2): 621-629.

Adenle, V.O. and Cardwell, K.F. (2000). Seed transmission of maize downy mildew (*Peronosclerospora sorghi*) in Nigeria. *Plant Pathology.*, **49**: 628–634.

Anahosur, K.H. and Patil, S.H. (1980). Chemical control of sorghum downy mildew in India. *Pl. Dis.*, **64**: 1004–1006.

Anaso, A.B. (1989). Survival of downy mildew pathogen of maize in Nigerian, Guinea Savana. App. *Agric. Res.*, **4**: 258–263.

APS, (1973). *A Compendium of Corn Diseases*. American Phytopathological Society, St Paul, Minnesota, USA.

Balasubramanian, K.A. (1976). Chemical control of downy mildew of sorghum. *Curr. Sci.*, **45**: 416–417.

Bigirwa, G, Adipala E. and Esele, J.P. (1998). Occurrence of *Peronosclerospora sorghi* in Uganda. *Pl. Dis.*, **82**: 757–760.

Bock, C.H. (1995). Studies on the epidemiology, variability and control of sorghum downy mildew (*P.sorghi*) (Weston and Uppal) C. G. Shaw, in Africa. U.K. Univ., Department of Agriculture. Ph. D Thesis.

Bock, C.H., Jeger, M.J., Mughogho, L.K., Mtisi, E. and Cardwell K.F. (1998). Production of conidia by *Peronosclerospora sorghi* on sorghum crops in Zimbabwe. *Pl. Pathol.*, **47:** 243–251.

Bonde, M.R. (1982). Epidemiology of downy mildew diseases of maize, sorghum and pearl millet. *Tropical Pest Management*, **28**: 49–60.

Bonde, M.R., Peterson, G.L. and Duck, N.B. (1985). Effects of temperature on sporulation, conidial germination, and infection of maize by *Peronosclerospora sorghi* from different geographical areas. *Phytopathology.*, **75**: 122–126.

Broyles, J.W. (1956). Observations on time and location of penetration in relation to amount of damage and chemical control of *Physoderma maydis*. *Phytopathology.*, **46**: 8.

Bruck, R.I., Goodeng, A.V. and Main, C.S. (1982). Evidence for resistance to metalaxyl in isolates of *Peronospora hyoscyami*. *Pl. Dis.*, **66**: 44-45.

Chaube H.S. and Pundhir V.S. (2005). Crop Diseases and Their Management. Prentice Hall of India, Pvt. Ltd., New Delhi., pp. 372.

Chincholkal, S.B. and Mukerji, K.G. (2007). Biological Control of Plant Diseases: Haworth Food and Agricultural Products, Press TM, New York, London and Oxford., pp. 223-228

Christ, B.J., Person, C.O. and Pope, D.D. (1987). The genetic determination of variation in pathogenicity. (Eds. Wolfe, M.S., Caten, C.E.). Populations of Plant Pathogens: Their Dynamics and Genetics. Blackwell Scientific, Oxford, UK. pp. 719.

CMI. (1988). *Peronosclerospora sorghi* (Weston and Uppal) C.G. Shaw. Commonwealth Mycological Institute: Distribution Maps of Plant Diseases No. 179.

Crute, I.R. (1987). The occurrence, characteristics, distribution, genetics and control of metalaxyl resistant pathotypes of *Bremia lactucae* in the United Kingdom. *Pl. Dis.*, **71**: 763-767.

Dalmacio, S. (2000). Importance and growing concern for maize diseases in the Asian region. (Eds. Vasal, S.K., Ceniceros, F.G., Ming, F.X.). Proceedings of the Seventh Asian Regional Maize Workshop. PCARRD, Los Banos, Philippines., pp. 267-276.

De Leon, C., Granados, G., Wedderburn, R.N. and Pandey, S. (1993). Simultaneous improvements of downy mildew resistance and agronomic traits in tropical maize. *Crop Sci.*, **33**: 100-102.

Dekker, J. (1987). Build-up and persistence of fungicide resistance (Eds. Brent, E.S., Atkin, E.K.). Rational pesticide use, Cambridge University Press. Cambridge, UK. pp. 153-168.

El-Assiuty, E.M., El-Hmahmy, A. and Sharkawy, A.Y. (1991). *Bucillus subtilis, Psudomonas fluorescens* and *Verticillium tricorpus* as biological agents against late-wilt of maize. *Egypt J. Appl. Sci.*, 6(12): 824-829.

El-Fiki, A.I.I., Mohamed, F.G., El-Deeb, A.A. and Khalifa, M.M.A. (2004). Some applicable methods for controlling sesamum charcoal rot disease (*Macrophomina phaseolina*) under greenhouse conditions. *Egypt. J. Microbiol.*, **38** (4): 351-361

El-Kazzaz, M.K., Hassan, M.A., Goniem, K.E. and El-Zahaby, H.M. (2000). Biological control contrd of sugar beet root rot caused by certain soil-born fungi.Proceedings of the 8 th Cong. of Egyptian Phaytopathol. Society.

El-Mersawy, E.M. (2000). Biological control of maize downy mildew disease, caused by *Peronosclerospra sorghi* in Egypt. *J. Agric. Sci. Mansoura Univ.*, **25** (11): 6787-6794.

El-Shehawy, A.E. (2009). Studies on control of downy mildew disease of grain sorghum. Ph.D. Thesis, Tanta Univ. Egypt.

Exconde, O.R. and Molina, A.B. (1978). Ridomil a seed dressing fungicide for the control of Philippine corn mildew. *Philippine J. Crop Sci.*, **3**: 60-64.

Fredriksen, R.A., Bockholt, A.J., Clark, L.E., Cosper, J.W., Craig, J., Johnson, J.W., Jones, B.L., Matocha, P., Miller F.R., Reyes, L., Rosenow, D.T., Tuleen, D. and Walker, H.J. (1973). Sorghum downy mildew: A disease of maize and sorghum. Research Monograph 2. The Texas Agril. Expt. Sta. Texas A and M Univ., College Station, Texas. U.S.A., p.32.

Frederiksen, R.A. (1980). Sorghum downy mildew in the United States: Overview and outlook. *Pl. Dis.*, **64**: 903-908.

George, M.L.C., Regalado, E., Li, W., Cao, M., Dahlan, M., Pabendon, M., Warburton, M.L., Xia, X. and Hoisington, D. (2004). Molecular characterization of Asian maize inbred lines by multiple laboratories. *Theor. Appl. Genet.*, **109**: 80-91.

Guest, D.I. (1984). Modification of defence responses in tobacco and capsicum following treatment with fosetyl-Al. *Physiol. Plant Pathol.*, **25**: 125-134.

Hau, B.W., Drepper, J., Prasertkit, O., Kranz, J. and Renfro B. L. (1995). Temporal and spatial aspects of the epidemiology of sorghum downy mildew on maize. *Plant Pathology.*, **44**: 897-908.

Isakeit, T, Odvody, G, Jahn R and Deconini, L. (2003). *Peronosclerospora sorghi* resistant to metalaxyl treatment of sorghum seed in Texas. *Phytopathology.*, **93**: S39.

Jeffers, D., Cordova, H., Vasal, S., Srinivasan, G., Beck, D. and Barandiaran, M. (2000). Status in breeding for resistance to maize diseases at CIMMYT (Eds. Vasal, S.K.,

Gonzalez Cenicros, F. and XingMing, F.). Proc. Seventh Asian Regional Maize Workshop. PCARRD, Los Banos, Philippines., pp. 257-266.

Jeger, M.J., Gilijamse E, Bock, C.H. and Frinking H.D. (1998). The epidemiology, variability and control of the downy mildews of pearl millet and sorghum, with particular reference to Africa. *Plant Pathology*., **47**(5): 544-569.

Jones, B.L. Leeper, J.C. and Frederiksen, R.A. (1972). *Sclerospora sorghi* in corn: its location in carpellate flowers and mature seeds. *Phytopathology*, **62**: 817–819

Kings, S.B. and Webster, O.J. (1970). Downy Mildew of Sorghum in Nigeria. Indian *Phytopathology*, **23**: 342-349.

Kling, J.G., Cardwell, K.F. and Kim, S.K. (1994).Advances in screening methods and breeding for downy mildew resistance (*Peronosclerospora sorghi*) of maize. Paper presented at the Fourth Eastern and Southern Africa Regional Maize Conference: Maize Research for Stress Environments. Harare, Zimbabwe. March, 28-April, 1, 1994.

Krishnappa, M., Naidu, B.S. and Seetharam, A. (1995). Inheritance of resistance to downy mildew in maize. *Crop Improvement*., **22**: 33-37.

Kutama, A.S., Aliyu, B.S. and Emechebe, A.M. (2010). State of sorghum downy mildew in maize in the Sudan and Sahel Savanna agro-ecological zones of Nigeria. *Bayero J. Pure and App. Sci.*, **3**(1): 233 - 237.

Kutama, A.S., Aliyu, B.S. Nuraddin, A. and Kiyawa, S. A. (2008) Green house evaluation and epidemiology of sorghum downy mildew in some varieties of sorghum and their chemical control. *In:* Proceedings of International Conference on Research and Development.1 (13). November 25-28, Institute of African Studies, Univ. of Ghana, Accra, Ghana., pp. 63-67.

Lal, S, Saxena, S.C. and Upadhayay, R.N. (1980). Control of brown stripe downy mildew of maize. *Plant Dis. Rept*., **64**: 874-876.

Magarey, P.A., Wick, T.J. and Wachtel, M.F. (1989). Field evaluation of phosphoric acid for the control of grapevine downy mildew (Abstr.). *In:* 8th Conf. of Australian Pl. Path. Soc. 7. pp. 113.

McDonald, B.A. and Linde, C. (2002). Pathogen population genetics, evolutionary potential and durable resistance. *Ann. Rev. Phytopath*., **40**: 349-379.

Morsy, S.M, Abdel-Kawi, K.A. and Khalil, M.N.A. (2009). Efficiency of *Trichoderma viride* and *Bacillus subtilis* as biocontrol agents against *Fusarium solani* on tomato plants. *Egypt. J. Phytopath*., **38** (1): 47-57.

Nair, S.K., Prasanna, B.M., Rathore, R.S., Setty, T.A.S., Kumar, R. and Singh, N.N. (2004). Genetic variability in the Indian maize germplasm for resistance to sorghum downy mildew (*Peronosclerospora sorghi*) and Rajasthan downy mildew (*P. heteropogoni*). *Field Crop. Res.*, **89**: 379-387.

Nair, S.K., Setty, T.A.S., Rathore, R.S., Kumar, R., Singh, N.N., Prasanna, B.M. (2001). Towards molecular marker mapping of genes conferring resistance to sorghum downy mildew (*Peronosclerospora sorghi*) in maize. *Maize Genet. Coop. News Letter*., **75**: 47-48.

Odvody, G.N. and Frederiksen, R.A. (1984). Use of systemic fungicides metalaxyl and fosetyl-Al for control of sorghum downy mildew in corn and sorghum in South Texas. I: seed treatment. *Pl. Dis.,* **68**: 604–607.

Odvody, G.N. (2002). Resurgence of sorghum downy mildew on the upper coast of Texas: Preventing Future Epidemic. Corpus Christi Research and Extension Center., pp. 1-2.

Olanya, O.M. and Fajemisin, J.M.(1993) Host range and survival studies of *P. sorghi* at different locations in Nigeria. *International J. Pest Management*., **39**: 13-18.

Panicker, S and Gangadharan, K. (1999). Controlling downy mildew of maize caused by *Peronosclerospora sorghi* by foliar sprays of phosphonic acid compounds. *Crop Protection*., **18:** 115-118.

Pawar, M.N. Frederiksen, R.A. Mughogho, L.K. and Bonde, M.R. (1985). Survey of virulence in *Peronosclerospora sorghi* isolates from India, Ethiopia, Nigeria, Texas (USA), Honduras, Brazil, and Argentina. *Phytopathology*., **75**: 1374.

Payak, M.M. (1973). Progress in the control of downy mildew diseases of maize: An Inventory of Resistant Sources. Proc. 1st IACP Planning Session, Kualalumpur.

Pertot, I., Zosso, R., Amsalem, L., Baldessari, M., Angeli, G. and Elad, Y. (2008). Intergrating biocontrol agents in strawberry powdery mildew control strategies in high tunnel growing systems. *Crop Protection*., **27**: 622-631.

Perumal. R., Nimmakayala, P., Erattaimuthu, S.R., Reddy, U.K., Prom, L.K., Odvody, G.N., Luster, D.G. and Magill, C.W. (2008). Simple sequence repeat markers useful for sorghum downy mildew (*Peronosclerospora sorghi*) and related species. BMC Genetics, 9. doi: 10.1186/1471-2156-9-77.

Pingali, P.L. (2001). World Maize Facts and Trends, Meeting World Maize Needs: Technological Opportunities and Priorities for the Public Sector. CIMMYT, Mexico, D.F.

Pratt, R.K. and Janke, G.D. (1978). Oospores of *Sclerospora sorghi* in soils of south Texas and their relationships to the incidence of downy mildew in grain sorghum. *Phytopathology*., **68:** 1600-1605.

Raid, R.N. and Datno, L.E. (1991). Efficacy of aliette and ridomil combinations for control of lettuce downy mildew. *Fungic. Nematic. Tests*, pp.46: 133.

Rao, B.M., Shetty, S.H. and Safeeulla, K.M. (1984). Production of *Peronosclerospora sorghi* oospores in maize seeds and further studies on the seed-borne nature of the fungus. *Indian Phytopath*., **37**: 278-283.

Rao, B.M., Prakash, H.S., Shetty, H.S. and Safeeulla, K.M. (1985). Downy mildew inoculum in maize seeds: Techniques to Detect Seed-borne Inoculum of *Peronosclerospora sorghi* in maize. *Seed Science and Technology*., **13**: 593-600.

Rashid, Z., Zaidi, P.H., Vinayan, M.T., Sharma, S.S. and Srirama Setty, T.A. (2013). Downy mildew resistance in maize (*Zea mays* L.) across *Peronosclerospora* species in lowland tropical Asia. *Crop Protection*., **43** : 183-191.

Rathore, R.S. and Siradhana, B. (1987). Estimation of losses caused by *Perenosclerospora heteropogoni* on Ganga-5 maize hybrid. *Phytoparasitica.*, **19**: 119-120.

Rathore, R.S., Trivedi, A. and Mathur, K. (2002). Rajasthan downy mildew of maize: The problems and management perspectives. *In*: Proceedings of the 8th Asian Regional Maize Workshop, Bangkok, Thailand. August 5-8. 366-378

Raymundo, A.D. (2000). Downy mildew of maize in Asia: New perspectives in resistance breeding (Eds. Vasal, S.K., Ceniceros, F.G., Ming, F.X.). Proceedings of the Seventh Asian Regional Maize Workshop. PCARRD, Los Banos, Philippines., pp. 277-284.

Rifin, A. (1983). Downy mildew resistance of single cross progenies between Indonesian and Philippine corn inbred lines. *Penelit. Pertan.*, **3**: 81-83

Sadoma, M.T., El-Sayed, A.B.B. and El-Moghazy, S.M. (2011). Biological control of downy mildew disease of maize caused by *Peronosclerospora sorghi* using certain biocontrol agents alone or in combination. *J. Agric. Res. Kafer El-Sheikh Univ.* **37**(1): 1-11.

Schuh, W., Jeger, M.J. and Frederiksen, R.A. (1987). The influence of soil temperature, soil moisture, soil texture, and inoculum density on the incidence of sorghum downy mildew. *Phytopathology.*, **77**: 125-128.

Schuh, W., Jeger, M.J., Frederiksen, R.A. (1988). Comparisons of spatial patterns of oospores of *Peronosclerospora sorghi,* in the soil and of sorghum plants with systemic downy mildew. *Phytopathology.*, **78**: 432-434.

Shaarawy, M.A., Abdel Monem, A.M. and El-Wakil, A.A. (2002). A modified method for detecting of oospores of *Peronosclerospora sorghi*, the causal organism of sorghum downy mildew in Sudan grass seeds (*Sorghum sudanense*). *Egypt J. Agric. Res.*, **80**(3): 1033-1042

Singburaudom, N. and Renfro, B.L. (1982). Heritability of resistance in maize to sorghum downy mildew (*Perenosclerospora sorghi*). *Crop Prot.*, **1**: 323-332.

Singh, R.S. and Chanbe, H.S. (1968). The occurrence of *Sclerospora sacchari* and its oospores on Maize in India. *Labdev J. Sci. Technol.*, **6** (B): 197-200.

Singh, S.D. (1995). Downy mildew of pearl millet. *Plant Dis.*, **79**: 545-555.

Singh, S.D. and Shetty, H.S. (1990). Efficacy of systemic fungicide metalaxyl for the control of downy mildew of pearl millet. *J. Agric. Sci.*, **60**: 575-581.

Siradhana, B., Dange, S.R.S., Rathore, R.S. and Singh, S.D. (1980). A new downy mildew of maize in Rajasthan, India. *Curr. Sci.*, **49**: 316-317.

Sommartaya, T., Pupipat, U., Intrama, S. and Renfro, B.L. (1975). Seed transmission of *Sclerospora sorghi* (Weston and Uppal), the downy mildew of corn in Thailand. *Kasetsart Journal.*, **9**: 12-25.

Spencer, M.A. and Dick, M.W. (2002). Aspects of graminicolous downy mildew biology: Perspectives for tropical plant pathology and Peronosporomycetes phylogeny (Eds. Watling, R., Frankland, J.C., Ainsworth, A.M., Isaac, S. and

Robinson, C.H.). Tropical Mycology, Micromycetes, CAB International, Wallingford, UK. Vol. 2. pp: 63-81.

Tarr, S.A.J. (1962). Diseases of Sorghum, Sudan grass and Brown corn. Commonwealth Mycological Institute, Kew, Survey, pp. 171-181.

Thakur, R.P., Rao, V.P. and Sharma, R. (2009). Temporal virulence change and identification of resistance in pearl millet germplasm to diverse pathotypes of *Sclerospora graminicola*. *J. Pl. Pathol.*, **91**: 629-636.

Thakur, R.P., Rao, V.P., Wu, B.M., Subbarao, K.V., Shetty, H.S., Singh, G., Lukose, C., Panwar, M.S., Sereme, P., Hess, D.E., Gupta, S.C., Datta, V.V., Panicker, S., Pawar, N.B., Bhangale, G.T. and Panchbhai, S.D. (2004). Host resistance stability to downy mildew in pearl millet and pathogenic variability in *Sclerospora graminicola*. *Crop Prot.*, **23**: 901-908.

Wang, E., Ryley, M. and Meinke, H. (2000). Prediction of sorghum downy mildew risk in Australia using daily weather data. *Australasian Plant Pathology.*, **29**: 108-119.

White, D.G. (1999). Downy mildews. In: Compendium of Corn Diseases (3rd Edn.) *American Phytopathol. Soc., USA*, pp.25-32.

Wicks, T., Hall, B. and Pezzaniti, P. (1993). Fungicidal control of downy mildew on lettuce. *Aust. J. Exp. Agric.*, **33**: 381-384.

Wicks, T.J., Hall, B. and Pezzaniti, P. (1994). Fungicidal control of metalaxyl insensitive strains of *Bremia lactucae* on lettuce. *Crop Protection.*, **13**: 617-623.

Wicks, T.J., Magarey, P.A., Wachtel, H.F. and Frensham, A.B. (1991). Effect of post infection application of phosphonic acid on *Plasmopara citricola* on grapevine. *Plant Dis.*, **75**: 40-43.

Williams, R.J. (1990). Downy mildews of Tropical cereals: *Advances in Plant Pathology.*, **2**: 1-9.

Yao, C.L., Frederiksen, R.A. and Magill, C.W. (1992). Length heterogeneity in ITS 2 and the methylation status of CCGG and GCGC sites in the rRNA genes of the genus *Peronosclerospora*. *Curr. Genet.*, **22**: 415-420.

Yao, C.L., Magill, C.W., Frederiksen, R.A., Bonde, M.R., Wang, Y. and Wu, P.S. (1991). Detection and identification of *Peronosclerospora sacchari* in maize by DNA hybridization. *Phytopathology.*, **81:** 901-905.

Yen, T.T.O., Rathore, R.S., Setty, T.A.S., Kumar, R., Singh, N.N., Vasal, S.K., Prasanna, B.M. (2001). Inheritance of resistance to sorghum downy mildew (*P. sorghi*) and Rajasthan downy mildew (*P. heteropogoni*) in maize in India. *Maize Genet. Coop. Newslett.* **75**: 48-49.

Ziv, O. and Frederiksen, R.A. (1983). Control of foliar diseases with epidermal coating materials. *Plant Disease.*, **67**: 212-214.

2015, Major Constrains and Verdict of Crop Productivity *Pages 117–136*
Editor: **Dr. U.N. Bhale**
Published by: **ASTRAL INTERNATIONAL (P) LIMITED, NEW DELHI**

Chapter 7

Biotechnological Interventions for Crop Adaptation to Climate Change

Sanjay Jha[1] and Rajkumar[2]

[1] Associate Professor,
Gujarat Agriculture Biotechnology Institute (GABI),
[2]Assistant Research Scientist,
Main Cotton Research Station (MCRS),
Navsari Agricultural University (NAU), Athwa Farm,
Ghod Dhod Road, Surat – 395 007, Gujarat, India

Introduction

Climate change is unequivocal, particularly in terms of increasing temperature, increasing CO_2 concentration, widespread melting of snow and ice, and rising global average sea level, and increases in the frequency of drought. It would bring greater flooding of low-lying lands and increased flooding and runoff from tropical storms resulting in salinity changes and water logging. Climate changes are not new, eg. drastic decrease of CO_2 concentration about 400 million years ago leads to appearance of leaves and gave birth to agriculture after the end of the last ice age (about 11,000 years ago). In addition, the late Holocene droughts between 5000 and 1000 years ago resulted in the collapse of civilizations. The climate change, which is occurring now will have – and is already having – an adverse effect on food production and food quality especially with the poorest farmers and the poorest countries most at risk.

* E-mail: 1: sanjayjha14@gmail.com; 2: raj_11bk@rediffmail.com

The adverse effect is either a consequence of the expected or likely increased frequency of some abiotic stresses such as heat and drought, and of the increased frequency of biotic stresses (pests and diseases), and climate change is also expected to cause losses of biodiversity, mainly in the more marginal environments.

Throughout history, farmers have adopted new crop varieties and adjusted their practices in accordance with changes in the environment. However, as global temperature continues to rise, the pace of environmental change will likely be unprecedented. More frequent and intense precipitation events, elevated temperatures, drought, and other types of damaging weather are all expected to impact crop yield and quality (Hatfield *et al.*, 2011), making the challenge of feeding an estimated 9 billion people by 2050 exceedingly difficult. For example, after a 10-year drought, Australia experienced catastrophic floods during fall 2010 and winter 2011, leading to loss of an estimated $6 billion in grain harvests. Unpredictable and severe weather can also leave the most volatile regions of the world even more vulnerable to instability due to greater hunger, poverty, and food insecurity (CNA, 2007). It has been estimated that rising global temperatures between 1981 and 2002 reduced the yields of major cereals by $5 billion per year (Lobell and Field, 2007). Heat waves and drought in Europe in 2003 (Ciais *et al.*, 2005) significantly reduced productivity (*e.g.* maize) yield in northern Italy by 36 per cent. In 2009-10, heat waves also affected wheat production in central Asia, and extreme flooding affected agricultural production in south Asia. In addition to the challenge of temperature extremes (hot and cold) and drought or water stress as well as flooding associated with climatic variability, the incidence and severity of biotic stresses such as pests, diseases and the invasion of alien weed species are also likely to be greater. Insect ecology, epidemiology, and distribution mainly depend on temperature (Coakley *et al.*, 1999), while plant pathogens will be highly responsive to humidity, rainfall, and temperatures. Further, CO_2 may promote the rapid establishment of invasive insect species (Zavala *et al.*, 2008).

It is anticipated that by 2030, India will be most severely affected due to insufficient production of major food crops like cereals (rice, wheat and maize) and pulses because of massive impact of climate change on the tropics and sub-tropics, huge population growth and dependency of more than half of this population on agriculture (Reynolds and Ortiz, 2010). To meet the dietary demand of the fast-increasing population in India, it is thus imperative to develop and grow high-yielding durable stress tolerant cultivars ('smart' crop varieties). Crop wild relatives (CWRs), underutilized crops, and neglected species, that are conserved ex-situ, on-farm, and in-situ, are veritable repositories of the beneficial heritable traits lost in the course of domestication, including those for adapting to climate change (Gur and Zamir, 2004); with the help of appropriate biotechnological approaches those genomic segments precisely where these 'landmarks' are located on the genome can be identified and assembled into the envisaged 'smart' crop varieties.

Crop improvement has always addressed both abiotic and biotic stresses, and strategies of adaptation to climate changes may include a more accurate matching of phenology to moisture availability using photoperiod temperature responses, increase access to a suite of varieties with different duration to escape or avoid predictable occurrences of stress at critical periods in crop life cycles, improved water use efficiency, and re-emphasize population breeding. However, conventional crop breeding for tolerance to these effects has had considerable success, but is slow and largely limited to exploiting the existing genetic variation in crop plants and their very close relatives. Most of these yield, and abiotic (drought, salinity, cold, high temperature, acidity and sodicity) and biotic (evolution of new races of pathogens due to climate change) stress component agronomic traits are complex and quantitative in nature and multiplicatively governed by many major and several small effect genes/QTLs (quantitative trait loci) (Wang *et al.*, 2008). Henceforth, understanding the molecular mechanism and dissection of such complex quantitative traits is the prime objective in current plant genomics and molecular breeding research.

Genome sequences are available now for many crop species such as rice, sorghum, maize and soybean. Furthermore, the advent of 'next-generation sequencing' had made it possible to sequence the transcriptomes or genomes of any species relatively quickly and cheaply. These genome or transcriptome sequences coupled with genetic approaches can be used for identifying suitable genes conferring stress tolerance that can be deployed in crop improvement either by using molecular breeding or genetic engineering approaches. Superior genes or alleles have been identified in the same species could be transferred into elite genotypes through molecular breeding. Moreover, by using genetic engineering/transgenic approaches gene(s) responsible for an advantageous trait could be used to develop stress tolerant crops, even from non-plant species.

Recognizing the negative impact of global warming and climatic change on crop production and to counter it, researchers could use biotechnological approaches like molecular breeding and genetic engineering to provide a buffer against increased unpredictability of climate change.

Molecular Breeding

Molecular plant breeding aims to improve crop variety in context to its yield, quality and resistance by the means of latest innovations made in genetics and genomics. Our understanding about the association between genotype and phenotype has been growing with the help of genomics tools. One of the major applications of genomics directly related to breeding has been identification of molecular markers associated with the trait of interest for breeders. Such markers have greatly helped the breeding communities for several crops in overcoming the constraints of phenotypic selection which sometimes is unable to identify individual with highest breeding value. This section presents some selected breeding methodologies that use genomics tools to facilitate breeding for developing the superior lines or genotypes plants for yield and stress tolerance.

Marker Development and Mapping

Molecular markers and genetic maps are pre-requisites in several markerassisted selection (MAS) and genomicsassisted breeding programmes for crop improvement (Varshney *et al.*, 2010). Availability of large amount of sequence data from genome/ transcriptome because of advent in next generation sequencing (NGS) technologies and high-throughput genotyping platforms, together with advances bioinformatics tools and databases, allows discovering new genes and regulatory sequences and their positions, and makes available large collections of molecular markers. Various molecular markers such as restriction fragment length polymorphism (RFLP), randomly amplified polymorphic DNA (RAPD), microsatellite or simple sequence repeat (SSR), amplified fragment length polymorphism (AFLP), single nucleotide polymorphism (SNP) markers have been developed in a range of crops, SSR and SNP markers have emerged as the markers of choice (Gupta and Varshney, 2000; Varshney *et al.,* 2007). Further, genomic approaches like TILLING (target induced local lesions in genomes) and EcoTILLING (Eco type TIILING), which examines natural genetic variation in populations and has been successfully utilized in animals and plants to discover SNPs including rare ones.

The main goal of genetic mapping is to detect neutrally inherited markers in close proximity to the genetic causatives or genes controlling the complex quantitative traits. Genetic mapping can be done mostly in two ways

1. Using the experimental populations (also referred to as "biparental" mapping populations) that are called QTL-mapping as well as "genetic mapping" or "gene tagging,"
2. Using the diverse lines from the natural populations or germplasm collections that is called "LD-mapping" or "association mapping."

Both linkage analysis and association studies rely on co-inheritance of functional polymorphism and neighboring DNA variant. The difference is that in linkage analysis (panel a, using F_2 design as an example), there are only a few opportunities for recombination to occur within families and pedigrees with known ancestry, resulting in relatively low mapping resolution; whereas in association mapping (panel b, showing only haplotype) historical recombination and natural genetic diversity were exploited for high resolution mapping.

QTL-Mapping

1. The experimental populations such as F_2, back cross (BC), double haploid (DH), recombinant inbred line (RIL), and near isogenic line (NIL) populations, derived from the genetic hybridization of two parental genotypes with an alternative trait of interest, need to be developed.
2. These experimental populations including a large number of progenies or lines are measured for the segregation of a trait of interest in the different environmental conditions.
3. Set of polymorphic DNA markers, differentiating the parental genotypes and segregating in a mapping population need to be identified and genotyped. For that, usual practice is that, first, the parental genotypes are screened, and if markers are polymorphic over the parents, then, all individuals of a mapping population are genotyped with these polymorphic molecular markers. Once genotypic data of a mapping population is ready, marker data is used to construct the framework linkage maps, representing the order (position) and linkage (a relative genetic distance in cM) of used molecular markers along the linkage groups or segments of particular

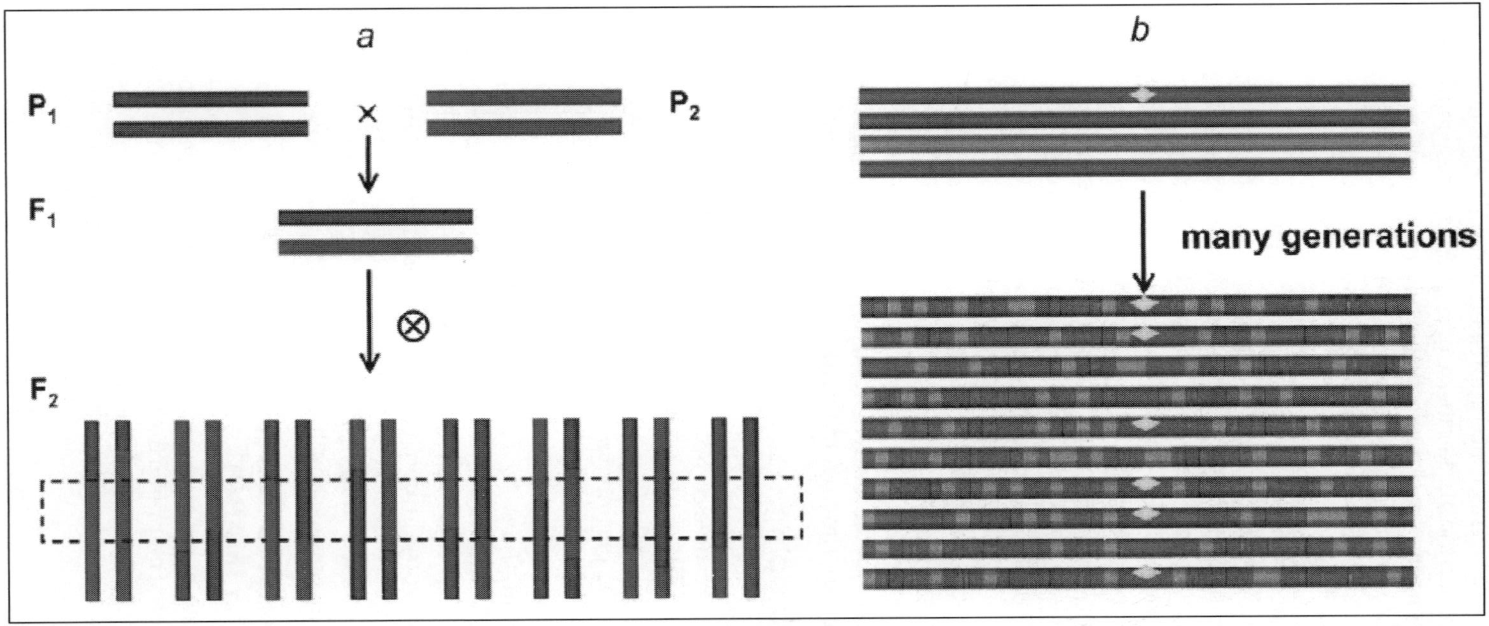

Figure 7.1: Schematic Comparison of Linkage Analysis with Designed Mapping Populations and Association Mapping with Diverse Collections (*Source*: Zhu *et al.*, 2008).

chromosomes. This is accomplished through assessing of recombination rates between the marker loci. Consequently, these markers ordered along the linkage map are statistically correlated with phenotypic characteristics of individuals of a mapping population, and QTL regions affecting a trait of interest, along with closely positioned marker tags to that QTL, are identified.

The precision of QTL-mapping largely depends on the genetic variation (or genetic background) covered by a mapping population, the size of a mapping population, and a number of marker loci used. Once QTLs affecting a trait of interest accurately tagged using above-outlined approach, marker tags are the most effective tools in a crop improvement that allows the mobilization of the genes of interest from donor lines to the breeding material through marker-assisted selection (MAS). Although traditional QTL-mapping will continue being an important tool in gene tagging of crops, it is a "now classical approach" and overall is very costly, and has low resolution with simultaneous evaluation of only a few alleles in a longer research time scale. Linkage analysis in plants typically localizes QTLs to 10 to 20 cM intervals because of the limited number of recombination events that occur during the construction of mapping populations and the cost for propagating and evaluating a large number of lines. In linkage mapping, the major limitation, hampering the fine mapping, is associated with the availability of only a few meiotic events to be used that occurred since experimental hybridization in a recent past.

Association Mapping

Association mapping, also known as linkage disequilibrium (LD) mapping, has emerged as a tool to resolve complex trait variation down to the sequence level by exploiting historical and evolutionary recombination events at the population level. Conventional linkage analysis with experimental populations derived from a bi-parental cross provides pertinent information about traits that tends to be specific to the same or genetically related populations, while results from association mapping are more applicable to a much wider germplasm base. The ability to map QTLs in collections of breeding lines, landraces, or samples from natural populations has great potential for future trait improvement and germplasm security. Association mapping is complementary to advanced backcross QTL (AB-QTL) and introgression library (IL) in that it is an additional tool for evaluating extant functional diversity in each crop species on a much larger scale. Association mapping has a number of advantages over linkage mapping including the potential for increased QTL resolution, and an increased sampling of molecular variation.

1. Marker-Assisted Selection (MAS) and Marker-Assisted Backcrossing (MABC)

Once the markers associated with a trait of interest are identified through linkage mapping or association mapping, the next step is to use these markers in the breeding programs. In this context, the selection of one or a few genes (QTLs) through molecular markers using backcrossing is a very efficient technique (Utomo and Linscombe, 2009).

There are three levels of MABC: (i) foreground selection (Hospital and Charcosset, 1997) which includes screening of target gene or QTL using molecular markers, this step can also be used for selection of recessive allele for backcrossing as recessive alleles require one generation of selfing for its expression, (ii) recombinant selection involves selection of the BC progeny containing the target gene and recombination events (between the target locus and linked flanking markers). The purpose of this selection step is to minimize the 'linkage drag' by using markers that flank the target gene. This linkage drag poses a big problem during selection through conventional breeding methods. Furthermore this recombination selection event is usually carried out using two backcross (BC) generations (Frisch *et al.*, 1999) (iii) background selection involves use of markers that are unlinked to the target locus for the selection of BC progeny containing highest proportion of recurrent parent (RP). In summary, the MABC employs linked markers to select the target gene/QTL from the donor parent and the unlinked markers to recover RP. Traditional approaches of recovery of RP genome take upto six BC generations but the use of markers enables to achieve the same in even in BC2. MABC approach has been very popular approach in molecular breeding community. While MABC approaches are in routine in many multi-national companies, MABC has been successfully utilized in some breeding programmes in public sector. Success stories of MABC leading the development of improved varieties and superior genotypes have been detailed in crops, like bell pepper (Thabuis *et al.*, 2006) and Tomato (Robert *et al.*, 2001). In recent years, MABC approach is becoming more popular in developing countries like India, China, Bangladesh and Thailand where public sector in collaboration with the international agriculture research centres have been working towards developing the improved varieties in their targeted environments.

In marker- assisted backcrossing, a particularly critical step is the identification of major QTLs with consistent effects across different genetic backgrounds. A notable exception is *Sub1*, a major QTL that has been widely targeted *via* MABC in order to improve submergence tolerance in rice (Septiningsih *et al.*, 2009). Further, the genetic improvement of Basmati rice for yield, quality and resistance to bacterial leaf blight and blast diseases was achieved by pyramiding the multiple genes through MAS and MABC (Joseph *et al.*, 2004; Sundaram *et al.*, 2008; Singh *et al.*, 2011). Moreover, the improvement of hybrid performance and bacterial blight, blast and brown plant hopper disease resistance of Indian rice varieties have been well executed through MAS using fertility restorer and disease resistant genes, respectively (Basavaraj *et al.*, 2010; Singh *et al.*, 2011).

2. Advanced-Backcross (AB-QTL) Analysis

Although LD mapping offers the possibility of utilizing the potential of exotic germplasm for crop improvement, advanced backcross QTL (AB-QTL) and introgression library (IL) are well-known strategies for mining alleles from exotic germplasm to improve the productivity, adaptation, quality, and nutritional value of crops. Further, though MAS has been quite successful, it has always been a difficult task to tackle linkage drag especially when a QTL or a gene is to be introgressed from wild/exotic species. Furthermore, in MAS, QTL/gene discovery and variety development are two separate processes. To deal with this problem and to harness the potential of the wild/unadapted germplasm in breeding programs, Tanksley and Nelson (1996) proposed the advanced backcross QTL (AB-QTL) analysis, which integrates QTL analysis with variety development, by simultaneously identifying and transferring favorable QTL alleles from unadapted (*e.g.* land races, wild species) to cultivated germplasm. In this context, a superior cultivar/variety is crossed with a wild species leading to the production of a backcross population (BC2, BC3) and molecular markers are used to monitor the transfer of QTLs by conventional backcrossing. In the advanced backcross (AB) design, QTL analysis will be delayed until a later generation like the BC2 or BC3 generation. The logic behind this approach is that the effect of individual QTLs can be more precisely measured because undesirable effects of wild species on the elite background are reduced since later generation progenies such as BC2 or BC3 carry a smaller number of genes from the donor parent. The advanced backcross approach has already been successfully utilized in different crops including rice (Manosalva *et al.*, 2009), barley (Schmalenbach *et al.*, 2008), maize (Mano and O Mori, 2008), tomato (Tanksley *et al.*, 1996; Stevens *et al.*, 2007), and wheat (Ibrahim *et al.*, 2010).

3. Marker-Assisted Recurrent Selection (MARS)

There are cases where quantitative variation is controlled by many genes (QTLs) with minor effect; in such cases the previous approach (MABC) has certain limitations in the introgression of the target trait. Moreover, the markers identified linked with a trait to be used in MABC are generally identified biparental mapping populations. This limits the study of allelic diversity and genetic background which are very essential in crop breeding program. Limited statistical tools for studying polygenic traits controlled by many small effect loci are yet another drawback of MAS (Boer *et al.*, 2007, Cooper *et al.*, 2007). Furthermore, minor QTLs show an inconsistent QTL effect. Even though the effect of these minor QTLs is consistent, introgression of these QTLs through MABC approach becomes extremely difficult, as a larger number (sometimes unmanageable) of progenies, depending on the number of QTLs, are required to select appropriate lines in MABC. In cases as mentioned above, marker assisted recurrent selection (MARS) can be used for pyramiding of several genes/ QTLs (of minor effect) in a single genotype (Hospital and Charcosset 1997, Edward *et al.*, 1994). MARS is based on adhoc significance test which include the identification of trait associated markers and estimation of their effect. The approach involves multiple cycles of marker based selection that includes, (i) identification of F2 progeny which contain favorable alleles for most if not all QTLs, (ii) recombination of the selected progenies to the selfed ones, and (iii) repetition of these cycles. However, development of publicly available appropriate statistical tools for calculating the selection indices to select the progeny lines for the selection cycles is still an issue. According to the recent studies, the response of MARS is larger in case of prior knowledge of the QTLs and the response decreases as the knowledge of the number of minor QTL associated with the trait decreases (Bernardo and Charcosset, 2006). In sweet corn, MARS was employed to fix six marker loci in two different F_2 populations which showed an increase in the frequency of marker allele from 0.50 to 0.80 (Edward **et al.,** 1994). Similarly in a separate study, enrichment of rust resistance gene (Lr34/

Yr18) with the increase in frequency from 0.25 to 0.60 was reported in wheat BC1 through MARS (Kuchel ***et al.***, 2007). MARS, becoming a popular approach, can thus be effectively utilized for selection of traits associated with multiple QTLs by increasing the frequency of favourable QTLs or marker alleles. Several multinational companies such as Syngenta and Monsanto use MARS in the breeding programmes of several crops such as maize, soybean etc. (Ragot *et al.*, 2000, Ribaut and Ragot, 2006). Recently, some international agricultural research centres in collaboration with Generation Challenge Program (GCP) have also initiated MARS in crops like chickpea, sorghum, rice, cowpea, etc for pyramiding favorable drought tolerant alleles.

Genome-Wide Selection (GWS)

In addition to MARS, the genome-wide selection is another approach that can be used to pyramid favourable alleles for minor effect QTLs at whole genome level. The main advantage of GWS is that it does not require any a prior knowledge about the loci influencing the target traits. Unlike MABC or MARS, the GWS calculates the marker effects across the entire genome that explains entire phenotypic variation. The genome wide marker data (marker loci or haplotypes) available or generated on the progeny lines, therefore, are used to calculate genomic estimated breeding values (GEBV) (Meuwissen *et al.*, 2001). It is important to note that the GEBVs are calculated for individuals based on genotyping data using a model that was 'trained' from individuals having both phenotyping and genotyping data. These GEBVs are then used to select the progeny lines for advancement in the breeding cycle. In summary, the GWS provides a strategy for selection of an individual without phenotypic data by using a model to predict the individual's breeding value. However, to maximize the GEBV accuracy in GWS, it is very critical to select the appropriate training population (used to develop the model for calculating GEBVs) that is representative of selection candidates in the breeding programme to which GWS will be applied. Although, to the best of our knowledge, there is no published report on deploying GWS in a crop breeding programme, availability of high-throughput genotyping platforms in several crops makes it feasible to generate genome wide marker data and undertake the GWS in breeding pogramme. It is anticipated that at least a few breeding programmes for major crops like rice, sorghum, soyabean, tomato, potato, etc (where draft genome sequence is available) should be using GWS soon.

Genetic Engineering

Genetic engineering approach involve the incorporation of exogenous DNA or ribonucleic acid (RNA) sequences, using either biolistics or vectors, into the genome of the recipient organism which, as a result, expresses novel and agronomically useful traits. Thus, genetic engineering tools focus on providing the ability to directly detect and transfer genes of interest from other plant lines or organisms into the crop of interest without the continuing need to use the appearance or stress response of the plant (its phenotype). With the advent of genetic transformation techniques, it has become possible to engineer plants to suit the environment better through incorporation of genes for tolerance to biotic and abiotic stress (Ortiz, 1998). Over the past few decades, breeding possibilities have been broadened by genetic engineering and gene transfer technologies, as well as by gene mapping and identification of the genome sequences of model crop plants, and which resulted in efficient transformation and generation of transgenic lines in a number of crop species. Further, pyramiding of desirable genes with similar effects can also be achieved by using these approaches. Genetic transformation by *Agrobacterium* mediated approaches has been successfully demonstrated for economically important traits like biotic resistance (Wani and Sanghera 2010), abiotic stress tolerance (Gosal *et al.*, 2009), in different crop plants.

Climate change is generating increase in atmospheric CO_2 concentrations (rising from about 270 µmol.mol"1 in 1750 to current concentrations larger than 385 µmol.mol"1). This increase in atmospheric CO_2 has been accompanied by a coincident increase in the even more potent forcing gases methane, ozone, and nitrous oxide such that combined ambient greenhouse gas concentrations are now expected to exceed concentrations of 550 µmol.mol"1 by 2050. Genetically modified (GM) crops offers a solution to decrease green house gases and therefore mitigates climate change. Since, the first approvals in 1996 for commercialization biotech crops, for the last 18 years GMOs have been contributing to the reduction of CO_2 emissions. They allow farmers to use less and environmentally friendly energy and fertilizer, and practice soil carbon sequestration. For example, herbicide tolerant biotech crops such as soybean and canola facilitate zero or no-till, which significantly reduces the loss of soil carbon (carbon sequestration) and CO_2 emissions, reduce fuel use, and significantly reduce soil erosion. Insect resistant biotech crops require fewer pesticide sprays which results in savings of tractor/fossil fuel and thus less CO_2 emissions.

Since the recorded commercialization of GM crops in 1996 up to 2013, millions of farmers in ~30 countries worldwide contributed to an unprecedented 100-fold increase in the global area of transgenic crops (1.6 billion hectares). Biotech crops have reduced the amount of pesticides used by 497 million kilograms. In 2012 alone, fewer insecticide sprays reduced CO_2 emissions by 26.7 billion kilograms, equivalent to taking 11.8 million cars off the road for a year (ISAAA report, 2013, www.isaaa.org).

Tremendous success in terms of pest control and yield improvements has been seen globally witnessed with genetically modified versions of soy, maize and cotton produced for insect pest resistance and/or herbicide tolerance. More results that are moderate were seen with transgenic alfalfa, canola, papaya and squash. To date these commercialized genetic modifications have involved genetically simple (single or double gene) traits. A major reason for the relatively slow progress in conventional breeding responses to the stresses related to climate change arises from the fact that plant adaptation, for example to drought or salinity effects, are not likely to be single gene changes. Whole metabolic pathways or cascades of pathways are likely to be involved. Strategies for the use of selected genes to improve tolerance to abiotic stresses in crops include gain- and loss-of-function approaches that target single genes at various levels. These genes could be enzymes, proteins, or regulatory genes such as transcription factors. Tissue-specific, constitutive, or stress inducible-promoters have been used to express the selected genes in order to achieve maximum efficiency in stress protection with as few as possible negative effects on growth and productivity. Lusser *et al.* (2011) summarized following techniques for development of GM/ transgenic crops *viz.*, i) Zinc finger nuclease (ZFN): single mutations or short indels are generated or new genes are introduced into pre-determined target sites of the genome ii) Oligonucleotide directed mutagenesis (ODM): targeted mutations of one or a few nucleotides are induced iii) Cisgenesis and intragenesis: GMOs are produced by the insertion of hereditary materials derived from the species itself or from a cross-compatible species and are contiguous and unchanged (cisgenesis) or the inserted DNA may be a new combination of DNA fragments but must still be from the species itself or from a cross-compatible species iv) RNA-dependent DNA methylation (RdDM): Still being refined, modified gene expressions are epigenetic with the new phenotypes inherited only over a few generations v) Grafting (on GM rootstock): desired improvements are achieved by the grafting of non-transgenic scions onto GM rootstock vi) Reverse breeding: combination of recombinant DNA techniques and cell biology procedures is used to generate suitable transgene-free homozygous parental lines rapidly for reconstituting elite heterozygous genotypes vii) Agro-infiltration: used mostly in research settings, for example to study plant-pathogen interaction in living tissues, to select parental lines or to evaluate the efficacy of transgenes, a liquid suspension of *Agrobacterium* sp. containing the desired gene(s) is used to infiltrate plant tissues, mostly leaves, so that the genes are locally and transiently expressed at high levels viii) Synthetic genomics: Large functional DNA molecules that are synthesized without any natural templates are used for constructing viable minimal genomes which can serve as platforms for the biochemical production of chemicals such as biofuels and pharmaceuticals

Generation of huge genomic information and the use of bioinformatic tools have led to the identification of signaling pathways and regulatory genes and networks controlling complex traits related to environmental stresses. However, drought together with soil salinity, stresses that threaten worldwide crop productivity under climate change scenario. In terms of physiology, both drought and salt stresses, as well as heat or cold stress, produce overlapping responses in plants, including accumulation of osmoprotectants and heat shock proteins, some of which are mediated by the actions of the phytohormone abscisic acid (ABA).

Salinity Tolerant Crops

Plants try to cope with salinity by inducing various metabolic changes like production of osmolytes, antioxidative enzymes and upregulating various genes involved in stress response like ion transporters, ion channels, transcriptional factors and various signaling pathway components. NtAQP1, a gene encoding for aquaporin was identified in tobacco and shown to provide protection against salinity stress in transgenic tomatoes (Sade *et al.*, 2010). *NtAQP1* plays a key role in preventing root/shoot hydraulic failure, enhancing water use efficiency and thereby improving salt tolerance. Transgenic rice plants overexpressing Arabidopsis *CBF3/DREB1A* or *ABF3* TF showed improved tolerance to drought and salinity without growth retardation (Oh *et al.*, 2005). In Australia, field trials of 1,161 lines of GM wheat and 1,179 lines of GM barley modified to contain one of 35 genes obtained from wheat, barley, maize, thale cress, moss or yeasts are in progress since 2010 and will run till 2015. Some of the genes are expected to enhance tolerance to a range of abiotic stresses including drought, cold, salt and low phosphorous. Sugarcane that contains transcription factor (OsDREB1A) is also under field trial from 2009 to 2015 (Tammisola, 2010).

Drought Resistant Crops

Transgenic plants carrying genes for water-stress management have been developed. Structural genes (key enzymes for osmolyte biosynthesis, such as proline, glycine/betaine, mannitol and trehalose, redox proteins and detoxifying enzymes, stress-induced LEA proteins) and regulatory genes, including dehydration–responsive, element-binding (DREB) factors, zinc finger proteins, and NAC transcription factor genes, are being used. Transgenic crops carrying different drought tolerant genes are being developed in rice, wheat, maize, sugarcane, tobacco, Arabidopsis, groundnut, tomato, potato and papaya. Phospholipase C1 gene (ZmPLC1) cloned from maize encodes phosphatidylinositol-specific phospholipase C (PI-PLC) and the up-regulated the expression of ZmPLC1 was recorded in maize roots under dehydration (Zhai *et al.*, 2005). Research results show that an enhanced expression of ZmPLC1 improves drought tolerance in transgenic maize (Wang *et al.*, 2008). Under drought stress, the transgenic maize had higher relative water content, better osmotic adjustment, increased photosynthesis rates, lower percentage of ion leakage, less lipid membrane peroxidation, and higher grain yield than the control.

The expression of the Arabidopsis' *HARDY* (HRD) gene in transgenic rice improved water-use efficiency and the ratio of biomass produced to the amount of water used, by enhancing photosynthetic assimilation and reducing transpiration (Karaba *et al.*, 2007). Rice plants that consumed less water exhibited increased shoot biomass under irrigated conditions and an adaptive increase in root biomass under drought stress. This result shows the usefulness of research undertaken in a model plant such as Arabidopsis for improving water-use efficiency in a main staple such as rice. Similarly correlation of drought tolerance with root architecture (spread, depth and volume) has been examined in cowpea (South Africa, West Africa and India), rice (India) and beans (Central and South America). Transgenic maize that expresses *Escherichia coli's* glutamate dehydrogenase (gdhA) gene (Lightfoot *et al.*, 2007) is another another potential avenue for breeding drought tolerance. Germination and grain biomass production were increased in gdhA transgenic maize in the field during seasons with significant water scarcity. Water deficit tolerance under controlled conditions was also increased.

Further, GM crop with signaling components and transcription factors (TFs) leads to the expression of their target transcriptome that consists of several genes involved in stress adaptation. For example, the Dehydration-Responsive Element Binding gene 1 (DREB1) and DREB2 are transcription factors that bind to the promoter of genes such as the dehydration-responsive Arabidopsis gene rd29A, thereby inducing expression in response to drought, salt, and cold (Ortiz *et al.*, 2007). The enhanced production of the signaling hormone Abscisic acid (ABA) by the overexpression of the LOS5/ABA3 gene encoding a Molybdenum Cofactor Sulfurase, required for ABA synthesis, conferred enhanced drought tolerance in transgenic rice plants under field conditions (Xiao, B.Z. *et al.*, 2009). Nelson *et al.* (2007) showed that an orthologous maize transcription factor (ZmNF-YB2) confers improved performance under drought. Transgenic maize plants with increased ZmNF-YB2 expression show tolerance to drought as measured by chlorophyll content, stomatal conductance, leaf temperature, reduced wilting, and maintenance of photosynthesis under limiting water. Such stress adaptations will contribute to grain yield, when this maize grows in drought-prone environments. Similarly, the overexpression of the rice AP37 (an APETALA2-type TF) gene resulted in the enhanced expression of several target genes and produced 16–57 per cent higher grain yield under field drought stress conditions (Oh *et al.*, 2009). Hence, transcriptome engineering seems to be promising for the development of abiotic stress-tolerant crops.

Drought accelerates leaf senescence, leading to a decrease in canopy size and loss in photosynthesis, thereby reducing crop yields. A delay in drought-induced leaf senescence could therefore enhance crop tolerance to water scarcity. Rivero *et al.* (2007), using tobacco as the model plant, showed that the suppression of drought-induced leaf senescence led to outstanding drought tolerance as measured by the vigorous growth of the transgenic plants after a long drought period that killed the control plants. Furthermore, the transgenic tobacco plants maintained high water contents and retained, although reduced, photosynthetic activity during the drought period. Moreover, there was minimal yield loss in the transgenic plants when they were watered with only 30 per cent of the amount of water used for the control plants. The new generation of traits to counter drought includes the use of RNA interference in which defense to stress will still be maintained, but without impairing the growth of the plant.

Biotech Crops for Cold Tolerance

By using genetic and molecular approaches, a number of relevant genes for cold tolerance have been identified. The genes controlling the CBF cold-responsive pathway and together with DREB1 genes, integrate several components of the cold acclimation response to tolerance low temperatures (Sanghera *et al.*, 2011). In 2012, a genetically modified drought tolerant maize MON 87460 that expresses cold shock protein B has been approved in the US for release in the market (http://www.aphis.usda.gov/newsroom/2011/12/brs_actions.shtml). Cold tolerant GM eucalypti are being developed, which is currently being field tested in the US by Arborgen LLC since 2010. Thale cress has been improved to contain the DaIRIP4 from *Deschapsia antarctica*, a hairgrass that thrives in frosts down to -30 C, and sugarcane are being introgressed with genes from cold tolerant wild varieties (Tammisola, 2010).

Monsanto (http://www.monsanto.com/) researchers showed that bacterial cold shock proteins (Csps) can confer improved stress adaptation in multiple plant species. CspB codes for and is responsible for an RNA chaperone, which is a commonly occurring protein molecule that binds to RNAs and facilitates their function. In maize and rice, CspB works by helping the plant maintain growth and development during times of inadequate water supply (Castiglioni *et al.*, 2008). They further stated that, expression of CSP proteins in maize is not associated with negative pleiotropic effects, indicating that stress tolerance does not come at a cost to crop yield under limiting water.

Biotech Crops for Heat Stress

Expression of heat shock proteins (HSPs) has been associated with recovery of plants under heat stress. HSPs bind and stabilize proteins that have become denatured during stress conditions, and provide protection to prevent protein aggregation. In GM chrysanthemum containing the DREBIA gene from *Arabidopsis thaliana*, the transgene and other heat responsive genes such as the HSP70 were highly expressed when exposed to heat treatment. The transgenic plants maintained higher photosynthetic capacity and elevated levels of photosynthesis-related enzymes (Hong *et al.*, 2009). Similarly cultivars are now being developed which are tolerant to excess heat during pollination for cowpea and corn, and flooding early in the growing season for soybean and rice (Hall, 2004; Van Toai *et al.*, 2010; Bailey-Serres *et al.*, 2010). Maize hybrids are now being developed that have a better synchronization of pollination and flowering under heat and water stress (Ribaut *et al.*, 2009). Cultivar differences for heat tolerance exist in some crops such as rice, cowpea, and peanut, but knowledge about the effects of extremely high temperatures is very limited because diverse germplasm has not been extensively screened. Continued work in this area will provide new germplasm for plant breeders to incorporate into productive varieties for farmers.

Biotech Crops for High CO_2 and Better N Use

The main challenge for crop genetic enhancement will be to breed crops that maximize the advantages of rising CO_2. Ainsworth *et al.* (2008) suggest that substituting current Rubisco for Rubisco from other species, particularly non-green algae, which has a markedly lower specificity and higher catalytic rate, could dramatically increase C gain at current and elevated CO_2 levels. Improving the nitrogen-use efficiency (NUE) of crops will be a key factor for reducing N fertilizer pollution as well as for improving yields in N-limiting environments. Shrawat and Good (2008) give a brief summary of crop genetic engineering for enhancing NUE. Engineering plants with transport gene systems, glutamine synthetase and glutamate synthase gene systems, gene systems regulating N metabolism or manipulating N remobilization are among the ongoing NUE research undertakings. Research suggests it could be possible to enhance or manipulate N metabolism and crop growth. Improved NUE in crops should lead to reduced fertilizer application and thereby lower emissions of greenhouse gases into the atmosphere.

Furthermore, to tackle the issue of tolerance to multiple stresses at different stages of plant growth and development. A combination of genes is required to offset the adverse impact of climate variability on plants. Apart from genes encoding effector proteins, signaling proteins and/or TFs, there is a need for a repertoire of promoters to drive transgene expression in a precise and predetermined fashion in specific tissues or plant organs. Transgenic rice plants overexpressing OsNAC10 under the rootspecific promoter RCc3, but not under the control of the constitutive GOS2 promoter, conferred a yield advantage under drought stress conditions in the field (Nelson, D.E. *et al.*, 2007 and Jeong *et al.*, 2009, 2010).

In spite of the low numbers of commercial GM crops (that is herbicide tolerance and insect resistance or their combinations) and the transformation events that confer the modified agronomic traits. The increased cultivation of GM crops globally, signals a trend in crop improvement that can no longer be ignored. Technically, the drawbacks to more widespread development of GM varieties include the lack of efficient genotype independent regeneration systems for most crops and the stacking of transformation events. However, the successful stacking of genes conferring insect resistance and herbicide tolerance is indicative of progress in addressing this constraint. Another hindrance to wider adoption of the GM technology is the absence of biosafety regulatory frameworks as specified by the Cartagena Protocol on Biosafety to the Convention on Biological Diversityin many countries. Nevertheless, the research and development in biotechnology across the globe has shown considerable scale and momentum in recent past.

Conclusion

In conclusion, the advances in genome sequencing, development of high throughput omic technologies, quantitative genetics, and bioinformatics allowed for fast and accurate identification of adaptive traits expressed in different environments. Development cultivars (using molecular and genetic engineering approaches) adapted to climate change (smart or designer crops) with improved features that can use natural resources such as water, soil nutrients, atmospheric carbon and nitrogen with a far greater efficiency has enhanced the crop production. However, molecular breeding - derived products have been accepted and adopted, genetic engineering-derived crops still have a long way to go to gain universal acceptance and reach farmers' fields.

References

Ainsworth, E., A. Rogers and Leakey, A.D.B. (2008). Targets for crop biotechnology in a future high-CO_2 and high-O_3 world. *Plant Physiol.*, **147**: 13-19.

Bailey-Serres, J., F. Fukao, P. Ronald, A. Ismail, S. Heuer, and D. Mackill. (2010). Submergence tolerant rice: SUB1's journey from landrace to modern cultivar. *Rice*, **3**: 138–147.

Basavaraj, S. H., V. K. Singh, A. Singh, A. Singh, A. Singh, S. Yadav, R. K. Ellur, D. Singh (2010). Marker-assisted improvement of bacterial blight resistance in parental lines of Pusa RH10, a superfine grain aromatic rice hybrid. *Mol. Breed.*, **2**: 293–305.

Bernardo, R. and A. Charcosset (2006). Usefulness of gene information in marker-assisted recurrent selection: A simulation appraisal. *Crop Sci.*, **46:** 614–621.

Boer, M. P., D. Wright, L. Feng, D. W. Podlich, L. Luo, M. Cooper and van Eeuwijk F. A. (2007). A mixed-model quantitative trait loci (QTL) analysis for multiple-environment trial data using environmental covariables for QTL-by-environment interactions, with an example in maize. *Genetics.*, **177**: 1801–1813.

Castiglioni, P., D.Warner, R. J.Bensen, D. C.Anstrom and Harrison and Stoecker, M. *et al.* (2008). Bacterial RNA chaperones confer abiotic stress tolerance in plants and improved grain yield in maize under water-limited conditions. *Plant Physiol.*, **147:** 446–455.

Ciais, Ph., M. Reichstein, N. Viovy, A. Granier, J. Oge´e, V. Allard, M. Aubinet and Buchmann N. *et al.* (2005). Europe-wide reduction in primary productivity caused by the heat and drought in 2003. *Nature.*, **437**: 529–533.

CNA (2007). National Security and the Threat of Climate Change. Available at http: //security and climate. cna. org/(Verified 9 June 2011). Alexandria, VA.

Coakley, S.M., H. Scherm, and Chakraborty, S. (1999). Climate change and plant disease management. *Annual Review of Phytopathology*, **37:** 399-426.

Cooper, M., D. W. Podlich and Luo, L. (2007). Modeling QTL effects and MAS in plant breeding. In RK Varshney, R Tuberosa, eds, Genomics-Assisted Crop Improvement, Vol 1: Genomics Approaches and Platforms. Springer, Dordrecht, The Netherlands, pp. 57–96.

Edward, M. and Johnson, L. (1994). In Proc Joint Plant Breeding Symp Series of CSSA and ASHA, Corvallis, OR. Am Soc Hort Sci, Alexandria VA, (1994) pp. 33-40.

Frisch, M., M. Bohn and Melchinger, A. E. (1999). Minimum sample size and optimal positioning of flanking markers in marker-assisted backcrossing for transfer of a target gene. *Crop Sci.*, **39**: 967-975.

Gosal, S. S., S. H. Wani and Kang, M. S. (2009). Biotechnology and drought tolerance. *J. Crop Improvement*, **23**: 19–54.

Gupta, P. K., and Varshney, R. K. (2000). The development and use of microsatellite markers for genetics and plant breeding with emphasis on bread wheat. *Euphytica.*, **113:** 163-185.

Gur, A. and Zamir, D. (2004). Unused natural variation can lift yield barriers in plant breeding. *PLoS Biol.*, **2**: 1610–1615.

Hall, A. E. (2004). Breeding for adaptation to drought and heat in cowpea. *Eur. J. Agron.*, **21**: 447–454.

Hatfield, J., K. Boote, B.A. Kimball, R. Izaurralde, D. Ort, A. Thomson, and D. Wolfe. 2011. Climate Impacts on Agriculture: Implications for Crop Production. *Agron. J.*, **103**: 351–370.

Hong, B., C. Ma, Y. Yang, T. Wang, K. Yamaguchi-Shinozaki and Gao J. (2009). Over-expression of AtDREB1A in chrysanthemum enhances tolerance to heat stress. *Plant Mol Biol.*, **70**(3): 231-240.

Hospital, F. and Charcosset, A., 1997, Marker-assisted introgression of quantitative trait loci. *Genet.*, **147**: 1469-1485.

Ibrahim, S. E., A. Schubert, K. Pillen and Léon, J. (2010). Quantitative trait loci analysis for drought tolerance in an advanced backcross population of spring wheat. *Sudan J. Agricul. Res.*, **15**: 1-18.

ISSSA report (2013). Beyond Promises: Top 10 Facts about Biotech/Gm Crops in 2013. (available at http: //www.isaaa.org/resources/publications/ biotech_booklets/top_10_facts/download/Top per cent 2010 per cent 20Facts per cent 20Booklet.pdf).

Jeong, J. S., Y. S. Kim, K. H. Baek, H. Jung, S. H. Ha and Choi, Y. D., *et al.* (2010a). Root-specific expression of osnac10 improves drought tolerance and grain yield in rice under field drought conditions. *Plant Physiol.*, **153**: 185–197.

Jeong, J. S., Y. T. Park, H. Jung, S. H. Park, and Kim, J. K. (2009). Rice NAC proteins act as homodimers and heterodimers. *Plant Biotechnol. Rep.*, **3**: 127–134.

Joseph M, S. Gopala Krishnan, R. K. Sharma, V. P. Singh, A. K. Singh, N. K. Singh and Mohapatra, T. (2004) Combining bacterial blight resistance and Basmati quality characteristics by phenotypic and molecularmarker-assisted selection in rice. *Mol. Breed.*, **13** : 377–387.

Karaba, A., S. Dixit, R. Greco, A.Aharoni, K. R. Trijatmiko, N. Marsch-Martinez, A. Krishnan, K. N. Nataraja, Udayakumar, M. and Pereira, A. (2007). Improvement of water use efficiency in rice by expression of HARDY, an Arabidopsis drought and salt tolerance gene. *Proc. Natl. Acad. Sci. U.S.A.* **104**: 15270–15275.

Kuchel, H., R. Fox, J. Reinheimer, L. Mosionek, N. Willey, H. S. Bariana, and Jefferies, S. (2007). The successful application of a marker-assisted wheat breeding strategy. *Mol. Breed.*, **20**: 295-308.

Lightfoot, D.A., Mungur, R., Ameziane, R., Nolte, S., Long, L., Bernhard, K., Colter, A., Jones, K., Iqbal, M.J., Varsa, E. and Young, B. 2007. Improved drought tolerance of transgenic *Zea mays* plants that express the glutamate dehydrogenase gene (gdhA) of *E. coli*. *Euphytica*, **156**: 103-116.

Lobell, D. and Field, C. (2007) Global scale climate-crop yield relationships and the impacts of recent warming. Environ. *Res. Lett.*, **2**(1): 1-7.

Lusser, M., C. Parisi, D. Plan, and Rodríguez-Cerezo E.(2011).New plant breeding techniques. State-of-the-art and prospects for commercial development. Luxembourg: Publications Office of the European Union.

Mano, Y. and Omori, F. (2008). Yoshiro Mano and Fumie Omori. Verification of QTL controlling root aerenchyma formation in a maize × teosinte "Zea nicaraguensis" advanced backcross population. *Breed. Sci.*, **58:** 217-223.

Manosalva PM, R. M. Davidson, B. Liu, X. Zhu, S. H. Hulbert, H. Leung, and Leach, J. E. (2009). A germin-like protein gene family functions as a complex quantitative trait locus conferring broad-spectrum disease resistance in rice. *Plant Physiol.*, **149**: 286–296.

Meuwissen, T.H.E., B. J. Hayes and Goddard M. E. (2001) Prediction of total genetic value using genome-wide dense marker maps. *Genetics.*, **157:** 1819–1829.

Nelson, D. E., P. P. Repetti, T. R. Adams, R. A. Creelman, J. Wu, D. C. Warner and Anstrom, D. C, *et al.* (2007) Plant nuclear factor Y (NF-Y) B subunits confer drought tolerance and lead to improved corn yields on water limited acres. *Proc. Natl. Acad. Sci. U.S.A.* **104**: 16450–16455.

Oh, S.J. Y. S. Kim, C.W. Kwon, H. K. Park ans Jeong, J. S. (2009) Overexpression of the transcription factor AP37 in rice improves grain yield under drought conditions. *Plant Physiol*. **150**: 1368–1379.

Oh, S.J., S. I. Song, Y. S. Kim, H. J. Jang, S. Y. Kim, M. Kim, Y. K. Kim, B. H. Nahm, and Kim, J. K. (2005) Arabidopsis CBF3/DREB1A and ABF3 in transgenic rice increased tolerance to abiotic stress without stunting growth. *Plant Physiol.*, **138**: 341–351.

Ortiz, R., M. Iwanaga, M. P. Reynolds, H. Wu, and J. Crouch (2007). Overview on crop genetic engineering for drought prone environments. *Journal of Semi-Arid Tropical Agricultural Research.*, 4 http: / / www.icrisat.org / jornal / SpecialProject / sp3.pdf.

Ragot M., G. Gay, J. P. Muller. and Durovray J. (2000). In molecular approaches for the genetic improvement of cereals for stable production in water-limited environments (JM Ribaut, D Poland, Editors), Mexico, DF: CIMMYT, 128 pp.

Reynolds, M. P. and Ortiz, R. (2010) Adapting crops to climate change: A summary; In MP Reynolds (eds) Climate change and crop production. Earthscan London., pp. 1–8.

Ribaut, J.M., F. J. Betrán, P. Monneveux, and T. Setter. 2009. Drought tolerance in maize. *In* S.C. Hakeand J.L. Bennetzen (ed.) Hand book of maize: its biology. Springer, Netherlands., pp. 311–344.

Rivero, R.M., M. Kojima, A. Gepstein, H. Sakakibara, R. Mittler, S. Gepstein and Eduardo, B. (2007). Delayed leaf senescence induces extreme drought tolerance in a flowering plant. *Proc. Natl. Acad. Sci.,* U.S.A. **104**: 19631-19636.

Robert V. J. M., M. A. L. West, S. Inai, A. Caines, L. Arntzen, J. K. Smith and St Clair, D. A. (2001). Marker-assisted introgression of black mold resistance QTL alleles from wild *Lycopersicon cheesmanii* to cultivated tomato (*L. esculentum*) and evaluation of QTL phenotypic effects. *Mol Breed.,* **8**: 217–233.

Sade, N., M. Gebretsadik, R. Seligmann, A. Schwartz, R. Wallach and Moshelion, M. (2010).The role of tobacco Aquaporin1 in improving water use efficiency, hydraulic conductivity, and yield production under salt stress. *Plant Physiol.,* **152**: 245–254.

Sanghera, G. S., S. H. Wani, W. Hussain, and Singh, N. B. (2011). Engineering cold stress tolerance in crop plants. *Curr Genomics.,* **12** (1): 30-43.

Schmalenbach, I., N. Koerber and Pillen, K. (2008). Selecting a set of wild barley introgression lines and verification of QTL effects for resistance to powdery mildew and leaf rust. *Theor. Appl. Genet.,* **117**: 1093–1106.

Septiningsih, E. M., A. M. Pamplona, D. L. Sanchez, C. N. Neeraja, G. V. Vergara, S. Heuer, A. M. Ismail and David J. M. (2009). Development of submergence-tolerant rice cultivars: the Sub1 locus and beyond. *Annals of Botany.,* **103**: 151–160.

Shrawat, A. K. and Good, A. G. (2009) Genetic engineering approaches for improving nitrogen use efficiency (NUE) in plants. *Information Systems for Biotechnology.,* pp.1-5.

Singh, A. K, S. Gopalakrishnan, V. P. Singh, K. V. Prabhu, T. Mohapatra, N. K. Singh, T. R. Sharma and Nagarajan, M. *et al.* (2011). Marker assisted selection: a paradigm shift in Basmati breeding. *Indian J. Genet.,* **71** : 120–128.

Stevens, R., M. Buret, P. Duffé, C. Garchery, P. Baldet, C Rothan and Causse. M. (2007). Candidate genes and quantitative trait loci affecting fruit ascorbic acid content in three tomato populations (2007). *Plant Physiol.,* **143**: 1943–1953.

Sundaram, R. M., M. R. Vishnupriya, S. K. Biradar, G. S. Laha, G. A. Reddy, N. Shoba

Rani, N. P. Sarma, and Sonti, R. V. (2008). Marker assisted introgression of bacterial blight resistance in Samba Mahsuri, an elite indica rice variety. *Euphytica.,* **160** : 411–422.

Tammisola, J. 2010. Towards much more efficient biofuel crops – can sugarcane pave the way? *GM Crops.,* **1**(4): 181-198.

Tanksley, S. D. and Nelson, J.C. (1996). Advanced backcross QTL analysis: a method for the simultaneous discovery and transfer of valuable QTLs from unadapted germplasm into elite breeding lines. *Theor. Appl. Genet.,* **92**: 191-203.

Tanksley, S. D., S. Grandillo, T. M. Fulton, D. Zamir, Y. Eshed, V. Petiard, Lopez, J. and Beck-Bunn, T. (1996). Advanced backcross QTL analysis in a cross between an elite processing line of tomato and its wild relative *L. pimpinellifolium. Theor. Appl. Genet.,* **92**: 213-224.

Thabuis, A., Palloix, A., Servin, B., Daubèze, A. M., Signoret, P., Hospital, F. and Lefebvre, V. (2004). Marker-assisted introgression of 4 *Phytophthora capsici* resistance QTL alleles into a bell pepper line: validation of additive and epistasis effects. *Mol. Breed.,* **14**: 9-20.

Utomo, H. S. and Linscombe, S. D. (2009). Current patents and future development underlying marker-assisted breeding in major grain crops. *Recent Pat DNA Gene Seq.,* **1**: 53-62.

VanToai, T., T. T. Cuc Hoa, N. T. Ngoc Hue, H. T. Nguyen, J. G. Shannon and Rahman, M. A. (2010). Flooding Tolerance of Soybean [*Glycine max* (L.) Merr.] Germplasm from Southeast Asia under Field and Screen-House Environments. *Open Agric. J.,* **4**: 38–46.

VanToai, Tara T.; Tran Thi Cuc Hoa; Nguyen Thi Ngoc Hue; Nguyen, Henry T.; Shannon, J. Grover; Rahman, Mohammed.

Varshney, R. K., K.C. Bansal, P.K. Aggarwal, S. Datta and Craufurd P.Q. (2011). Agricultural biotechnology for crop improvement in a variable climate: hope or hype? *Trends Plant Sci.,* **16**(7): 363-371.

Varshney, R. K., P. Langridge and Graner, A. (2007). Application of genomics to molecular breeding of wheat and barley. *Advances Genet.,* **58**: 121-155.

Wang, C. R., A. F. Yang, G. D. Yue, Q. Gao, Yin, H. Y. and Zhang, J. R. (2008). Enhanced expression of phospholipase C 1 (ZmPLC1) improves drought tolerance in transgenic maize. *Planta.,* **227**: 1127-1140.

Wani, S. H. and Sanghera, G. S. (2010). Genetic engineering for viral disease management in plants. *Notulae Scientia Biologicae.,* **2** (1): 20-28.

Xiao, B. Z., X. Chen, C. B. Xiang, N. Tang, Q. F. Zhang and Xiong, L.Z. (2009). Evaluation of seven function-known candidate genes for their effects on improving drought resistance of transgenic rice under field conditions. *Mol. Plant.,* **2**: 73–83.

Zavala, J. A., C. L. Casteel, E. H. DeLucia, and Berenbaum, M.R. (2008). Anthropogenic increase in carbon dioxide compromises plant defense against invasive insects. *Proc. Natl. Acad. Sci. U.S.A.* **105**: 5129-5133.

Zhai, S. M., Z. H. Sui, A. F. Yang and Zhang, J. R. (2005). Characterization of a novel phosphoinositide-specific phospholipase C from *Zea mays* and its expression in *Escherichia coli*. *Biotechnology Letters*, **27:** 799-804.

Zhu C., M. Gore, E. S. Buckler. and Yu, J. (2008). Status and prospects of association mapping in plants. *Plant Genome.*, **1**(1): 5-20.

2015, Major Constrains and Verdict of Crop Productivity *Pages 137–148*
Editor: **Dr. U.N. Bhale**
Published by: **ASTRAL INTERNATIONAL (P) LIMITED, NEW DELHI**

Chapter 8

Role of Biofertilizers in Plant Nutrient Management under Present Scenario

M.D. Khunt, V.A. Solanki, A.N. Sabalpara and Lalit Mahatma*

Department of Plant Pathology,
N.M. College of Agriculture, Navsari Agricultural University,
Navsari – 396 450, Gujarat, India

Introduction

Scientific intervention including improvement of genetic architecture, precise crop production technologies, accurate protection technologies and need based nutrients management in the crop production has changed the phenology and productivity of domesticated plant species. Significance of the improved genetic architecture of the wheat and nutrient management ushered green revolution in the late sixties of the last century. Success of the green revolution led farmers to use more and more chemical fertilizers in the agriculture. Injudicious use of chemical fertilizers in soil nutrient management has caused irreversible damage to the soil, water and environment. Therefore, there is an urgent need to device a technology which can minimize the use of chemical fertilizers and overcome hazardous effect of chemicals without compromising production. Organic amendments like farm yard manure, composted urban waste, vermicompost etc. are conventional and excellent source of nutrients; however, all these are in short supply. Biofertilizers offers alternatives of

* *Corresponding Author.* E-mail: mahatmalalit@yahoo.co.in

conventional as well as chemical source of nutrients. However, there is a wide gap in the prosperity and popularity of biofertilizers which need to address properly.

Methodological Approach of the Development of Chemicals

Different organic substances are being used for the supply of nutrients for the crop production since prehistoric period. Justus von Liebig (1840) recognized the significance of nitrogen and formulated the law of the minimum. John Bennet Lawes in 1842, patented a manure formed by treating phosphates with sulphuric acid and thus initiated the artificial manure industry. Before the development of ammonical fertilizers, Chilean nitrates (15 per cent N) was the most concentrated nitrogen however, adequate and easily available. The first synthetic products, ammonium sulphate was prepared from synthetic cyanamide ($CaCN_2$) was available only in limited quantities and was expensive. Fritz Haber (1909) developed the method of synthesizing ammonia from its elements with the help of an iron based catalyst. For the same he was awarded Nobel Prize of Chemistry in 1918. Within four years of this laboratory bench demonstration, BASF, Germany's under the leadership of Carl Bosch (1913) developed a commercial process (Haber-Bosch Process) and made it possible to mass produce cheaper nitrogenous fertilizers and awarded the Nobel Prize of Chemistry in 1931. Soon the fertilizer becomes popular in crop production. Influenced by the Justus von Liebig's law of minimum, Arnon and Stout (1939) proposed the criteria of essentiality of nutrients. According to the criteria an element is considered essential if, (1) the plant is unable to complete the vegetative or reproductive stage of its life cycle in its deficiency (2) The deficiency can be prevented or corrected only by supplying the specific element causing the deficiency (3) That element is directly involved in the nutrition of the plant. With time, Nicholas (1961) suggested that there is an additional fourth criterion: that the essentiality of any element is proved in all plants tested. Seventeen nutrients *viz.*, carbon (C), hydrogen (H), oxygen (O), nitrogen (N), potassium (K), calcium (Ca), magnesium (Mg), phosphorus (P), sulphur (S), chloride (Cl), iron (Fe), boron (B), manganese (Mn), zinc (Zn), copper (Cu), molybdenum (Mo), and nickel (Ni) are essentially required by the plants in balance proportion for the economic and optimum crop production. Other elements that have been identified as beneficial, but not necessarily essential to all plants, include cobalt (Co), selenium (Se), silicon (Si), sodium (Na), vanadium (V), and aluminium (Al). Among this list, N, P, and K are considered as the primary macronutrients; Ca, Mg, and S are considered secondary macronutrients; and the remaining are micronutrients. Green revolution realized in the late sixties of the last century is an outcome of efficient use of improved varieties and chemical fertilizers. Primary macronutrients (NPK) shares most of the chemical fertilizers production and consumption. Ammonium Sulphate (20.6 per cent N), Calcium Ammonium Nitrate (25 per cent N), Ammonium Chloride (25 per cent N) and Urea (46 per cent N) are straight nitrogenous fertilizers. Single Super Phosphate (16 per cent P_2O_5), Triple Super Phosphate (46 per cent P_2O_5) are Straight Phosphatic fertilizers. Murate of Potash (60 per cent K_2O) and Sulphate of Potash (60 per cent K_2O) are straight potash fertilizers. Fertilizers which supply more than one primary macronutrient are termed as complex fertilizers. Diammonium Phosphate (18-46-0 N-P-K), Mono Ammonium Phosphate (11-52-0), Nitro Phosphate (20-20-0 and 23-23-0), Ammonium Phosphate Sulphate(16-20-0 and 20-20-0), Urea

Ammonium Phosphate (24-24-0, 28-28-0 and 14-35-14) are the commonly used complex fertilizers.

Harmful Effects of Chemical Fertilizers

Use of chemical fertilizers and has increased 233 per cent after the green revolution. Excessive use of the fertilizers has seriously affected the soil physical, chemical and biological properties. Chemical fertilizer used per hectare in India is 121.4 kg which is very less than many developed and developing countries *viz.*, The Netherlands (665.5 kg); Egypt (624.8 kg); Japan (373.2 kg); China (301.5 kg); Britain (287.5 kg); Germany (205.4 kg); France (180.1 kg); USA (160.8 kg) and Italy (126.4 kg) (Anon, 2009). Ideal ration of NPK is 4:2:1, however, in India it is 6.5:2.9:1 indicating the injudicious and excessive use of chemical fertilizers mainly nitrogenous fertilizers (Anon, 2012). Over application of nitrogenous chemical fertilizers leads to softening of plant tissues and thus increases rate of disease. Fertilizer industry is considered to be source of natural radionuclides and heavy metals as a potential source. It contains majority of the heavy metals like Hg, Cd, As, Pb, Cu, Ni, and Cu; natural radionuclide like ^{238}U, ^{232}Th, and ^{21o}Po. The amount of phosphate and nitrate is increased in drinking water near to field having excessive use of chemical fertilizers. Nitrogen fertilizers consist of carcinogenic substances such as nitrosamines. The concentration of nitrous oxide (N_2O) which is one of the potent greenhouse gases is increasing constantly. In the process of nitrification (during which ammonia is oxidized to nitrate), microbes release NO and N_2O, as intermediates. *Nitrosomonas eutropha* is a nitrifying proteobacteria found in strongly eutrophic environments due to its high tolerance for elevated ammonia concentrations. Nitrogenous fertilizers increase ammonia concentration, causing *N. eutropha* to release more NO and N_2O in the process of oxidizing ammonium ions. Continuous and excessive use of acid-forming nitrogen fertilizers causes a decrease in soil pH. Therefore, liming is recommended in such field otherwise it decline efficiency of field. Use of basic fertilizers leads to increase in pH. Excess potash in the soil can lead to a calcium and magnesium deficiency in plants, since plants absorb calcium, magnesium and potassium largely in the ratio in which they are present in the soil. Muriate of potash contains about 50 per cent potassium and 50 per cent chloride. In the soil the chloride combines with nitrates to form chlorine gas which kills microbes. Applying 1 kg of potassium chloride to the soil is equivalent to applying 1 litre of Chlorine bleach. Approximately 2 ppm chlorine generally thought to be sufficient to sterilize drinking water. Potassium chloride application typically results in chloride levels as high as 50-200 ppm which adversely affects soil microflora leaving the soil biologically dead.

Biofertilizers and its Development

Schloesing in 1887 proved the bacterial origin of nitrification. Ulysse Gayon in 1885 isolated denitrifying bacteria that can reduce nitrates via NO and N_2O, return N_2 to the atmosphere (Payne, 1986). Hermann Hellriegel and Hermann Wilfarth in 1880 found that leguminous nodules are the sites where biofixation of nitrogen takes place (Galloway, *et al.*, 2013). Sergei Winogradsky, elucidated nitrogen cycle and the process by which diatomic nitrogen gas is converted into ammonium ions and becomes available to plants. Sergei Winogradsky devised enriched media and identified obligate anaerobe *Clostridium pasteurianum* capable of fixing atmospheric nitrogen. Martinus Beijerinck utilized the enriched media, however, deficit in nitrogen and isolated free living nitrogen fixating *Azotobacter chroococcum.* Soon the different microorganisms preparations *i.e.* biofertilizers were prepared for the use in agriculture.

According to the fertilizer Control Act-1985 the term Biofertilizer has been defined as "the product containing carrier based (solid or liquid) living microorganisms which are agriculturally useful in terms of nitrogen fixation, phosphorus solubilisation or nutrient mobilization, to increase the productivity of the soil and/or crop. Biofertilizers are considered to be the most safe, promising and economic component of sustainable agriculture. When biofertilizers are applied to seed, plant root, or soil, they colonizes the rhizosphere or the interior of the plant and promotes growth by providing and essential nutrients or increasing the supply or availability of primary nutrients to the host plant.

Types of Biofertilizers

Nitrogen Fixing Microorganisms

Atmosphere contains about 78 per cent nitrogen but none of the plant or animal can utilize it directly. Nitrogen fixing microorganisms can add atmospheric nitrogen into soil by the process known as Biological Nitrogen Fixation (BNF). Different microorganism may be grouped into symbiotic nitrogen fixers (*Rhizobium* spp., *Azorhizobium* spp., *Frankia* spp., *Azolla* spp., etc), Non-symbiotic nitrogen fixers (*Azotobacter* spp., *Klebsiella* spp., etc.) and associative symbiotic nitrogen fixers (*Azospirillum* spp. and *Gluconoacetobacter* spp.).

Symbiotic Nitrogen Fixers

Bacteria collectively known as Rhizobia are found to be associated with Leguminous plant, forms root nodules and occasionally stem nodules. Inside nodules, Rhizobia live as bacteroids, fix atmospheric nitrogen to ammonia. *Rhizobium* spp. inoculation is a famous practice to replace nitrogenous fertilizers in legumes (Gupta, 2004). Oxygen level inside nodules is usually regulated by specific protein called leg-haemoglobin, produced only after plant infection with *Rhizobium* spp. Plants root supply organic material for the growth of bacteriods, in turn, bacteria supplies nitrogen source.

It has been reported that reduction of nitrogenous chemical fertilizer by using nitrogen fixing bacterial inoculation is one of the efficient step in sustainable agriculture (Zahran *et al.,* 1999). Symbiotic association between *Rhizobium* and leguminous plant can fix nitrogen up to 100-200 kg N/ha/year in one crop season, can leave substantial nitrogen behind for subsequent crop (Tilak *et al.,* 2005). Association is also important in natural fallow, to reduce soil erosion and weeds, enrich soil organic nitrogen etc. (Carsky *et al.,* 2001). Application of *Pseudomonas* spp. and *Bacillus* spp. as co-inoculants with *Rhizobium* spp. in chick pea accelerates plant growth, nodulation and nitrogen fixation (Mahammadi *et al.,* 2010).

Another excellent case of symbiosis comprises several floating water fern (*Azolla* spp.) found in tropical and temperate ecosystems. There are seven important species have been reported *i.e. Azolla caroliniana, A. maxicana, A. microphylla, A. filiculoides, A. rubra, A. nilotica* and *A. pinnata.* It has the ability to fix atmospheric nitrogen though symbiosis with blue green algae (*Nostoc* spp. and *Anabaena* spp.). Therefore, it is considered an important potential source of nitrogen for wetland rice. The contribution of nitrogen from *Azolla* spp., to wetland rice plants has been found to be maximum when incorporated into the soil as green manure. Rice yield could be increased by 0.5-2 t/ha by applying *Azolla* (Gupta, 2004).

Frankia belongs to actinomycetes group, can fix atmospheric nitrogen symbiotically by forming stem nodules in *Casuarina* and *Alnus* plants.

Associative Symbiotic Nitrogen Fixers

Azospirillium is heterotrophic, associative symbiotic nitrogen fixer belongs to Spirilaceae family. *Azospirillim lipoferum, A. brasilense* and *A. amazonese* are important species in soil. *Azospirillum* sp. directly benefits plants improving shoot and root development and increasing the rate of water and mineral uptake by roots. It is recommended for maize, sugarcane, sorghum, wheat, pearl millet etc. Apart from 20-40 kg/ha nitrogen fixation, *Azospirillium* produces plant growth promoting substances. Corn inoculating with Azospirillum may be provide 10 to 20 per cent of nitrogen that is used for plant and with attention to the cultivation of this crop, annual will save millions of tons of nitrogen (Vose *et al.,* 1981).

Second associative nitrogen fixing, endophytic nitrogen fixing bacteria is belonging to Acetobacter genera. It was first describe in Brazil, associated with sugarcane plant, and first named as *Acetobacter diazotrophicus*. Then bacteria was renamed as *Gluconacetobacter diazotrophicus.* Although, Gluconacetobacter has also been found to be associated with non-sugar rich plants (Sevilla and Kennedy, 2000).

Non-Symbiotic Nitrogen Fixers

Azotobacter is aerobic, heterotrophic, cyst forming, capsulated, free living nitrogen bacteria, belongs to Azotobacteriaceae family. *Azotobacter chrooccum, A. vinelandii, A. insignis, A. beijerinckii* are most common found species in soil. However, number of *Azotobacter* rarely exceed in soil of 10^5 cfu/g due to presence of antagonistic organisms and limited nutrients. Apart from nitrogen fixation, *Azotobacter* produces antifungal antibiotics and plant growth promoting substances, thus promotes seed germination and plant vigour. According to field trial, *Azotobacter* treated (cfu 10^7/ml) melon and papaya seedlings were 68 per cent and 103 per cent heavier respectively than seedling without inoculation treatment (Berman- frank *et al.,* 2001). *Azotobacter* have been recommended as biofertilizer in number of crops including rice, maize, wheat, sugarcane, pearl millet, vegetable crops, horticultural crops, forest trees etc.

Phosphate Solubilizing Microorganisms

Phosphorous is the second most important nutrients of plants, absorbed as $H_2PO_4^-$ in acidic pH or as HPO_4^{2-} in alkaline pH. After the application of phosphorous fertilizers in soil chemical reactions converts available from of phosphorous in insoluble and unavailable form.

Highly alkaline and acidic soil reduces phosphate availability in soil. In alkaline soils, Ca is the dominant cation that will react with phosphate. A general sequence of reactions in alkaline soils is the formation of dibasic calcium phosphate dihydrate, octocalcium phosphate, and hydroxyapatite. The formation of each product results in a decrease in solubility and availability of phosphate. When pH is acidic, plant available phosphorus becomes increasingly tied up in aluminum phosphates. As soils become more acidic (pH below 5), phosphorus is fixed in iron phosphates. The amorphous Al and Fe phosphates gradually change into compounds that resemble crystalline variscite (an Al phosphate) and strengite (an Fe phosphate).

Fixed insoluble phosphate in the soil can be solubilized by a group of bacteria, known as phosphate solubilizing microorganisms (PSM), which have biochemical mechanism to convert insoluble unavailable phosphate to soluble available phosphate through organic acid production. Bacteria like *Bacillus, Pseudomonas, Rhizobium, Enterobacter* and fungi like *Aspergillus, Penicillium* are considered as potent PSM (Whitelaw, 2001). However, bacteria are more efficient in phosphate solubilization than fungi (Alam *et al.*, 2002). Therefore, application of PSB inoculants in agriculture has been preferred to save expensive chemical phosphatic fertilizers. PSB population in rhizosphere soil will always higher as compared to non-rhizospheric soil, suggesting synergistic effect of PSB and plant root. Application of PSB along with single super phosphate and rock phosphate reduce phosphatic fertilizer inputs by 25 per cent and 50 per cent respectively (Sundara *et al.*, 2002). PSB are recommended for all the crops.

Phosphate Absorbers

Mycorrhizae are considered as symbiotic association between fungi and plants root. Fungi absorbs nutrients like phosphate, zinc, calcium etc. from nearby soil and supplies to plant root while in turn, plant provides shelter and food to fungi. Mycorrhizae forms symbiotic association in majority of agricultural crops, except plants belonging to families of *Chenopodiaceae, Amaranthaceae, Caryophyllaceae, Polygonaceae, Brassicaceae, Commelinaceae, Juncaceae* and *Cyperaceae* (Mishra *et al.*, 2013).

Potash Solubilizers/Mobilizers

Potassium is the third most important nutrient for the plant and fourth most abundant nutrient constituting about 2.5 per cent of the lithosphere. Three forms of potassium found in the soil *viz.* Soil minerals, non exchangeable and available form. Soil contains only 1-2 per cent available potassium, rest of 98-99 per cent are in soil minerals or non exchangeable form that is unavailable for plants growth. Total pool of soil K is extremely complex and this can be solubilized by bacteria through production of organic acids.

Zinc Solubilizers

Zinc is considered as micronutrients for plant growth, may be present as available or unavailable form in the soil. Several bacterial species like *B. subtilis, Thiobacillus thioxidans* and *Saccharomyces* sp. can be used as biofertilizer as zinc solubilizers, *Bacillus* can be applied as zinc solubilizer biofertilizers along with cheaper insoluble zinc oxide, zinc carbonate and zinc sulphide to replace costly zinc sulphate (Mishra *et al.*, 2013).

Silicate Solubilising Bacteria (SSB)

Silicate is another micronutrient, required in minute quantity for plant growth. SSB have capacity to degrade aluminium silicates and silicates by producing organic acids like citric acid, oxalic acid, keto acid etc. Application of SSB with rice straw in field increased dissolution of silica and other nutrients in the soil.

Plant Growth Promoting Rhizobacteria (PGPR)

Plant growth-promoting rhizobacteria (PGPR) are group of bacteria that stimulate plant growth and increases crop yield. Due to increasing awareness of PGPR in agricultural production, scientists are trying to explore PGPR microorganisms for increasing plant growth and success rate of field applications (Burr *et al.*, 1984). PGPR belongs to several bacterial genera *e.g. Bacillus, Pseudomonas, Rhizobium, Bradyrhizobium, Enterobacter, Aclaligenes, Arthrobacter, Azotobacter, Cellulomonas, Flavobacterium, Streptomycetes, Xanthomonas* etc. (Waller, 1988). PGPR affects plant growth by nutrient uptake by direct or indirect mechanisms. Even, some of the bacteria have the capacity to secrete plant growth promoting chemicals; however, efficiency depends upon soil physical and chemical status. Surprisingly bacteria showed better activity in nutrient deficient soil as compare to nutrient rich soil (Egamberdiyeva, 2007). The bacterial inoculation into cotton plant significantly increases the seed yield, plant height and microbial population in soil (Anjum *et al.*, 2007). Use of PGPR with Phosphate enriched compost in an integrated manner improves the growth, yield and nodulation in chickpea (Shahzad *et al.*, 2008). Several PGPR isolated from farm waste compost show synergistic effect on growth of pearl millet (Hameeda *et al.*, 2006).

Multiple Bio-inoculants (Consortium)

Rhizosphere region is very complex and centre of plant-microbe, microbe-microbe interactions. Further, microbe- microbe negative interaction can be explained by taking example of interaction between fungi and bacteria for competition of space and nutrients as well as modification of rhizosphere habitats. In case of positive relationship between bacteria and fungi, *Paenibacillus* accelerates root colonization of mycorrhizae fungi and bacteria may also provide PGPR effect. Above association may positively benefits to plant by increasing nutrient uptake and transport (Bonfante, 2001). Further, co-inoculation of *Frankia* and AM fungi in *Hippophae tibetana* increases nitrogen fixation and AM colonization as compared to individual inoculation (Tian *et al.*, 2002).

Limitations and Reasons of Poor Share of Biofertilizers in Nutrient Management

Unfortunately, although biofertilizers are effective in the laboratory and in the field experimentation, they could not be more popularized in the field at a community level. New microorganisms added to the soil of a field cannot compete with the existing microflora and cannot maintain themselves for very long. Also, soil amendments, so far, have not been selective enough to support and build up only the populations of the introduced or existing antagonists. Thus, their potential is quite limited. There is a dying need to focus our attention on refinement of strategies for making the tool more effective. For this a multidimensional attention need to be paid. The potential cause of failure and their remedial measures may be as follows:

Lack of Native Strains of Microorganisms

It has been often noticed that the exotic bioinoculants are not easily acclimatize in the new ecosystem to balance the population dynamics. Particularly soil ecosystem is highly diverse providing entirely different environment for the growth of different microorganisms. Attempts should be made to decentralize our research efforts to isolate and popularize native strain than finding a single strain at national or state level.

Poor Soil Organic Amendments to Support the Growth of Potential Microorganism

Success of the biofertilizers depends on the availability of organic carbon in the soil. Inadequate supplementation of organic amendments, injudicious use of chemical fertilizers and intensive crop cultivation has eroded soil. Therefore, biofertilizers when added in the soil does not get the essential nutrients for the initial buildup of their population. Presently our efforts to add organic amendments are limited to supply essential nutrients for the crop growth. There is an urgent need to incorporate other soil amendments which not only improves soil physical and chemical properties but also helps in improvement of soil biological properties.

Inadequate Formulation of Commercially Available Products

Presently most of the biofertilizers are available in the form of powder or liquid. These formulations have their own limitation and viability of the biofertilizers have been questioned many times. The minimum requirement of cfu/g of formulation is around 10^8 to be passed through the laid quality parameters. Recently many technologies for the mass production of biofertilizers have come; however, an effective and reliable formulation has not been prepared. To make the technology more effective there is an urgent need to prepare formulations which have more shelf-life, viable cfu in good number till the product last and acceptable formulation.

Deprived Delivery System of Biofertilizers

Research efforts to improve delivery of the biofertilizers were never given due attention. To be effective, the biofertilizers must be applied near to the root system. Under the circumstances, the biofertilizers having potentiality to induce resistance in the plant can only show some effect. Presently nanotechnology is most widely used technology for the improvement of delivery system and has showed significant results in many fields. Efforts must be focused on the use of nanoparticles prepare the new formulation and improve the delivery system.

Lack of Consortium of Microorganisms

Many of the products available in the market as biofertilizers can be termed as "straight biofertilizers" as they are having strain or species of one microorganism. For almost all the ecosystem more than one microorganism have been identified and in may research, synergic effect of mix application of more than one microorganisms have been shown. However, availability of such product in the market is very rare. Efforts should be made to prepare such product and at the same time quality parameters for such product must be standardized so that the furious product cannot ditch farmers.

Less Attention of Progressive Stack Holder Towards the Biofertilizers Industries

Biofertilizers production and marketing work is being carried out by the small and medium entrepreneurs. Their risk bearing capacity is also accordingly limited; therefore, they do not invest heavily in the business to further refine the technology and initiate research and development work to further take-up new challenges. Involvement of large player is being encouraged by the governments in different capacities; however, outstanding involvement of the stack holder could not be seen.

Lack of Awareness among the Farming Communities

Development of awareness among the farming communities is most important factor to gain popularity of the technology. Adoption of the chemical fertilizers and pesticides were quite high therefore, the spectacular results of the Green revolution could be seen. Recently many models and channels of communication of effective technologies have been invented. At field level use of these channels could be seen for chemical input, however, campaigning of biofertilizers is not as intense as required to give a massive boost.

Government Policies and its Monitoring

Government policies are very strong motivator of entrepreneurs, researchers and farmers. Government should device policies to encourage stack holder to invest, researchers to develop new product and farmers to adopt newer technologies. Policies laid and channelization of entire product must be monitored thoroughly by the government to prevent substandard products. A good monitoring system will encourage farmers to adopt and will also help in popularization of technology.

Summary

Use of proper nutrients in agriculture is the most important input for the economic and sustainable crop production. Biofertilizer is an organic product, can be amended in soil to improve soil and plants health. Biofertilizers can be broadly classified into: Nitrogen fixers (symbiotic, non-symbiotic and associative symbiotic), phosphate solubilizers, phosphate mobilizers, potash mobilizers and microbes that supply micronutrients (zinc, silicate etc.). Presently, use of biofertilizers in sustainable agriculture is limited due to several hurdles like lack of native strains, poor organic nutrient status of the soil, poor awareness among farming community, poor interest of stack holders, improper methods of application adopted by farmers etc. Use of biofertilizers should be promoted by solving these constraints and popularizing it in farming community.

References

Alam, S., Khalil, S., Ayub, N and Rashid, M. (2002). *In vitro* solubilization of inorganic phosphate by phosphate solubilizing microorganism (PSM) from maize rhizosphere. *Intl J Agric Biol,* **4:** 454-458.

Anjum, M.A., Sajjad, M.R., Akhtar, N., Qureshi, M.A., Iqbal, A., Jami, A.R and Mahmud-ul-Hasan (2007). Response of cotton to plant growth promoting Rhizobacteria (PGPR) inoculation under different levels of nitrogen. *J Agril Res.,* **45:** 135-143.

Anonymous (2009) FAO, ResourceSTAT-Fertilizer. *Food and Agriculture Organization of the United Nations.* [Online]. Available: http: //faostat.fao.org/site/575/ Desktop Default.aspx? PageID=575#ancor, 12.03.2009.

Anonymous (2012). Indian Fertilizer Scenario, Department 0f Fertilizers, Ministry of Chemicals and Fertilizers, Government of India pp. 215.

Arnon, D. I. and Stout, P. R. (1939). The essentiality of certain elements in minute quantity for plants with special reference to copper. *Plant Physiol.,* **14**: 371-375.

Berman-Frank, I., Lundgren, P., Yi-Bu Chen, Küpper, H., Kolber, Z., Bergman, B and Falkowski P. (2001). Segregation of nitrogen fixation and oxygenic photosynthesis in the marine *Cyanobacterium trichodesmium. Science.,* **294**(16): 1534-1537.

Bonfante, P and Ruiz-Lozano, J.M. (2001). Intracellular *Burkholderia* strain has no negative effect on the symbiotic efficiency of the arbuscular mycorrhizal fungus *Gigaspora margarita. Plant Growth Regul.,* **34:** 347-352.

Bosch, Carl (1913) "The Nobel Prize in Chemistry 1931". *Nobelprize.org.* Nobel Media AB 2013. Web. 18 Jun 2014. http: //www.nobelprize.org/nobel_prizes/ chemistry/laureates/1931/

Burr, T.J., Caesar, A.M and Schrolh, N. (1984). Beneficial plant bacteria. Critical Review. *Plant Sci.,* **2:** 1–20.

Carsky, R.J., Becker, M and Hauser, S. (2001). Mucuna Cover Crop Fallow Systems: Potential and Limitations, In G. Tian, F. Ishida, and D. Keatinge, eds., Sustaining Soil Fertility in West Africa, Soil Science Society of America Special Publication No **58,** Madison, WI, pp. 111–135.

Egamberdiyeva, D. (2007). The effect of plant growth promoting bacteria on growth and nutrient uptake of maize in two different soils. *Appl Soil Ecol,* **36:** 184–189.

Elmerich, C. and Newton, W.E. (2007) Associative and endophytic nitrogen-fixing bacteria and cyanobacterial associations. Heidelberg, Germany: *Springers.* (doi: 10. 1007/1-4020-3546-2_1).

Galloway, J.N., Leach, A.M. Bleeker, A and Erisman, J.W. (2013) A chronology of human understanding of the nitrogen cycle. *Phil Trans R Soc., B* 2013 368: 20130120. http: //dx.doi.org/10.1098/rstb.2013.0120.

Gupta, A.K. (2004). The complete technology book on biofertilizers and organic farming. National Institute of Industrial Research Press. India.

Haber, F. (1909). Biographical". *Nobelprize.org*. Nobel Media AB 2013. Web. 17 Jun 2014. http: //www.nobelprize.org/nobel_prizes/chemistry/laureates/1918/haber-bio.html.

Hameeda, B., Rupela, O., Reddy, G and Satyavani, K. (2006). Application of plant growth-promoting bacteria associated with composts and macrofauna for growth promotion of Pearl millet (*Pennisetum glaucum* L). *Biol Fertil Soils*, **43:** 221-227.

Mishra, D.J., Singh, R., Mishra, U.K and Shahi Sudhir Kumar (2013). Role of Bio-Fertilizer in Organic Agriculture: A Review. *Res J Recent Sci.*, **2:** 39-41.

Mohammadi, K., Ghalavand, A., Aghaalikhani, M., Heidari, G.R and Sohrabi, Y. (2010). Introducing the sustainable soilfertility system for chickpea (*Cicer arietinum* L.). *Afr J* Biotechnol., 10(32): 6011-6020.

Nicholas, D.J.D. (1961). Minor Mineral Elements. *Annual Rev Plant Physiol*, **12:** 63-90.

Payne, W.J. (1986). Centenary of the isolation of denitrifying bacteria. *ASM News.*, **52:** 627–629.

Schloesing, J. and Muntz, A. (1876). Sur la Nitrification par les Ferments Organises. Comptes Rendus de l'Academie des Sciences, *Paris*, **LXXXIV:** 301-303.

Sevilla, M. *and* Kennedy, C. *(2000)* Colonization of rice and other cereals by *Acetobacter diazotrophicus*, an endophyte of sugarcane. *In The Quest for Nitrogen Fixation in Rice. Proceedings of the third working group meeting on assessing opportunities for nitrogen fixation in rice ed.* Ladha, J.K. *and* Reddy, P.M. *pp.*151–165, *Makati City (Philippines): International Rice Research Institute.*, pp. 151-165.

Shahzad, S.M., Khalid, A., Arshad, M., Khalid, M and Mehboob, I. (2008). Integrated use of plant growth promoting bacteria and Penriched compost for improving growth, yield and nodulating of Chickpea. *Pak J Bot.*, **40:** 1735-1441.

Sundara, B., Natarajan, V and Hari, K. (2002). Influence of phosphorus solubilizing bacteria on the changes in soil available phosphorus and sugarcane yields. *Field Crops Res.*, **77:** 43-49.

Tian, C., He, X., Zhong, Y and Chen, J. (2002). Effects of VA mycorrhizae and *Frankia* dual inoculation on growth and nitrogen fixation of *Hippophaë tibetana. For Ecol Manage.*, **170:** 307-312.

Tilak, K.V.B.R., Ranganayaki, N., Pal, K.K., De, R., Saxena, A.K., Nautiyal, C.S., Mittal, S., Tripathi, A.K and Johri, B.N. (2005). Diversity of Plant Growth and Soil Health Supporting Bacteria. *Curr Sci.*, **89:** 136–150.

Vose, P.B and Ruschel, A.P. eds (1981). Associative N_2–Fixation. Vol. I. pp. 179-184. CRC press, Florida.

Weller, D.M. (1988). Biological control of soil borne plant pathogens in the rhizosphere with bacteria. *Annu Rev Phytopathol.*, **26:** 379-407.

Whitelaw, M.A. (2001). Growth promotion of plants inoculated with phosphate solubilizing fungi. *Adv Agron.*, **69:** 99-151.

Zahran, H.H. (1999). *Rhizobium*-legume Symbiosis and Nitrogen Fixation under Severe Conditions and In an Arid Climate. *Microbiol Mol Biol Rev.*, **63:** 968-989.

2015, Major Constrains and Verdict of Crop Productivity Pages 149–160
Editor: Dr. U.N. Bhale
Published by: ASTRAL INTERNATIONAL (P) LIMITED, NEW DELHI

Chapter 9

Biopesticidal Approaches for the Management of Sorrel (*Rumex acetosa* L.) Diseases

U.N. Bhale*

Research laboratory, Department of Botany,
Arts, Science and Commerce College,
Naldurg, Tq. Tuljapur., Dist. Osmanabad – 413 602, M.S., India

ABSTRACT

Organic vegetable farming involves a widespread approach to soil health, crop health, and the maintenance of a biologically dynamic and diverse agro-ecosystem. Eco-friendly management of sorrel (Rumex acetosa L.) fungal diseases were done by using different leaf extracts, plant latex, essential oils and Trichoderma species as biological control agent. Antifungal activity of leaf extract of Azadirachta indica, Adhathoda vassica, Aegle mormelos, Ocimum gratissimum and Lantana camera showed significant inhibitory results. Among tested plant latex, of Jatropha curcas and Calotropis gigantean was restrict the growth of pathogens. In case of essential oils, Castor oil, Sunflower oil, Eucalyptus oil, Clove oil and Mustard oil were found suitable against pathogens. Antagonistic activities of all the dominant fungi were studied against biocontrol agent i.e. Trichoderma species (Trichoderma viride, T. virens, T. koningii. T. pseudokoningi and T.harzianum). It was found that growth of fungi were restricted in presence of Trichoderma species. All Trichoderma species showed more than 50 per cent antagonistic activity.

Introduction

Vegetables are the most important component of a balanced diet and we can now, grow varieties of different vegetables round the year. India is the world's second

* E-mail: unbhale2007@redffmail.com

largest producer of vegetables next to China. Vegetables are more susceptible to insects' pests and diseases due to their tenderness and softness as compared to other crops and virtual absence of resistance characters because of intensive hybrid (Chiranjeevi *et al.*, 2002). Effective management of a plant disease is a key to save plants from diseases caused from microbes, since plants are significant as they are both economical and aesthetic. Plants have always faced with potential deleterious organisms such as fungi, bacteria, viruses and nematodes responsible for crop losses worldwide. Pathogenic fungi alone cause 20 per cent reduction in the yield of major food and cash crops (Agrios, 2000). To avoid the implication of yield losses due to plant diseases, variety of control measures presently are in use. The chemical compounds are most commonly used for the controlling of plant diseases. No doubt the use of chemicals has been found very effective in controlling plant fungal diseases but some major problems threaten to limit the continued use of fungicides. The toxic effect of synthetic chemicals can be overcome, only by persistent search for new and safer pesticides accompanied by wide use of pest control methods, which are eco-friendly and effective (Mohana *et al.*, 2011). Biopesticides can be particularly effective when incorporated into integrated pest and disease management programs and when used in conjunction with conventional pesticides. Therefore the present investigation was made to evaluate biopesticidal materials against sorrel diseases.

Materials and Methods

Isolation of Pathogens

Infected parts of selected leafy vegetables *viz. Rumex acetosa,* was brought into the laboratory and cut into small pieces (1-2 mm) by sterilized blade then surface sterilized with 1 per cent mercuric chloride ($HgCl_2$) for 1 min. The pieces were washed with sterilized distilled water thrice. The pieces were incorporated with Potato Dextrose Agar medium (PDA) kept at 27±2°C in BOD incubator and pathogens were isolated and identified by manuals (Ellis, 1971; Subramanian, 1971).

I) Plant Extracts

Collection of Plant Materials

Fresh leaves of *Azadirachta indica, Ocimum gratissimum, Adhatoda vassica, Santalum album, Aegle mormelos, Datura stramonium, Nerium indicum, Lantana camara, Citrus limon* and *Catharanthus roseus* were collected and washed in distilled water.

Soxhlet Extraction

In a conventional Soxhlet extraction, 100gm of fresh sterilized plant leaves were placed in a thimble holder. The Soxhlet extraction was placed onto a flask containing the extractor solvent *i.e.* distilled water (100 mL). The Soxhlet was equipped with a condenser. The distilled water was heated to reflux. The water vapors were traveled up in distillation arm and floods into the thimble chamber of solid. The condenser ensures that water vapors cools and dripped back down into the thimble holder. The thimble containing solid material was filled slowly with warm water. Some desired compound was dissolved. When the thimble was fulfilled then the thimble chamber is automatically emptied through siphon side arm and extraction running back down to the distillation flask. The operation is repeated until complete extraction is achieved (Wang and Weller, 2006).

Determination of Antifungal Activity

The toxicity of stock extracted extracts was determined against resistant isolates of *Alternaria tenuissima* and *Fusarium proliferatum,* by food poisoning technique (Mishra and Tiwari, 1992) at four (25, 50, 75 and 100 per cent) different concentrations.

The resistant isolates of test fungi were multiplied on CZA medium. Petriplates containing CZA supplemented with different plant leaf extracts at four concentrations with their three replications were inoculated with fresh 7 days old culture of test fungi (8 mm) cork borer discs and kept upside down. The above plates were incubated in BOD incubator (at 28±2° C). Plates without plant extracts were served as control. Radial growth of the fungus on leaf extracts was measured at regular intervals and calculated by Vincent method.

II) Plant Latex

Plant Material and Latex Collection

Fresh latex *i.e. Jatropa curcus, Calotropis gigentea, Ficus bengalensis* and *Ficus glomerata* were aseptically collected from the aerial parts of the healthy plants as described by Aworh *et al.* (1994) in clean glass tubes containing distilled water to yield a dilution rate of 5:5 (v/v). The latex mixture was gently handled to maintain homogeneity during transport to the laboratory where it was stored at (4°C) until further use.

Preparation of Latex Extract

The fresh latex was selectively decanted and centrifuged at 5000 rpm for 5min. The precipitated material showing rubber aspect (polyisoprene) was pooled apart and the supernatant was decanted carefully. Finally the samples were centrifuged as previously described and the clear soluble supernatant was collected and lyophilized. The stock solutions of latex extract was diluted suitably as required from stock solution (Juncker *et al.,* 2009).

Determination of Antifungal Activity

Plant latex aqueous extracts of each prepared with distilled water and condensed to serve as stock extract was determined by food poisoning technique (Mishra and Tiwari, 1992) against tested pathogens Petriplates containing Czapek Dox Agar (CZA) medium, supplemented with different plant latex extracts at four concentrations (25, 50, 75 and 100 per cent) with three replications were inoculated with fresh 7 days old culture of test fungi in 8 mm discs and kept upside down. The plates were incubated in BOD incubator at 28 ± 2 °C. Plates without plant latex extracts served as control. Rradial growth was recorded daily for 8 days or until the plates were overgrown.

III) Essential Oils

Antifungal activity of eight essential oils such as Groundnut oil, Coconut oil, Soybean oil, Castor oil, Sunflower oil, Eucalyptus oil, Clove oil and Mustard oil were collected. Hundred ml of oil directly mixed with sterilized hot 100 ml PDA medium *i.e.* considered as 100 per cent conc. and tested against pathogens. For this experiment four different concentrations (25, 50, 75 and 100 per cent) were evaluated. Petri plates containing PDA supplemented with different oils at four different concentrations with three replications were inoculated with fresh 7th days old culture of test fungi and (8mm) cork borer disc and kept upside down and incubated in BOD incubator at 27±1°C. Plates without oils were served as control. Radial growth of the tested pathogens was measured at regular intervals (Wood Ward and De Groot, 1999).

IV) Biocontrol Agent (*Trichoderma* spp.)

Dual Culture Experiment

The Antagonistic efficacy of different species of *Trichoderma. T. viride, T. harzianum, T. virens,* T. *koningii* and *T. pseudokoningii* were tested against the resistant pathogenic fungi by dual culture experiment (Morton and Stroube, 1955). *Trichoderma* sp and test fungi were inoculated at 6 cm apart. Three replicates were maintained for each treatment and incubated at 28 ± 2°C for 7 days. Monoculture plates of both served as control. Seven days after incubation (DAI) radial growth of test fungi and *Trichoderma* isolates were measured. Colony diameter of test fungi in dual culture plate was observed and compared with control. The growth inhibition was calculated by using the formula: 100 X C - T/C, Where C = growth in control and T = growth in treatment (Vincent, 1947).

Statistical Analysis

Statistical analyses of the experiments were performed by using the book of Mungikar (1997).

Results and Discussion

I) Plant Extracts

Plant leaf extracts was tested against three pathogenic fungi and determined their antifungal activity. Different concentrations of plant leaf extracts (25, 50, 75 and 100 per cent) were tested against pathogenic fungi. Among tested leaf extracts, *Adhatoda vassica* showed significant reduction of radial growth of *Alternaria tenuissima,* (67.77 per cent) at 100 per cent conc., *Santalum album* showed significant reduction of radial growth of *Fusarium proliferatum* (80.00 per cent) at 100 per cent conc (Table 9.1).

II) Plant Latex

Plant latex extracts was tested against three pathogenic fungi and determined their antifungal activity. Different concentrations of plant latex (25, 50, 75 and 100 per cent) were tested against pathogenic fungi. Minimum inhibitory concentration (MIC) was measured. *Jatropa curcus* latex extract showed more than 75 per cent reduction of radial growth of *Alternaria tenuissima, Fusarium proliferatum* at 50 per cent conc (Table 9.2).

III) Essential Oils

Essential oils *viz.*, Groundnut oil, Coconut oil, Soybean oil, Castor oil, Sunflower oil, Eucalyptus oil, Clove oil and Mustard oil were tested against nine pathogenic fungi and determined their antifungal activity. Different concentrations of essential oils (25, 50, 75 and 100 per cent) were tested against pathogenic fungi. Minimum inhibitory concentration (MIC) was measured (Table 9.3).

Table 9.1: Antifungal Activity of Leaf Extracts against Pathogenic Fungi of *Rumex acetosa*

Plant Species	*Conc (per cent)*	*Radial Growth of A. tenuissima (mm)*	*Inhibition (per cent)*	*Radial Growth of F. proliferatum (mm)*	*Inhibition (per cent)*
Azadiracta indica	25	48	46.66	88	2.22
	50	41	54.44	86	4.44
	75	37	58.88	82	8.88
	100	35	61.11	77	14.44
Ocimum gratissimum	25	80	11.11	70	22.22
	50	77	14.44	65	27.77
	75	65	27.77	59	34.44
	100	59	34.44	52	42.22
Adhatoda vassica	25	32	64.44	44	51.11
	50	28	68.88	39	56.66
	75	25	72.22	32	64.44
	100	20*	67.77	28	68.88
Santalum album	25	90	00.00	27	70.00
	50	87	3.33	21	76.66
	75	72	20.00	19	78.88
	100	66	26.66	18*	80.00
Aegle mormelos	25	66	26.66	67	25.55
	50	61	32.22	58	35.55
	75	55	38.88	51	43.33
	100	45	50.00	48	46.66
Datura stramonium	25	59	34.44	85	5.55
	50	43	52.22	81	10.00
	75	31	65.55	79	12.22
	100	29	67.77	69	23.33
Nerium indicum	25	67	25.55	59	34.44
	50	59	34.44	55	38.88
	75	55	38.88	48	46.66
	100	40	55.55	39	56.66
Lantana camera	25	37	58.88	32	64.44
	50	35	61.11	28	68.88
	75	32	64.44	27	70.00
	100	29	67.77	25	72.22

Contd...

Table 9.1–*Contd...*

Plant Species	*Conc (per cent)*	*Radial Growth of A. tenuissima (mm)*	*Inhibition (per cent)*	*Radial Growth of F. proliferatum (mm)*	*Inhibition (per cent)*
Citrus limon	25	80	11.11	49	45.55
	50	77	14.44	45	50.00
	75	72	20.00	42	53.33
	100	68	24.44	38	57.77
Catharanthus roseus	25	33	63.33	90	00.00
	50	29	67.77	88	2.22
	75	28	68.88	78	13.33
	100	27	70.00	69	23.33
Control		90.00	–	90.00	–

* Significantly reduced mycelial growth.

Table 9.2: Antifungal Activity of Plant Latex Extracts against Pathogenic Fungi of *Rumex acetosa*

Plant Species	*Family*	*Conc. (per cent)*	*Radial Growth of A.tenuissima (mm)*	*Inhibition (per cent)*	*Radial Growth of F. proliferatm (mm)*	*Inhibition (per cent)*
Jatropa curcas	Euphorbiaceae	25	16	82.02	14	84.44
		50	15	83.14	06	93.33*
		75	12	86.51*	00	100.00*
		100	00	100.00*	00	100.00*
Calotropis gigantea	Asclepiadaceae	25	39	56.17	59	34.44
		50	28	68.53	40	55.55
		75	24	73.03	35	61.11
		100	22	75.28	30	66.66
Ficus bengalensis	Moraceae	25	18	79.77	30	66.66
		50	17	80.89	28	68.88
		75	15	83.14	24	73.33
		100	13	85.39	20	77.77
Ficus glomerata	Moraceae	25	36	59.55	31	65.55
		50	30	66.29	29	67.77
		75	26	70.78	22	75.55
		100	22	75.28	20	77.77
Control			89.11	–	90.00	–
CD (P=0.05)			–	6.24	–	8.60

*Significantly reduced mycelial growth.

Table 9.3: Antifungal Activity of Essential Oils against Pathogenic Fungi of *Rumex acetosa*

Plant Species	*Conc (per cent)*	*Radial Growth of A. tenuissima (mm)*	*Inhibition (per cent)*	*Radial Growth of F. proliferatum (mm)*	*Inhibition (per cent)*
Groundnut oil	25	26	71.11	22	75.55
	50	25	72.22	19	78.88
	75	23	74.44	18	80.00
	100	22	75.55	16	82.22
Coconut oil	25	42	53.33	30	66.66
	50	39	56.66	26	71.11
	75	37	58.88	24	73.33
	100	35	61.11	22	75.55
Soybean oil	25	36	60.00	28	68.88
	50	34	62.22	26	71.11
	75	31	65.55	18	80.00
	100	28	68.88	16	82.22
Castor oil	25	34	62.22	39	56.66
	50	29	67.77	36	60.00
	75	27	70.00	33	63.33
	100	25	72.22	29	67.77
Sunflower oil	25	24	73.33	22	75.55
	50	22	75.55	20	77.77
	75	20	77.77	18	80.00
	100	18	80.00	16	82.22
Eucalyptus oil	25	14	84.44	41	54.44
	50	12	86.66	37	58.88
	75	10	88.88*	35	61.11
	100	00	100.00*	32	64.44
Clove oil	25	18	80.00	20	77.77
	50	16	82.22	16	82.22
	75	13	85.55	15	83.33
	100	11	87.77	13	85.55
Mustard oil	25	12	86.66	16	82.22
	50	11	87.77	14	84.44
	75	10	88.88*	12	86.66
	100	00	100.00*	10	88.88*
Control		90	–	90	–
CD (P=0.05)			11.92		07.22

* Significantly reduced mycelial growth.

IV) Biocontrol Agent (*Trichoderma* sp.)

The inhibitory effects of the essential oils on pathogenic fungi were represented in Table 9.3. Eucalyptus oil and mustard oil completely inhibited the radial growth of *Alternaria tenussima* at 100 per cent conc and mustard oil showed more than (88.88 per cent) reduction of radial growth of *Fusarium proliferatum* at 100 per cent conc (Table 9.4; Figure 1 & 2).

Alternaria tenuissima

Results indicated that, all *Trichoderma* species were showed antagonistic activity. But among them, *T.viride, T.koningii* and *T.pseudokoningii* showed 80 per cent antagonistic activity than others in case of sensitive isolate of *Alternaria tenuissima*. Resistant isolate of pathogen was restricting the antagonism in some extent. Overall, all *Trichoderma* species were found more than 50 per cent antagonistic nature.

Fusarium proliferatum

Tested five species of *Trichoderma i.e. T.virens, T. harzianum* against *Fusarium proliferatum* found effective. Among the *Trichoderma* species *T. viride* showed the best performance *in vitro* biological control of *F. proliferatum* (78.88 per cent) followed by others.

Table 9.4: Antagonistic Activity of *Trichoderma* Species against Pathogenic Fungi of *Rumex acetosa*

Trichoderma Species	*Radial Growth of A. tenuissima (mm)*	*Per cent of Antagonism*	*Radial Growth of F. proliferatum (mm)*	*Per cent of Antagonism*
T.viride	20	77.77	25	72.22
T.harzianum	30	66.66	33	63.33
T.koningii	21	76.66	40	55.55
T.pseudokoningii	25	72.22	32	64.44
T.virens	25	72.22	30	66.66
CD (P=0.05)	3.04	4.82		

* Significantly reduced the mycelial growth.

Eco-friendly Manegement of plant diseases is one of the viable alternatives in sustainable agriculture because it is safe and environmentally friendly (Newsham *et al.*, 1995). Biological control is an effective, eco-friendly and alternative approach for any disease management practice.*In vitro* effect of aqueous extract (1.5 per cent) of three medicinal plant parts namely rhizome of *Zingiber officinale,Cucurcuma longa* and bulb of *Allium sativum* on four plant pathogenic fungi (*viz. Helminthosporium oryzae, Alternaria solani,Fusarium solani* and *Sclerotium rolfsii*), *Allium sativum* showed maximum inhabitation at 15 per cent in four test fungi (Chaudhary, 2004). Similarly, Ramos *et al.* (2007) has reported 35 per cent growth reduction of mycelia of *Phytophthora* on neem leaf extract media. Saha *et al.* (2005) recorded 100 per cent inhibition of spore germination of *Pestalotiopsis theae* (Saw.) Stey, *Colletotrichum camelliae* Mess., *Curvularia*

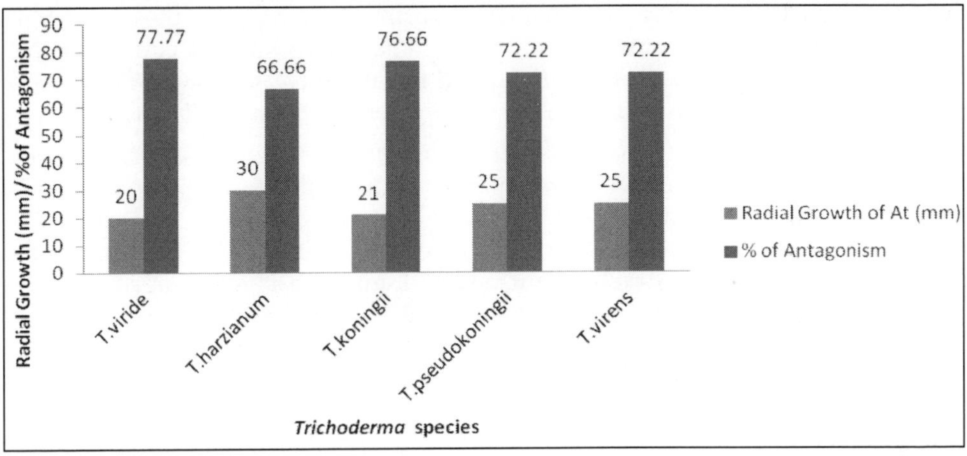

Figure 9.1: Antagonistic Activity of *Trichoderma* Species against *Alternaria tenuissima.*

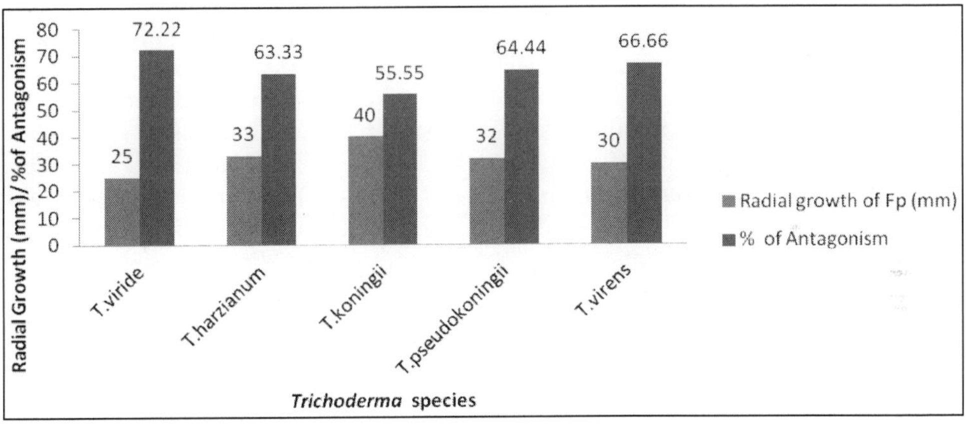

Figure 9.2: Antagonistic Activity of *Trichoderma* Species against *Fusarium proliferatum.*

eragrostidis (P. Hennings) Meyer and *Botryodiplodia theobromae* Patouiilard due to extracts of *A. indica*. Govindachari *et al.*(1999) reported that n-hexane fractions of leaf extracts of *A. indica* showed highly antifungal activity against spore germination of *Fusarium oxysporum* and *Colletotrichum Lindemuthianum.*Some plants and plant products were also found to be useful in controlling *Alternaria infection* (Neeraj, 2010). The neem leaf extract showed high efficacy to inhibit the radial growth of *A. solani* (43.3 and 26.7 per cent respectively at 0.1 per cent and 0.01 per cent) (Sharma *et al.*, 2007). Alcoholic extracts of *Semecarpus anacardium, Commiphora stocksiana, Azadirachta indica* and *Polyalthia longifolia* found maximum inhibition of *Fusarium oxysporum f. spinaciae* at 10 per cent and spore germination at 75 per cent and 100 per cent conc. respectively (Bhale *et al.*, 2005).

The antifungal potency of *C. gigantea* latex extract on the *C. albicans* showed a larger diameter of clearance than that of other fungal strains (Venkatesan and

Subramanian, 2010). Raghavendra and Gurumurthy (2011) reported the latex extract were screened *in vitro* against human pathogenic strains such as Gram positive; *Staphylococcus aureus, Bacillus subtilis,* Gram negative; *Salmonella typhi, Klebsiella phenonemia* and two fungal strains; *Aspergillus niger* and *Candida albicans.*

Among the antagonists, *Bacillus subtilis, Trichoderma viride* and *Gliocladium virens* inhibited mycelia growth of *A. solani.* In pot culture experiment, post inoculation spray application of palmarosa oil (0.1 per cent) or Neem 60 EC (3 per cent) at 75 DAS significantly reduced the tomato leaf blight disease incidence. In the field among the selected effective plant extracts, plant oils and antagonists spray application of palmarosa oil (0.1 per cent) at the initial appearance of the disease symptoms significantly reduced the leaf blight disease incidence besides increasing tomato yield (Vadivel and Ebenezar, 2006). The essential oil from the leaves of *H. suaveolens* has been reported earlier by Saxena *et al.* (1978). Earlier the antifungal and antimicrobial activity of *H. suaveolens* oil was reported by Iwu *et al.* (1990). Antifungal properties of plant essential oils have been reported by researchers throughout world (Bouchra *et al.*, 2003).

Antifungal activity of eucalyptus (*Eucalyptus camaldulensis* Dehnh.) essential oil was evaluated and suppressed the mycelial growth of post harvest pathogenic fungi *Penicillium digitatum, Aspergillus flavus, Colletotrichum gloeosporioides* and soilborne pathogenic fungi *Pythium ultimum, Rhizoctonia solani, Bipolaris sorokiniana* pathogenic fungi (Nafiseh *et al.*, 2011). Different isolates *i.e. Trichoderma viride* (12), *T. harzianum* (12), *T. koningii* (07), *T. pseudokoningii* (05) and *T. virens* (10) strains were isolated *and* used for antagonistic studyand results indicated that, among *T. viride* isolates, Tv1 *and* Tv5 showed maximum antagonism, but in case of Tv7 and Tv10 showed minimum antagonism against *A. spinaciae*. In case of *F. oxysporum* f. sp. *spinaciae* showed significant inhibition by Tv4 and reduced the per cent inhibition by Tv8 and Tv6 (Bhale, 2012). Recently, Waghmare and Kurundkar (2011) reported efficacy of *Trichoderma* species against *F. oxysporum* f. sp. *carthami* causing wilt of safflower and found that isolates no 29 *and*33 were found minimum growth of the pathogen as compared to others. The species of *Trichoderma* significantly inhibited the mycelial growth of plant pathogens (Rajkonda *et al.*, 2011).

Conclusion

The integrated management of fungal diseases of leafy vegetable crops, plant products and bio-control agents showed promising results with eco-friendly approach. Biological control offers, durable environmentally safe and cost effective alternative to chemical for the efficient management of plant disease. The plant extracts boost vegetable plants health by protecting from plant parasites. There is a need to sustain the biocontrol programs using *Trichoderma* whose beginnings were promising in order to maintain the sustainability and productivity.

Acknowledgement

Author thankfully acknowledged to UGC, New Delhi for the financial assistance of Major Research Project.

References

Agrios, G.N. (2000). Significance of plant diseases. In: *Plant Pathology*. Academic Press London, pp.25-37.

Aworh, O.C., Kasche, V., Apampa, O.O. (1994). Purification and properties of Sodom apple latex proteinases. *Food Chem.,* **50**: 359-362.

Bhale, U. N. M. G. Ambuse, V. S. Chatage and Rajkonda, J. N. (2012). Bioefficacy of *Trichoderma* isolates against pathogenic fungi inciting spinach (*Spinacea oleracea* L.), *JBiopest.*, **5**(2): 222-227.

Bhale, U. N., Kamble, S. S. and Gangawane, L. V. (2005). Effect of plant extracts on growth and spore germination in *Fusarium oxysporum f.* sp. *spinaciae* causing wilt of spinach. *Bioinfolet (Journal of Life Science)*, **2** (2): 122-125.

Bouchra, C., Achouri, M., Idrissi Hassani, L. M. and Hmamouchi, M. (2003). Chemical composition and antifungal activity of essential oils of seven Moroccan Labiatae against *Botrytis cinerea* Pers: Fr. *J. Ethnopharmacol.*, **89**: 165-169.

Chaudhary, H. C. and Raj, S. K. (2004). Antifungal property of some medicinal plants *invitro*. *J. Mycopathol. Res.*, **42**(1): 109-111.

Chiranjeevi, C., Reddy, I. P., Naryanamma, M. and Neerja, (2002). Effect of shoot clipping and insecticides on the incidence of fruit borer in brinjal. International conferece on Vegetables. Nov.11-14 Bangalore, India, pp.271.

Ellis, M. B. (1971). 'Dematiaceous Hyphomycetes' (Commonwealth Mycological Institute: Kew, UK), pp.20-72.

Govindachari, T. R., Suresh, G. and Masilamani, S. (1999). Antifungal activity of *Azadirachta indica* leaf hexane extract. *Fitoterapia,* **70:** 417-420.

Iwu, M. M., Ezeugwa, C. O., Okunji, C. O., Sanson, D. R. and Tempesta, M. M. (1990). Antimicrobial activity and terpenoids of essential oil of *H. suaveolens. Journal of Crude Drug Resistance,* **28:** 73-76.

Juncker, T., Schumacher, M., Dicato, M., Diederich, M. (2009). UNBS1450 from *Calotropis procera* as a regulator of signaling pathways involved in proliferation and cell death. *Biochem Pharmacol*, **78**(1): 1-10.

Mishra, M. and Tiwari, S. N. (1992). Toxicity of *Polyalthia longifola* against fungal pathogens of rice. *Indian Phytopath.,* **45**: 59-61.

Mohana, D.C., Pravin Prasad, Veena Vijaykumar and Raveesha, K.A. (2011). Plant extracts effect on Seed-borne pathogenic fungi from seeds of paddy grown in southern India. *Journal of Plant Protection Research*, **51**(2): 101-106.

Morton, D.J.and Stroube, W.H. (1955).Antagonistic and stimulating effects of soil microorganisms upon sclerotium. *Phytopathol*, **45**: 417-420.

Mungikar, A.M. (1997). An Introduction to Biometry. *Saraswati Printing Press, Aurangabad,* pp.57-63.

Nafiseh, K., Raheleh, M. and Seyed, E. R. (2011). Evaluation of eucalyptus essential oil against some plant pathogenic fungi. *Journal of Plant Breeding and Crop Science,* **3**(2): 41-43.

Neeraj and Shilpi Verma (2010). *Alternaria* diseases of Vegetable Crops and New Approaches for its Control. *Asian J. Exp. Biol. Sci.*, **1** (3): 681-692.

Newsham, K. K., Fitter, A. H. and Watkinson, A. R. (1995). Arbuscular mycorrhizas protect an annual grassfrom from root pathogenic fungi in the field. *Journal of Ecology*, **83**: 991-1000.

Raghavendra, R. and Gurumurthy, D. M. (2011). *In vitro* antimicrobial activity of various plant latex against resistant human pathogens. *International Journal of Pharmacy and Pharmaceutical Sciences*, **3**(4): 70-72.

Rajkonda, J. N., Sawant, V. S., Ambuse, M. G. and Bhale, U. N. 2011. Inimical potential of *Trichoderma* species against pathogenic fungi. *Plant Sciences Feed*, **1**(1): 10-13.

Ramos, A. R., Falcao, L. L., Barbosa, G. S., Marcellino, L. H. and Gander, E. S.(2007). Neem (*Azadirachta indica* A. Juss) components: Candidates for the control of *Crinipellis perniciosa* and *Phytophthora* sp. *Microbiol. Res.*, **162:** 238-243.

Saha, D., Dasgupta, S. and Saha, A. (2005). Antifungal activity of some plant extracts against fungal pathogens of tea (*Camellia sinensis*). *Pharmac. Biol.*, **43:** 87-91

Saxena, H. O., Dutta, D. K. and Brahman, M. (1978). A survey of Vetiver, Lemongrass and *Hyptis suaveolens* in Orissa for their essential oils. *Indian Perfumer.*, **22:** 298-299.

Sharma, A., Dass, A. and Pau, M. S. (2007). Antifungal effect of neem extract on some common phytopathogenic fungi. *Adv. Pl. Sci.*, **20** (2): 357-358.

Subramanian, C. V. (1971)."Hypomycetes an account of Indian species except *Cercospora*". Indian Council of Agricultural Research, New Delhi, p.810.

Vadivel, S. and Ebenezar, E.G. (2006). Eco-friendly management of leaf blight of tomato caused by *Alternaria solani, J. Mycol. Pl. Pathol.*, **36**(1): 79-83.

Vincent, J.M. (1947). Distortion of fungal hyphae in the presence of certain inhibitors. *Nature.* **150** : 850.

Venkatesan, S. and Subramanian, S. P. (2010). Evaluation of antifungal activity of *Calotropis Gigantea* latex extract: An *in vitro* study. *International Journal of Pharmaceutical Sciences and Research*, **1** (9): 88-96.

Waghmare, S. J. and Kurundkar, B. P. (2011). Efficacy of local isolates of *Trichoderma* spp. Against *Fusarium oxysporum* f. sp. carthami causing wilt of safflower. *Advances in Plant Science*, **24**(1): 37-38.

Wang, L. and Weller C. L. (2006). Recent advances in extraction of Nutraleuticals from wants. Trend in food science and technology (Elsevier), **17**: 300-312.

Woodward, B. and De Groot, R (1999). Tolerance of *Wolfiporia cocoa* isolates to copper in agar media. *Forest Prod. J.*, **49**(4): 87-94.

2015, Major Constrains and Verdict of Crop Productivity *Pages 161–169*
Editor: Dr. U.N. Bhale
Published by: ASTRAL INTERNATIONAL (P) LIMITED, NEW DELHI

Chapter 10

Drivers of Food Crops Productivity in Nigeria: Issues and Way Forwards

Olayiwola Olubodun Olaniyi*

Development Policy Centre (DPC), P.M.B.3073 Agodi GRA, Ibadan, Oyo State, Nigeria

ABSTRACT

Taking into consideration the importance of food crop productivity for farm planning and policy decisions at the micro level, the present study focused on major constraints associated with food crop productivity in Nigeria. Agricultural productivity is the key challenge being presently discussed by agricultural stakeholders as being imperatives to the attainment of food security, reducing the present high rate of unemployment and the diversification of Nigeria economy from being oil dependent. Panel data for 36 states were collected from the data bank in Nigeria and ordinal least estimate (OLE) method was used for the analysis. The results show the significant contributions of female farmers in food production. The input fertilizer use was positively related with crop productivity despite it is not significant. Based on the findings, it is good to suggest policy measure that allows adoption of technologies that will support productivity gains in crop production through less of land expansion practices.

Keywords: *Drivers, Food crops, Productivity and Nigeria.*

Introduction

Agricultural productivity may be defined in general terms as an index of the ratio of the value of total farm output to the value of total inputs used in farm production

* E-mail: olubodun_2008@yahoo.com

(Olayide and Heady (1982). Since one of the main objectives of any society is the attainment of an optimal high level of living with a given amount of effort, any increase in productivity of resources employed in farm production amounts to progress. Increase in agricultural productivity will therefore contribute to the well-being of the economy as a whole. The ultimate objective of the interest of economists in productivity should be to find ways of increasing output per unit of input and of attaining desirable inter-firm, intra-firm and inter-sector transfers of production resources, thereby providing the means of raising standard of living. The input-output process of farm production according to Olayide and Heady (1982) is important in at least four major problem areas. These are distribution of income, the allocation of resources, the relation between stocks and flows and the measurement of efficiency or productivity. Within the concept of productivity, a meaningful assessment will depend upon a clear and precise definition of input and output in such a way that their movements over time are not equal. Also the need to seek to determine which inputs and outputs are consistent with the particular productivity concept in question. Thus, being faced with separate and distinct conditions efforts are directed out to measurement of labor or capital or land productivity. In other words, resource productivity is definable in terms of individual resource inputs or in terms of combination of them. In this write-up, the concept of labor/land productivity or yield shall be defined as the ratio of total output of a particular crop to labor/land inputs as a bench-mark, change in productivity over time will depend upon changes in types and quantities of inputs. Maximum resource productivity will imply obtaining the maximum possible output from the minimum possible set of input. In this context, optimal productivity of resources implies an efficient utilization of resources in the production process. This means that productivity and efficiency are synonymous in this context. An increase in farm output will result from one of three forces. First, it will result from an increased quantity of inputs, with no change in output per unit of input. Second, it will result from increased productivity of inputs with no change or a decrease in quantity of inputs. Thirdly, it will result from a combination of changes in inputs and productivity. This situation makes the concept of efficiency a central issue in production economics. It therefore becomes imperative to mention notable efficiency theoretical frameworks developed by Farrell, Coelli and Battesse and as well discuss at least one of them in detail. Coelli and Battesse (1996) defined efficiency couched in three-related terms. First, they define 'technical' efficiency as the measure of a firms' success in producing maximum output from a given set of inputs. This indicates all those undisputed gains that can be obtained by simply organizing management better. Second, "price" efficiency was defined as the measure of a firm's success in choosing an optimal set of inputs. This is an indication of the gains that can be obtained by varying the input ratios on certain assumptions about the future price structure. Third, they define "overall" efficiency as the simple product of the technical and the price efficiencies. In summary, the three efficiency measures of Farrell are defined as follows from Equation 2.1.

$$\text{Technical Efficiency (TE)} = \frac{OA}{OP} \quad (1.0)$$

$$\text{Price Efficiency (PE)} = \frac{OR}{OA} \tag{1.1}$$

$$\text{Economic Efficiency (EE)} = \frac{OA}{OP} \times \frac{OR}{OA} = \frac{OR}{OP} \tag{1.2}$$

Productivity in this paper was however, assessed using land *i.e.* the quantity of crop produced per unit area. The analysis was done across the three agro ecological zones in order to identify areas where each crop is best suited. Also, food crop Production is concerned with the relative performance of the process used in transforming input into output. The analysis of efficiency is generally associated with the possibility of farms producing a certain optimal level of output from a given bundle of recourses or certain level of output at least-cost. Farrell (1957) distinguishes between three types of efficiency: (a) Technical Efficiency (TE), (b) Allocative or price Efficiency (AE) and (c) Economic Efficiency (EE).Technical efficiency in production is a measure of how close a firm is to the maximum output level as defined by the frontier given its input level. This definition of technical efficiency implies that differences in technical efficiency between firms exist.

The production function pre-supposes technical efficiency, whereby maximum output is obtained from a given level of input combination. Therefore, it is a factor-product relationship. An important assumption relating to efficiency is that firms operate on the outer bound production function, that is, on their efficiency frontier. When firms fail to operate on the outer bound production function, they are said to be technically inefficient. For such firms an improvement in technical efficiency maybe achieved in three ways (Heady 1960). Firstly, technical efficiency can be enhanced through improved production techniques. This may imply a change in factor proportions through factor substitution under a given technology. Hence, it may represent a change along the given production function. Secondly, technical efficiency can also be improved through an improvement in the production technology. This represents a change in the production itself such that the same amounts of resources produce more output, or alternatively, the same amount of output is derived from smaller quantities of resources than before. Thirdly, technical efficiency can be improved through an improvement in both production technique and technology. While technical efficiency is only concerned with the physical relationship between input and output, Allocative efficiency takes into account price relationship in addition to the physical relationship. Thus, Allocative efficiency is concerned with choosing optimal sets of inputs. In the regard, a firm is allocatively efficient when production occurs at a point where the marginal value product is equal to the marginal factor cost (MVP= MPC).

Economic efficiency is a situation where there are both technical efficiency and Allocative efficiency. Therefore, the achievement of either of technical efficiency or Allocative efficiency is a necessary but not a sufficient condition to ensure economic efficiency (Ellis, 1988). The simultaneous achievement of both efficiencies however, provides the sufficient condition for economic efficiency. The sufficient condition according to Heady (1960), occurs when price relationships are employed to denote maximum profits for the firm or when choice indicators are employed to denote the maximization of other economic objectives. Thus, economic efficiency refers to the choice of the best combination for a particular level of output, which is determined by both input and output prices.

Resource productivity is defined in terms of individual resource inputs or in terms of combination of them (Olayide and Heady, 1982). For instance, labor productivity is defined as the as the ratio of total output to the input of land, fertilizer and seeds respectively. Maximum efficiency is achieved when it is impossible to reshuffle resources without decreasing the quantum or value of output. To achieve this, resources must be allocated in such a way that there is no difference between the marginal value productivity of the various resources and their respective prices. Heady (1960) has, however suggested the use of the net profit from resources, rather than resource prices, to obtain a single index of efficiency. However, Olayiwola (2013) has argued that, so long as the pricing system accurately reflects the value system and consumer choices, the value productivity of resources can serve as an index of production efficiency despite its limitations. For a multi-product firm, resources are said to be allocated optimally among feasible production enterprises when the marginal value product (MVP) of every input is equal to its price in all enterprises in which it is employed (Upton, 1973).

Concept Note on Major Agricultural Constraints in Nigeria

1. Poor Agricultural Pricing Policies

Fertilizer use is promoted mainly by the fertilizer subsidy policy in Nigeria. In spite of economic reforms in Nigeria, fertilizer subsidies have remained. There is renewed consideration of input subsidies, at least as a means to reduce attendant effects of market failures. Input subsidies were widely practiced in the 1960s through 1980s. The costs of subsidies became high and unsustainable.

2. Low Access to Agricultural Credit

Access to agricultural credit has been positively linked to agricultural productivity in several studies. Yet this vital input has eluded smallholder farmers in Nigeria. Cooperatives, friends, and family members dominate the sources of farm credit among the rural farmers surveyed in southwest Nigeria. Banks with large loan funds are generally difficult to access. Issues of collateral and high interest rates screen out most rural smallholders. Another problem associated with smallholder access to agricultural credit is that agricultural loans are often short term, with fixed repayment periods; this may not suit annual cropping, especially when loan release is not coordinated with growing cycles of crops. Short-term loans are also unsuitable for livestock production. For credit to be most effective, loan terms must flexibly relate to cash flows in the target business, the input demand/supply structure, and quantifiable business risk

3. Poor Funding and Coordination of Agricultural Extension

Specific constraints identified in the implementation of the training and visit (T and V) system in Nigeria included bureaucratic procedures, and location of crop and livestock extension staff in different departments and ministries, which tended to promote rivalry and duplication of? resources. Related to these issues was the fact that the extension system was implemented with a huge bias in favor of cropping activities.

4. Poor Market Access and Marketing Efficiency

Limited or poor-quality roads and rail transportation inhibit timely access to inputs, increase costs of inputs, and decrease access to output markets. Thus, investment in infrastructure contributes to agricultural productivity. The bulky nature of primary produce has discouraged production because rural farmers have limited access to markets and good feeder roads. Economic reforms in Nigeria have led to increased private-sector participation in the supply of most purchased inputs in Nigeria, but most suppliers are based in urban areas. End users of the inputs are in rural areas, which are poorly linked to urban suppliers. Transaction costs of inputs increase delivery costs to rural farmers. However, given the prevailing poor marketing infrastructure and the attendant high transaction costs, fertilizer subsidies in Nigeria may not be effective at this time.

5. Land Tenure System and Land Degradation

The communal system of land ownership prevails among most ethnic groups in the south, in which individual ownership of land is embedded in group or kinship ownership. Communal ownership of land in Nigeria has been associated with such problems as limited tenure security, restrictions on farmers' mobility, and the inevitable fragmentation of holdings among future heirs. In addition, group ownership restricts access rights of community members outside the owning group, a situation that limits the use of land as collateral for agricultural credit. But communal ownership has also been credited with preserving traditional land use practices such as bush fallowing, which has helped retard problems of land degradation.

Methodology

Nigeria is a country with population of 140 million above (National population Census Report). It is basically divided into six geopolitical zones. Nigeria lies on the west coast of Africa and occupies approximately 923,768 square kilometers of land and shares borders with Chad, Cameroun and Benin. The country made up of 38 states and Abuja is the state capital. The states and capital territory are divided

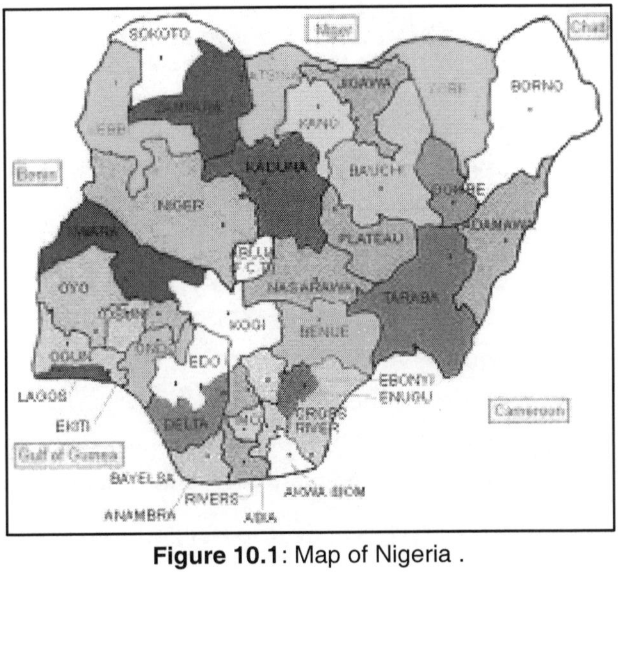

Figure 10.1: Map of Nigeria .

into approximately 774 local government areas. The spatial distribution of the population is uneven with majority of the population living in the rural areas, while approximately 36 per cent of the population lives in the urban areas. Map of Nigeria is show in Figure 10.1

The Secondary data were mainly collected from the published Ministry documents of Government of Nigeria and web site of National Bureau of Statistics, Abuja, Nigeria. The productivity estimate is expressed as:

$$RESOURCE = MVP_{xi} = \frac{\sum_{i=1}^{n} \bar{y} * b_i}{\sum_{i=1}^{n} b_i * P_{xi}}$$

The stochastic production frontier (SPF) representing Cobb –Douglas production technology characterized by variable return to scale is specified as In $Y_i = B_o + \Sigma B_y$ In $X_{ki} + V_i - U_i$ where i = 1, 2 - - - n The error term is made up of two components V_i and $U_{i.}$ V_I is a two-sided conventional error that is independent of $U_{i'}$, and is assumed to be distributed as N~ (O, Σ2v) (*i.e.* it is normally distributed with mean of zero and constant variance). This component is supposed to capture statistical noise (*i.e.* measurement error) and random exogenous shocks that disrupt production. The second component U_i is also a random variable, but unlike $V_{I'}$, it is only a one-sided variable taking non-negative values. This term captures technical inefficiency of production. There is no consensus among econometricians as to what specific distribution U_i should have. The technology adoption index is a catch all measure of technology adoption practices of the farmers. The technology adoption index was computed as:

$$TAI = \frac{1}{G}\left[\frac{AH_i}{CA_i} + \frac{FA_i}{FR_i} + \frac{IA_i}{IR_i} \pm \frac{CP_{Mi}}{PM_R}\right] \times 100$$

where,

i = Number of farmers 1, 2, 3. n and G= 4

AH_i = Area under modern varieties (ha)

CA_i = Total area

FA_i = Quantity of fertilizer (NPK) (kg/ha)

FR_i = Recommended dose of fertilizers (kg/ha)

IA_i = Number of irrigation

IR_i = Recommended irrigation

CPM_I = Quality of plant protection chemicals (liters/ha)

PM_{Ri} = Recommended plant protection chemicals

A trend of real prices of crops is expressed as:

$$Y = ab^t \qquad (2)$$

where,

Y= Trend value of dependent variable (area/total production/yield)

a = Constant

b = Trend coefficient (slope of line)

t = Time variable (Years)

The function takes the form of a linear equation in logarithmic and become log-linear as under:

$$\text{Log } Y = \log a + \log b \tag{3}$$

Results and Discussion

Variables	*OLE without Time factor*	*OLE with Time Factor*
Log crop area	0.072**(0.05)	0.027(0.04)
Share of female labor	0.562*** (0.31)	0.315***(0.34)
Log total labor	0.035 (0.22)	0.175(0.33)
Log fertilizer cost per ha.	0.017(0.06)	0.061(0.23)
Maize share	0.07(0.54)	0.05(0.62)
Cocoyam share	0.064(0.32)	–0.234(0.11)
Melon share	–0.543(0.22)	0.841** (0.33)
Rice Share	0.578*** (0.25)	0.321(0.46)
Cassava share	0.664*** (0.35)	–0.312(0.64)
Yam share	0.731***(0.44)	–0.452(0.33)
Beans share	–0.643**(0.22)	–0.998(0.45)
Groundnut share	–0.341(0.52)	–0.745(0.35)
Dry savannah	–0.421(0.56)	–0.346(0.55)
Wet savannah	0.005(0.23)	0.003(0.11)
No. of observations	350	350
t- values	0.002***	0.004***
Dublin Watson	2.02	2.01
R-square	0.78	0.65
Adjusted R-square	0.55	0.49
Adoption Index	200	150

Note: OLE (ordinal least estimate) and parenthesis indicate standard error.

1 per cent indicate by ***, 5 per cent ** and 10 per cent *.

The results show the differences when compare the productivity gain with respect to time variations. Ordinal least estimate model without time factor are compared with ordinal least estimate with time factor. The results are robust with the different specification in terms of with and without time factor. The productivity gain in food production is not too much achieved through land use area expansion. The share of the female labor is positively associated with productivity and it means that there is higher productivity of the female labor for stable food crops. This result is in line with Oni *et al.* (2008) findings that explained the significant role of female farmers in food production and ensuring food security.

Controlling for farm area and crop area, total labor did not affect crop productivity significantly for both ordinals least estimate model of with and without time factor. The reason behind the aforementioned point was as a result of gender disparity associated with agricultural policies and programmes in Nigeria. A similar finding was achieved by Olayiwola (2013) in a study related to poverty cycle disparity among the peri-urban farming household in Uyo, Nigeria. Fertilizer use was positively associated with productivity but the association was not significant in the model of ordinal least estimate of with and without time factor. The non- significant impact of fertilizer on crop productivity might be due to inadequate use of fertilizer during food production. This is one of the major constraints of crop productivity in Nigeria and the problem persists as a result of unclear fertilizer market linkage between the Government and the farmers.

Based on the competiveness of the selected crops, result shows that rice, cassava and yam recorded higher value of production per unit area than maize. Both rice and cassava are included in the presidential crop initiated in Nigeria but maize does not included. Farmers that plant the presidential crops in Nigeria have comparative advantage in terms of inputs subsidies and minimum support price initiated by the government. This is done to promote presidential crops and to increased investment but other crops such as maize does not have such benefit.

A double log model of ordinal least estimate for with and without time factor was used to address the skewed distributions that are common for all continuous variables. Test was conducted for the first order of auto correlation and was found to be significant at t value of 0.002 for ordinal least estimate without time factor and 0.004 for ordinal least estimate with time factor respectively. R-square explained the variability that exist in the model and R-square for the ordinal least estimate without time factor(0.78) is more better than R- square of ordinal least estimate with time factor(0.65).It indicates that 78 per cent of variability in inputs can be explained in ordinal least estimate without time factor while only 22 per cent can be explained by the disturbance term but in model of ordinal least estimate with time factor, only 65 per cent variability was caused by inputs factor and 35 was caused as a result of error term.

Conclusions

This research work has provided proof to show that agricultural crop production has grown in the past decade, mainly due to area land expansion and to a limited extent to improve agricultural productivity. Though the Nigerian government has designed a number of policies and strategies to reduce poverty and improve food security, achievement of such strategy requires facilitating adoption of improved crop production technologies. Results from the study suggest the share of female family labor increases agricultural productivity. This implies that female farmers are critical to productivity enhancement in crop production in Nigeria. The fact that fertilizer exhibit positive sign and was not significant makes this paper to conclude that fertilizer is a yield enhancing technology but that there exist high level of inadequate use. It therefore becomes imperative to either increase supply to farmers as at when due or enhances their capacity in terms of right quantity to apply. This study also shows that two of the presidential crops cassava and rice have higher returns than Maize. These results justify the inclusion of cassava and rice in the presidential initiative. This implies that there is a need to increase maize productivity as well as stepping up productivity of cassava and rice which are main drivers of agricultural crop productivity in Nigeria.

Way Forwards

The potential to solve the problem associated with crop productivity in Nigeria is simple if government and other related institutions pay more attentions to agriculture sector. Government must find solution to problem of land fragmentation through embarked on effective policies that will militate against this problem. Also, adequate farm inputs like agro- chemicals must be made available at cheap price to the farmers and government must follow a clear cut linkage supply system of inputs. Adoption of technologies that will bring about productivity gains in crop production through less of land expansion practices must be encouraged. It is also recommended that farmers' capacity in terms of right quantity use of fertilizer should be enhanced, while female farmers' participation in crop production should be encouraged.

References

Coelli, T.J, Prasada, R. and Battese, G. (1996). Identification of factor which influence the technical efficiency of Indian farmers. *Australia Journal of Agricultural Economics*. **40**(1): 103-128.

Ellis, I.A (1980). Productivity technology and efficiency.SIN working paper series, No. 117.

Farrell, J.M.(1957) The measurement of productivity efficiency. *Journal of the Royal Statistical Society Series* A 120, pp. 253-281.

Heady, E.O (1960). Economics of production and resource use. Prentice-Hall, New Jersey. National Bureau of Statistics Bulletin Report.

Olayide, S.O., and Heady, E.O (1982). Introduction to Agricultural Production economics, Ibadan. University Press, University of Ibadan, Nigeria.

Olayiwola, O.O.(2013). Poverty Cycle disparity among the peri-urban farming household in Uyo, Nigeria. Abhinav –*International Monthly Journal of Research in Management and Technology*., **2**(8): 17-24

Oni, I.A., Ojo, S.O. and Yusuf, S.Ass (2008). Study on food crop productivity in Nigeria. *Journal of Economics and Rural Development*., **18**(1): 36-45.

Upton, M (1973). Land Management system in south eastern Nigeria. Unpublished Ph.D. thesis, University of Ibadan.

2015, Major Constrains and Verdict of Crop Productivity *Pages 170–181*
Editor: Dr. U.N. Bhale
Published by: ASTRAL INTERNATIONAL (P) LIMITED, NEW DELHI

Chapter 11

Plant Growth Promoting Rhizobacteria (PGPR): Mediating and Facilitating Microorganisms for Plant Sustainable in Soil

H.C. Lakshman*

Microbiology Laboratory, P.G. Department of Studies in Botany Karnatak University, Pavate Nagar, Dharwad – 580 003, Karnataka, India

Introduction

In Indian soils, availability of Phosphorous (P) is a limiting factor in crop yields. A considerable portion of cultivated land in our country has turned into saline or alkaline soil, which is a serious environmental degradation. It is estimated that about 12 million ha of land has been affected by salinity condition in India (Gaind and Gour, 1999). Specifically, in Gujarat, a total of 58.41 lakh ha land is salt affected. The problematic (saline-alkaline) soils of Saurashtra exhibits relatively higher magnitude of P deficiency due to high calcareous nature of the soil. Some microorganisms dissolve mineral phosphate from the otherwise unavailable phosphorus sources in soil and make them available to plants. These microorganisms, termed phosphate solubilizing microorganisms, can be used as microbial inoculants in soil to ameliorate P nutrition of plants.

Phosphorus is an important element in a complete and balanced fertility program that can improve crop health (Armstrong, 1999). The concentration of phosphorus in

* *Corresponding Author.* E-mail: dr.hclakshman@gmail.com

the soil solution is low but plant requirements are high. The situation is further aggravated by the fact that P is readily fixed in the soil and average utilization efficiency of added fertilizer P by plants ranges from 15 to 25 per cent (Kamprath and Watson, 1986). The MPS property among soil bacteria is wide spread in pseudomonads. The fluorescent pseudomonads are a group of bacteria that have received special attention because of their plant growth promotional activity and ability to control plant diseases. Certain strains of fluorescent pseudomonads are known to possess the MPS activity (Dave and Patel, 1999; Lakshman, 2010). If we can obtain efficient P-solubilizing fluorescent *Pseudomonas,* in addition to promoting plant growth and suppressing plant pathogens, it would enhance the utility of single bio-inoculants.

Plant growth promoting rhizobacteria (PGPR) are a subset of free-living rhizobacteria that increase plant growth after inoculation onto seeds. PGPR offer disease control against diverse soil borne and foliar pathogens, promote vegetative growth and yield, promote legume nodulation and seedling emergence. The PGPR promote the plant growth by solubilisation of insoluble phosphates present in the soil, production of phytohormones such as auxins, cytokinins and gibberellins, and/or increased iron supply to the plant mediated by siderophores. PGPR increase plant growth, indirectly, by suppressing the soil-borne pathogens through competition for nutrients or pro-duction of antibiotics/hydrogen cyanide and/or siderophores. PGPR also induce systemic resistance in the host plant conferring resistance against a broad spectrum of diseases. Both fungal and bacterial inoculants are capable of increasing the vegetative growth of the plant and yield. Arbuscular mycorrhizal (AM) fungi establish a symbiotic association with plant roots and occur on all crop plants with few exceptions. AM fungi enhance the phosphorus nutrition of the host plant, since the hyphae of AM fungi extend beyond the phosphorus depletion zone which occur near the root as a result of phosphorus absorption by the plant, and translocate phosphorus to the plant. AM fungi also increase the plant uptake of Zn, Cu, S, K, Ca, Mg, Mn, Cl, Br and water, alter the plant's physiology to reduce stress response to soil drought, reduce plant response to high salt levels, toxicities associated with mine spoils, landfills and heavy metals. Mycorrhizal fungi produce metabolites that alter the root regeneration and morphology, thus greatly increasing absorptive surface area and feeder root longevity.

Pseudomonas fluorescens a Promising PGPR

The plant growth promoting rhizobacteria belonging to pseudomonas (*P. fluresecens, P. putida, P cepacia*) are no recognized as the early root colonizers and effective in suppression of a variety of root borne fungal pathogens, directly by producing antifungal compounds and indirectly by promoting plant health. The secondary metabolites identified from the pseudomonads which are antifungal in nature are the antibiotics phenazines, pyrroles, tropolone, siderophores and batalactums. The plant growth promoters are the phytohormones vitamins and cytokinins. The specific strains also induce systemic in the root system through elevation in phytoalexing levels in roots leading to overcome fungal infection.

These are several practical reasons for using *Pseudomonas fluorescens* rather than Bt to produce the toxin. For example, *Pseudomonas fluorescens* produces higher concentration of the toxins, the toxin remain within *the Pseudomonas fluorescens* cells (encapsulated) when the cells are killed, making collecting the toxins relatively easy; and *P.flurorescens* does not produce the large number of additional toxins made by Bt, which might harm non target species. The active ingredient can be applied by several methods including hand spraying, aerial spraying and ground based applications. It can also be applied through irrigation systems. Bt delta endotoxins encapsulated in killed *P. fluorescens* are non hazardous to humans as human gut lacks any receptors that would make the toxins hazardous, and this active ingredient also does not pose a risk to the environment, or to non target organism. Organism other than moths and butterflies appear to lack the gut receptors that would make the toxins hazardous.

Barber and Lynch (1977), have first made a simple but devastating calculation, in which they showed the amount of C released from the roots was unlikely to be sufficient for any useful fixation. If we take an estimate of C release as 20 per cent of total photosynthate, and efficiency of fixation as quoted above, it appears that some 1.5 per cent N of total dry weight could be fixed, if the released C were used for this purpose, at the same efficiency as in legumes. The symbiotic endophyte *Rhizobium* at least has good access to the C produced by the host. Associative N-fixers, such as *Azospirillum* in rhizosphere, are dependent upon the exudates from the root, and competition for this is intense because of the rapid increase of the general bacterial population. Free-living N-fixing bacteria are also inefficient because of their large need to use C for respiratory protection of the anaerobic condition of nitrogenase. Nye and Tinker (1977), have discussed these effects in more quantitative terms, in relation to the possible amounts of microbial compounds being produced. The over-riding factor is that these effects could mostly be produced directly by the plant itself, and in some cases such as pH change, to a much larger extent than any conceivable micro biological process (Tinker, 1980; Hedlley *et al.*, 1982). The burden of proof appears to rest firmly on those who propose that such mechanisms are important.

Plant Growth Promoting Rhizobacteria as Phosphate Solubilizers

The group of rhizosphere bacteria (rhizobacteria) that exerts a beneficial effect on plant growth is referred to as Plant Growth Promoting Rhizobacteria (PGPR) (Schroth and Hancock, 1981). PGPR belongs to several genera, *e.g., Actinoplanes, Agrobacterium, Alcaligenes, Amorphosporangium, Arthrobacter, Azotobacter, Bacillus, Bradyrhizobium, Cellulomonas, Enterobacter, Erwinia, Flavobacterium, Pseudomonas, Rhizobium, Streptomyces* and *Xanthomonas* (Weller, 1988). *Bacillus* sp. are appealing candidates as PGPR because their endospore producing ability which makes them ideal inoculants for dry areas. Currently *Pseudomonas* species are receiving much attention as PGPR, because of their multiple effects on plant growth promotion.

PGPR are believed to improve plant growth by colonizing the root system and pre-empting the establishment of or suppressing Deleterious Rhizosphere Micro Organisms (DRMO) on the root (Schroth and Hancock, 1981). Inoculating planting material with PGPR presumably prevents or reduces the establishment of pathogens (Suslow, 1982). Production of siderophore is yet another mechanism through which microbes influence plant growth. Siderophores are low molecular weight high affinity

Fe+++ chealtors that transport iron into bacterial cells and are responsible for increased plant growth by PGPR (Kloepper *et al.*, 1980). Under Fe-deficient conditions, fluorescent pseudomonads produce yellow-green fluorescent siderophore-iron complex (Hohnadel and Meyer, 1986) which creates an iron deficient environment deleterious to fungal growth.

Inoculations of legumes with root colonizing bacteria and *Rhizobium* has been demonstrated to affect symbiotic nitrogen fixation by enhancing the colonization and nodulation of the legume roots by rhizobia through multitudinous desirable characteristics. Attributes of *Rhizobium*, host and associated rhizosphere microflora influence the outcome of the competition for nodulation sites (Tilak and Saxena, 1994). Besides indigenous *Rhizobia*, other rhizosphere microorganisms and AM fungi contribute to the competitive success of an inoculant strain (Saxena *et at*, 1997).

Bacteria such as *Pseudomonas* and *Bacillus* excrete acids into the growth medium and hence solubilize bound phosphates (Gaur, 1985). These organisms are quite useful in the utilization of rock phosphates with low content of P_2O_5. Field experiments conducted with P-solubilizers like *Aspergillus awamori*, *P. striata* and *B. polymyxa* significantly increased the yields of various crops like wheat, rice, cowpea etc. In the presence of rock phosphate, a saving of 30 kg P_2O_5/ha with the use of phosphate solubilizing microorganisms has been reported by Gaur (1985).

Root Exudates as Structural Analogous

Often, the presence of structure analogs to the chemical of interest will enhances microbial degradation. Such was the case with polychlorinated biphenyl (PCB) degradation, where biphenyl was used to enhance the activity of PCB – degrading bacteria. Donnelly and co-workers (1994) (Missed in Ref List) have explored this idea further by using plants to provide the structural analong(s) in the form of root exudates could support growth of PCB-degrading bacteria while also maintaining their ability to degrade various PCB congeners. Additional studies revealed that, among others, red mulberry (*Morus rubra*) was capable of producing large amounts of root phenolics. The presence of exudates diverse in nutritional quality makes the rhizosphere a good place for co-metabolism of chemical contaminants. Rovira *et al.* (1979) discussed and define the nutrients present in the rhizosphere, including sloughed-off debris, contents of dying cells, as well as the soluble, low-molecular weight compounds more normal regarded as 'exudates' from a healthy root. The latter comprise very many compounds, namely sugars, organic acids, and amino acids. It is virtually certain that the amount and composition of exudates varies with growth conditions, so it is extremely difficult to generalize about the substrate available K to rhizosphere bacteria. If we assume that P and other elements cans be obtained as inorganic ions from the main body of the soil, the primary question concerns the total amount of combined C supplied.

Newman and Watson (1977) concluded that soluble exudates would normally be in the range of 1 – 10 per cent of total photosynthesis by the host plant. Measurements in axenic solution culture usually give 1-2 per cent at the most, and it may be that this figure is only modestly increased in non-sterile soil. However, larger amounts of solid root debris must be produced because work with C suggests that C equals to as much as 25 per cent of net photosynthesis may be lost from the root surface. N fixation and its C cost Rhizosphere organisms certainly include N/fixers of the genera *Azotobactor* and *Azospirillum*. There have been claims that important amounts of N can be fixed in the rhizosphere, or even in the root cortex, by such organisms. The principle is not in doubt, and the debate is simply about the probable importance of this process. In this context, the need for C by all N-fixers is important, because it sets the limits within which the process has to work. It is discussed here for both the free-living organisms and *Rhizobium*.

By providing a niche suitable to a diverse population of micro organism, vegetation may enhance microbial degradation because of a key group of organisms. Generally, the *rhizosphere* is colonized by a predominantly Gram-negative bacteria community. Interestingly, Gram-negative bacteria appear to have some important metabolic capabilities for degrading xenobiotic chemicals not found in Gram-positive microorganisms. Glutathione-S-transferase (an enzyme responsible for conjugation of xenobiotics in mammals, plants, and microorganisms) is particularly active among Gram-negative bacteria, including the genera *Pseudomonas* and *Enterobacter* (Brown and Rovira, 1999). For chemicals that are easily degraded, an increase in the numbers of microorganisms in the rhizosphere is likely the reason behind the enhanced degradation. In this simple scenario, the presence of 100-fold more microorganisms in rhizosphere soil compared with non/vegetated soil translates into an increase in the degradation rate for the contaminant. As good example of this mechanism is the recent research with vegetation and the primary component of de-icing fluid, ethylene glycol (EG) (Baath, 1990). Based on this study the positive impact that vegetation could have through simply increasing microbial biomass would appear to be limited to sites with readily degradable contaminants.

Increase or Synergistic Microbial Activity

The complex nature of most synthetic chemicals that are encountered by microorganisms can require the interaction of microbial communities at the molecular, physiological, and ecological levels to achieve bio-chemical transformation (Dhanya and Potter, 2007). The plant root zone fosters these types of interactions. Collectively, the microbial community may provide the spectrum of derivative enzymes, each of which may be required for mineralization but may not all be present in a microbial strain.

Bacillus thuringiensis delta endotoxins encapsulated in killed pseudomonas fluorescens is used for controlling moth larvae on a wide range of crops including vegetable, field crops (Corn and Cotton) fruits and nuts. The active ingredient *B. thuringiensis* delta endotoxins encapsulated in *Pseudomonas fluorescens* is created by taking the genetic material that codes for two protein from bacterium *Bacillus thuringiensis* (Bt) and inserting it into a different bacterium, *Pseudomonas fluorescens*. These two Bt toxins (Cryl AC and Cry IC) are toxic to the larvae of certain species of moth and butterflies. These Bt toxins cause death by attaching to specific receptors in the larval gut; eventually rupturing the gut and killing the larvae. Because only certain species of moths and butterflies contain the needed receptor, this active ingredient does not harm other organisms such as other insects' birds or humans.

Increased Phosphate Solubilisation

Next to nitrogen, phosphorus is the most important macronutrient required for plant growth. A large proportion of phosphorus present in the soil is in insoluble form rendering it unavailable for plant uptake. It is estimated that, in certain soils, up to 75 per cent of the applied phosphorus fertilizers may be converted into insoluble mineral complexes (Goldstein, 1986; Azcon, 1989) and farmers need to apply several-fold excess of phosphorus fertilizers. Phosphate solubilising microbes (PSM) convert inorganic unavailable P forms into orthophosphate anion and make them available for plant uptake and this process is referred to as mineral phosphate solubilisation (MPS). Addition of rock phosphate to soil along with PSM increased phosphorus uptake of the plant and showed positive growth responses. Among fungi, *Aspergillus* and *Penicillium* spp. and among bacteria *Bacillus, Enterobacter, Erwinia* and *Pseudomonas* spp. are the prominent phosphate solubilizers.

Gram negative bacteria secrete organic acids such as citrate, lactate, succinate etc. which are partly responsible for phosphate solubilisation (Kim *et al.*, 1997). Direct oxidation of glucose to glueonic acid by glucose dehydrogenase (GDH) located on the outer leaf of cytoplasmic membrane, leads to the release of protons that finally solubilize the insoluble phosphates. The conversion of glucose to glueonic acid by GDH transfers electron from aldose sugars to the electron transport chain and requires pyrroloquinoline quinone (pqq) as a cofactor (Duine, 1991). The glueonic acid is further oxidized to 2-keto-gluconic acid, which acidifies the microniche in the vicinity of the root system and could have a major role in solubilizing inorganic phosphates.

Transgenic tobacco plants expressing the bacterial citrate synthase, produced and secreted more citrate into the rhizosphere and conferred a distinct advantage when grown soil containing calcium phosphate. Similarly, PSM genetically manipulated to over produce organic acids may have increased MPS activity. Cloning of the genes encoding for pqq synthase from *Erwinia herbicola* and *Rahnella aquatilis* in *E. coli* conferred the MPS trait to *E. coli* (Kim *et al.*, 1998). The apo-GDH enzyme produced by E. *coli* utilizes pqq as a co-factor resulting in the production of holoenzyme. Transfer of pqq gene cluster from PSM into microbial strains producing apo-GDH and having better rhizosphere competence or disease control potential will be advantageous. Apo-GDH has also been reported in *Agrobacterium, Azotobacter* and *Rhizobium* spp. Transfer of the pqq gene cluster from PSM to *Azotobacter* or *Rhizobium* spp. without affecting their nitrogen fixing ability will enable to develop a single microbial inoculants to improve both nitrogen and phosphorus nutrition of the plant.

Bacteria scavenge inorganic phosphate (Pi) from organic phosphate by a process mediated by acid and alkaline phosphatases and the induction of these enzymes is regulated by *Pho*-regulon. Phosphate starvation induced both glucose dehydrogenase and alkaline phosphatase in *Enterobacter asburiae* (Gyaneshwar *et al.*, 1999). Identification of common regulatory controls between the *Pho*-regulon, and MPS genes could be helpful to develop strains with constitutive expression of these systems.

Role in Enhance the Nitrogen Fixation

Flavanoids, chalcones, conjugated isoflavanoids and betaines secreted by the host plant activate the *nod* genes of the microbial symbiont. The lipo-oligosachharide chitin molecules produced by rhizobia upon activation of *nod* genes induce specific gene expression in the cognate host plant. By altering the specificity of *nod* genes to plant signals and the structure of the *nod* factors the symbiotic host range of *Rhizobium* can be extended to non-legume plants. However, the oxygen sensitivity of nitrogenase poses a problem for its activity in non-legume plants which do not contain leghaemoglobin. Hence, simultaneous alteration of nitrogenase for its oxygen sensitivity is essential to extend the host range of *Rhizobium* to non-legume plants.

The nodulation genes of *Rhizobium leguminosarum* and *R. meliloti* are located on symbiotic (*sym*) plasmids. Introduction of sym plasmid from *R. leguminosarum* and *R. trifolii* into *Agrobacterium tumefaciens* conferred the ability to nodulate pea and clover, respectively but the nodules formed were ineffective without formation of bacteriods. Results from several similar studies also indicate that infection and nodule initiation genes could be expressed in heterlogous rhizobia leading to broadening of host range but effective bacteriod development leading to formation of nitrogen-fixing nodules is difficult to achieve (Azcon, 1989; Lakshman, 1996).

Role in the Production of Chitinases and β-Glucanases

Chitinases and β-glucanases produced by biocontrol agents are the key hydrolysing enzymes involved in lysing the cell wall of pathogenic fungi (Elad *et al.*, 1983; Haran, *et al.*, 1996). *Psuedomonas* spp. (Koby *et al.*, 1994) genetically engineered to produce chitinase was proved to be effective in control of plant diseases. Transformants of T. *harzianum* over expressing a gene *Chit33* encoding 33kDa chitinase, showed increased antagonistic activity against *Rhizoctonia solani*. Similarly, transformants of *T. longibrachiatum* overexpressing the β -1,4-endoglucanase gene *egl*1 were effective than the wild-type strain in protecting cucumber seedlings against P. *ultimum* infection. These two studies indicate the possibility of improving biocontrol strains by over expression of chitinases and β -glucanases.

A few attempts have been made to improve the production of chitinases by wild-type *Trichoderma* strains by cloning chitinase under the promoter of cellobiohydrolase gene (cbh 1) (Clark *et al.*, 1996), by cloning under the control of a constitutive viral promoter (Haran *et al.*, 1993), by homologous transformation (Giczey, *et al.*, 1998) and transformation with a constitutively expre-ssed heterologous fungal chitinase gene.

Role in the Antibiotic Production

In addition to lytic enzymes involved in mycoparasitism, antibiotics produced by bio-control strains play a major role in antagonism and disease control (Fravel, 1998). Several illustrations from literature indicate that mutations in antibiotic encoding gene(s) or regulatory gene(s) affecting the production of antibiotics also affected the disease supressibility of bio-control agent and complementation of these genes from wild strains restored the disease supressibility (Hill *et al.*, 1994). Transfer of genomic fragment required for biosynthesis of 2, 4-diacetylphloroglucinol (phl) from P. *fluorescens* F113 to phi P. *fluorescens* Ml 14 enhanced its antagonism towards *P. ultimum* (Fenton *et al.*, 1992).

The production of antibiotics pyrrolnitrin, pyoluteorin and 2, 4-diacetyl-phloroglucinol by *P. fluorescens* pf-5 is regulated by *rpoS* gene. The *rpoS*- mutant overproduced pyoluteorin and 2, 4-diacetyl phloroglucinol, and was superior to the wild type in suppression of seedling damping-off of cucumber caused by *Pythium ultimum*. Identification of such regulatory gene(s) and their manipulation to increase the antibiotic production may increase the disease control potential of biocontrol agents. Also, the spectrum of bio-control agents can be increased by transfer of gene(s) encoding for antibiotics with broad spectrum of activity.

Role in the Enhanced Siderophore Production

Siderophores are low-molecular weight iron chelating compounds produced by several Gram +ve, Gram -ve bacteria and fungi. Siderophores produced by microbial inoculants can effectively scavenge most of the Fe^{3+} available in their vicinity, thus making it unavailable for proliferation of fungal pathogens and other deleterious microorganisms. Siderophore produ-cing strains uptake its Fe^{3+}-siderophore complex through specific receptors in the outer membrane. Manipulation of these receptor proteins to recognize Fe^{3+}-siderophore complexes from other strains enhances the survival chances of microbial inoculants. In fluorescent pseudomonads both the biosynthesis of siderophore moiety and uptake of Fe^{3+}-siderophore complex are strongly repressed by iron. Complete understanding of the mechanism of repression enables constru-ction of strains that could produce siderophore (s) under a wider range of physio-logical conditions.

Role in Control of Plant Diseases

The major success in the field of biological control was the control of crown gall caused by *Agrobacterium tumefaciens* using *A. radiobacter* strain 84 (Kerr, 1972). Since then, several bacterial and fungal antagonists proved successful in control of plant and postharvest diseases. Among these biocontrol agents bacterial strains of *Bacillus* and *Pseudomonas* spp. and fungal strains of *Trichoderma* and *Gliocladium* spp. were potent antagonists against a wide range of plant pathogens (Lakshman *et al.*, 2012). The factors contributing to biological control of plant diseases include: parasitism of the fungal pathogen by the bio-control agent and PGPR acts antagonism against the pathogen mediated by antibiotics produced by bio-control agent. Experiments proved that, PGPR lysis of the fungal pathogens by chitinases, β-1,3-glucanses and other hydrolytic enzymes produced by bio-control agent and aids in competition for nutrients between the pathogen and bio-control agent and it secretes the induction of phytoalexins and PR-proteins in the host by the bio-control agent.

Conclusion

A biotechnological goal is to use a combined inoculation of selected rhizosphere microorganisms to minimize fertilizer application and to maximize plant growth and nutrition. Selected combinations of microbial inocula enhanced the positive effect achieved by each microbial group, improving plant development in infertile soils. A free living bacterium producing certain stimulating metabolites enhances plant growth. Plant growth promoting rhizobacteria (PGPR) are beneficial bacteria colonizing plant roots or niches close. Several mechanisms have been proposed which account for the stimulating effect, inclusing a range of metabolites (amino acids, vitamins, phytohormones), solubilization or mineralization process. AM fungi differing in their ability to enhance nutrient uptake and the different physiology of AM colonized roots may alter conditions for rhizosphere microbial groups. Therefore, growth enhancement induced by bacterial/fungal inoculation may also be due to the ability of the beneficial bacteria to protect plants from certain deleterious rhizosphere associated microorganisms and/or to reduce the population of the latter.

Acknowledgement

Author is very thankful to DST-SERB, New Delhi for sanctioning major research project " Experimental studies on additive effects of PGPR and hydrolytic enzymes on Niger plants infected with AM fungi" and financial support.

References

Armostrong, D.L. (1999). Phosphorous nutrition improves plant disease resistance. In: Better crops with plant food. Potash and phosphate institute, Georgia, USA., pp: 26-27.

Asaka, O. and Shoda, M. (1996). Biocontrol of *Rhizoctonia solani* damping-off of tomato with *Bacillus subtilis* RBI4. *Appl. Env. Microbiol.*, **62**: 4081-4085.

Azcon, R. (1989). Selective interaction between free living rhizosphere bacteria and AM fungi. *Soil Biol. Biochem*. 21: 639-644.

Baath, E. (1990). Thymidine incorporation into soil bacteria. *Soil Biol. Biochem*., **22**: 803-810.

Bagyaraj, D.J., Krishnaraj, P.U. and Khanuja, S.P.S. (2000). Mineral phosphate solubilization: Agronomic implications, mechanism and molecular genetics. *Proc. Ind. Natl. Sci. Acad.*, **66:** 69-82.

Barber, D.A. and Lynch, J.M. (1977). Microbial growth in the rhizosphere. *Soil Biol Biochem.*, **9**: 305-308.

Brown, G.D. and Rovira, A.D. (1999). The rhizosphere and its management to improve plant growth. *Adv. Agronomy.*, **66**: 1-102.

Clark, M.E., Harman, G.E. and Penttila, M. (1996). Enhanced expression of endochitinase in *Trichoderma harzianum* with the cbhl promoter of *Trichoderma reesei. Appl. Env. Microbiol.*, **62**: 2152-2155.

Dava, A. and Patel, H.H. (1999). Inorganic phosphate solubilizing soil *Pseudomonas. India J. Microbiol.*, **30**: 305-310.

Dhanya, M.K. and Potty, V.P. (2007). Siderophore production by *Pseudomonas fluorescens* isolated from rhizosphere of *Solenostemon rutundifolias. J. Root Crops.*, **33**(2): 138-140.

Duine. J.A. (1991). Quinoproteins: enzyme containing quinoid cofactor pyroloquinoline quinone topaquinone or tryptophan-tryptophan quinone. *Eur J. Biochem.*, **200**: 271-284.

Elad. Y., Chet, I., Boyle, P. and Henis. Y. (1983). Parasitism of *Trichoderma* spp. on *Rhizoctonia solani* and *Sclerotium rolfsii* - scanning electron microscopy and fluorescence microscopy. *Phytopathology.*, **73**: 85-88.

Fenton. A.M. Stephens. P.M., Crowley. J., O'Callaghan.M. and O'Gara. F. (1992). Exploitation of gene(s) involved in 2, 4-diacetylphloroglucinol biosynthesis to confer a new biocontrol capability to a *Pseudomonas* strain. *Appl. Env. Microbiol.*, **58**: 3873-3878.

Fravel, D.R. (1998). Role of antibiosis in the biocontrol of plant diseases. *Annu. Rev. Phytopathol.*, **26:** 75-91.

Gaind, S. and Gaur, A.C. (1999). Microbial phosphate solubilization as influenced by sodium chloride. *Indian J. Expt Biol.*, **37**: 209-210.

Gaur, A.C. (1985). Phosphate solubilizing microorganisms and their role in plant growth and crop yields, proc. Soil Biol. Symp. Hisar, -India (eds. M.M. Mishra and K.K. Kapoor), pp. 125-138.

Giczey, G. Kerenyi, Z., Dallmann, G. and Hornok, L. (1998). Homologous transformation of *Trichoderma hamatum* with an endochitinase encoding gene, resulting in increased levels of chitinase activity. *FEMS Microbiol. Lett.*, **165**: 247-252.

Goldstein, A.H. (1986). Bacterial solubilization of mineral phosphates: historical perspectives and future prospects. *Am. J. Altern. Agric.*, **1**: 51-57.

Gyaneshwar, P., Parekh, L.J., Archana, G., Poole, P.S., Collins, M.D., Hutson, R.A. and Naresh Kumar, G. (1999). Involvement of a phosphate starvation inducible glucose dehydrogenase in soil phosphate solubilization by *Enterobacter asburiae. FEMS Microbiol Lett.*, **171**: 223-229.

Haran, S., Schickler, H., Oppenheim, A. and Chet, I. (1996). Differential expression of *Trichoderma harzianum* chitinases during mycoparasitism. *Phytopathology.*, **86**: 980-985.

Haran, S., Schickler, H., Pe'er, S., Logemann, S., Oppenheim, A. and Chet, I. (1993). Increased constitutive chitinase activity in transformed *Trichoderma harzianum*. *Biological Control.*, **3**: 101-108.

Hedly, M.J., Nye, O.H. and White, R.E. (1982). Plant induced changes in the rhizosphere of rape (Brassica napus var. Emeral) seedlings II Origin of the pH change. *New Phytol.*, **91**: 31-44.

Hill. D.S., Stein, J.K., Torkewitz, N.R., Morse, A.M., Howell, C.R., Pachlatko. J.P., Becker, J.O. and Ligon, J.M. (1994). Cloning of genes involved in the synthesis of pyrrolonitrin from *Pseudomonas fluorescens* and role of pyrrolonitrin synthesis in biological control of plant disease. *Appl. Env. Microbiol.*, **60**: 78-85.

Hohandel, O.B. and Meyer, J.M. (1986). Pyoverdine-facilitated iron uptake among fluorescent pseudomonades. In : Iron Siderophoes and Plant Diseases (Swinburne, T.R. ed.). Plenum Press, New York, USA, pp. 119-129.

Kamprath, E.J. and Watson, M.E. (1980). Conventional soil and tissue tests for assessing the phosphorous status of soils. In: The role of phosphrous in Agriculture (Eds) Khaswneh, T.E., Sample, E.C. and Kamprath, E.J. American Society of Agronomy, Madison, WI, pp. 433-464.

Kerr, A. (1972). Biological control of crown gall: Seed inoculation. *J. Appl. Bacterial.*, **35**: 493-497.

Kim, K.Y., Jordan, D. and Krishnan, H.B. (1998). Expression of genes from *Rahnella aquatilis* that is necessary for mineral phosphate solubilization in *Escherichia coli*. *FEMS Microbiol. Lett.*, **159**: 121-127.

Kim, K.Y., McDonald, G.A. and Jordan, D. (1997). Solubilization of hydroxyapatite by *Enterobacter agglomerans* and cloned *Escherichia coli* in culture medium. *Biol. Fertil. Soils.*, **24**: 347-352.

Kloepper, J.W., Leong, J., Teintze, P., Arayangkool, T. Sintwongse, P., Siripaibool, C, Wadisirisuk, P. and Boonkerd, N. (1988). Nitrogen fixation (^{15}N dilution) with soybeans under Thai field conditions. II. Effect of herbicides and water application schedule. *Plant and Soil.*, **108**: 87-92.

Kloepper. J.W., Leong, J., Teintze, M. and Schroth, M.N. (1980). Enhanced plant growth by siderophores produced by plant growth promoting rhizobacteria. *Nature.*, **286:** 885-886.

Koby. S., Schickler, H., Chet, I. and Oppenheim, A.B. (1994). The chitinase encoding Tn-7 based chiA gene endows *Pseudomonas fluorescens* with the capacity to control plant pathogens in soil. *Gene.*, **147**: 81-83.

Lakshman, H.C. (1996). VA-Mycorrhizal studies in some important timber tree species. Ph.D. thesis Karnatak University, Dharwad-580 003, India., p. 349.

Lakshman, H.C. (1999). Aggregation of sand dune soil by arbuscular mycorrhizal fungi and its use in revegetaion practices. *India J Envorn Ecoplan.*, **2**(3): 247-252.

Lakshman, H.C. (2003). Synergistic effect between PSB and AM fungi on growth and P uptake of *Clitoria ternatea* L. In: Mineral phosphate solubilization. Eds. A.R. Alagawadi, P.U. Krishnaraj and M.S. Kuruvinashetti. U.A. S. Dharwad-580 005., pp: 159-162.

Lakshman, H.C. and Nadagouda, M.G. (2010). Microbial solubilisation of PSB and AM fungal use for yield, phosphate uptake in improvement of nodulation and yield of Vicia faba L. (Broad bean). *Inter. J. of Agric. Sci.*, **6**(1): 319-321.

Lakshman, H.C., Kavatagi, P.K. and Kurandawad, J.M. (2012). Arbuscular mycorrhizal fungi and plant growth promoting microorganisms as disease control agents. In: Glimpses of arbuscular mycorrhizae fungal research. Lambert Academic publishers, Germany., pp: 49-56.

Newman, E. and Watson, A. (1977). Microbial abundance in the rhizosphere: a computer model. *Plant and Soil.*, **48**: 17-56.

Nye, P.H. and Tinker, P.B. (1977). Solute movement in the soil root system. Oxford Blacwell Scientific Publications., p. 342.

Pathak, V.N. (1997). Post-harvest fruit pathology - present status and future possibilities. Indian *Phytopath.*, **50:** 161-185.

Rovira, A.D., Poster, R.C. and Martin, J.K. (1979). Note on geominology: organisms, nature and nomenclature of the organic materials in the rhizosphere. In: soil/

root interface (Eds.) J.L. Harley and Prescot, R. London Academic press., pp.1-14.

Saxena, A.K., Rathi, S.K. and Tilak, K.V.B.R. (1997). Differential effect of various endomycorrhizal fungi on nodulation effects of greengram by *Bradyhizobium* sp. (Vigna) strain S 24. *Biol. Fertil. Soils.*, **24**: 175-178.

Saxena. A.K. and Tilak, K.V.B.R. (1998). Free-living nitrogen fixation: Its role in crop production. In : Microbes for Health, Wealth and Sustainable Environment (ed. A.K. Varma), Malhotra Publishing Co., New Delhi, pp. 25-64.

Schroth, M.N. and Hancock, J.G. (1981). Selected topics in biological control. *Annual Review of Microbiol.*, **35:** 453-476.

Tilak, K.V.B.R. (2002). Biofertilizers: Their role in integrated nutrient management. In: Bioinoculants for sustainable agriculture and forestry. Scientific Publishers (India) pp. 1-5.

Tilak, K.V.B.R. and Saxena, A.K. (1994). Competition among *Rhizobium* strains for nodulation of legumes. In: Biology and Application of Nitrogen-fixing organisms: Problems and Prospectes. (Eds A.B. Prasad and A. Vaishampayan), Scientific Publsihers, India, pp. 241-256.

Tinker, P. (1980). The role of rhizosphere microorganisms in phosphorus uptake by plants. In: The role of phosphorus in Agriculture, Eds. Wasanch, F.K. and E. Sample. Madison: American Society of Agronomy., pp.617-654.

Weller, D.M. (1988). Biological control of soil borne plant pathogens in the rhizosphere with bacteria. *Annu. Rev. Phytopathol.*, **26**: 379-407.

2015, Major Constrains and Verdict of Crop Productivity *Pages 182–197*
Editor: Dr. U.N. Bhale
Published by: ASTRAL INTERNATIONAL (P) LIMITED, NEW DELHI

Chapter 12

Recent Advances in Researches on Foot Rot and Leaf Rot of Betelvine (*Piper betle* L.) Caused by *Phytophthora* spp.: A Threat to Betelvine Crop

Basudeb Dasgupta*

Department of Plant Pathology,
Bidhan Chandra Krishi Viswavidyalaya,
Mohanpur – 741 252, Nadia, West Bengal

Introduction

Betelvine (*Piper betle* L.) is a perennial dioecious creeper, belonging to the family Piperaceae. It is an important commercial crop of India, Bangladesh, Sri Lanka and some extend of Malaysia, Singapore, Thailand, Philippines and Papua New Guinea. In India it is commercially grown over an area of 54,000 ha providing livelihood to hundreds of thousands of families engaged in its cultivation and trade. Out of the total cultivable area under betelvine West Bengal constitute about 1/3 *i.e.* 18,203 ha. In West Bengal, Betelvine is usually grown under artificially erected structure known as Baroj, Bareja or Bheet, which is a kind of hut whose sides and roof is made of jute slaths or straw on a light frame work of bamboo which provides high moist and humid condition favouring several diseases that in turn are prerequisites for good

* *Corresponding Author.* E-mail: dasgupta.basudeb824@gmail.com

harvest. The major constraint to cultivation of betelvine is its diseases that severally damage foot, stem, root and foliage. The serious diseases reported include a foot rot syndrome produced by a number of pathogens including *Phytophthora parasitica* var. *piperina*, *Phytophthora nicotianae* var *parasitica* species of *Rhizoctonia*, *Pythium* and *Sclerotium rolfsii* Sace, and foliage diseases like leaf rot by *Phytophthora parasitica*, *Phytophthora palmivora*, leaf spot and stem anthracnose caused by *Colletorichum capsici* and bacterial leaf spot and stem rot caused by *Xanthomonas campestries* pv. *betlicola*. Among the pathogens, *Phytophthora* spp. perhaps ranks first in its destructiveness under both field and storage conditions. The extent of losses may vary from 30 – 100 per cent in case of foot rot and 20 – 40 per cent in case of leaf rot, leading to almost total crop failure (Maiti and Sen, 1982; Dasgupta *et al.*, 2000). The present investigation was undertaken to study the recent advances of researches on foot rot and leaf rot of Betelvine caused by *Phytophthora* spp.

Phytophthora parasitica, the major fungal pathogen causes qualitative and quantitative losses of betelvine (*Piper betle* L.). Among the soil borne pathogens attacking betel vine, *Phytophthora* spp. is wide spread in occurrence and probably most devastating. Incidence of two rots (foot rot and leaf rot) has been recorded in almost all betelvine plantations.

The present investigation was undertaken to study the present scenario of researches on foot rot and leaf rot of Betelvine caused by *Phytophthora* spp.

History and Geographical Distribution of *Phytophthora* sp. in Betelvine

The pathogenic effect of the genus *Phytophthora* sp. came into light with the out burst of late blight of potatoes in Island. The genus was first reported by Anton de Bery in 1876. The history including ethno-history has been mentioned in detailed (Balasubrahmaniyam *et al.*, 1994; Dasgupta and Sen., 1997). The disease in focus has been reported from almost all betelvine growing countries in the world including Indonesia, Mayanmar (Su, 1937), Sri Lanka (Paul, 1939), Bangladesh (Roy, 1948; Turner, 1969) etc.

The most destructive fungus is the foot rot caused by *Phytophthora* spp. Dastur (1926, 27) in Madhya Pradesh, reported the occurrence of foot rot caused by *Phytophthora parasitica*. The disease was noticed to be of destructive nature in Tamil Nadu (Sundaram, 1927). McRae (1928) reported this disease from West Bengal and the disease was found to be commonly prevalent in Uttar Pradesh and Assam (Chowdhury, 1944) and Madhya Pradesh (Saksena, 1977) etc. Agarwal *et al.* (1959) reported it from the leaf rot in Gupteshwar and Johri *et al.* (1981) from the vines of betelvine in Mahoba, Uttar Pradesh. In West Bengal, Dasgupta and Sen (1997, 1999) recorded the highest intensity of foot rot and leaf rot from Midnapore and Nadia districts.

Symptomatology

Dastur (1935) gave an accurate description of foot rot and leaf rot caused by *Phytophthora* spp. In *Phytophthora* induced foot rot, wet rot associated with wilting of vines is common. In leaf rot circular, brown to black, water soaked spots developed on the leaves that enlarge fast extending the rot to the petiole and stems. Leaves subsequently are shed (Dastur, 1935; Khare *et al.*, 1988). Maiti and Sen (1977) reported that under intermittent wet and dry conditions clear zonations were seen on leaf giving a wavy appearance due to shrinkage or collapse of cells during intermittent dry phase. This disease was first identified by Dastur (1935) from Maharashtra. The pathogen is a soil inhabitant (Dastur, 1935) but it perpetuates well in host tissues as dormant mycelium. Foot rot showed rotting at soil level (Saksena, 1977; Maiti and Sen, 1979; Johri *et al.*, 1984). The foot rot is more prevalent during or just after rainy season while leaf rot occurs mainly during June to October because it is most common during or just after rains (Chattopathyay and Maiti, 1990). The observations of Meherotra (1981) and Balasubrahmaniyam *et al.* (1994) were also similar to the earlier observations of Dastur (1935).

Losses

It is difficult to assess the qualitative and quantitative losses caused by individual disease. However, the extent of loss by individual disease has been recorded by several workers.

The foot rot caused by *Phytophthora* sp. infection, the extent of loss varied from 30-100 per cent (Saksena, 1977) and from 20-40 per cent due to leaf rot (Maiti and Sen, 1982). The extent of loss varied from moderate to severe (8-100 per cent) was reported from West Bengal, Assam, Uttar Pradesh, Tamil Nadu, Maharashtra (Sen *et al.*, 1984). Dasgupta and Sen (1999) reported from West Bengal that the extent of loss may vary from 5-90 per cent due to foot rot and 5-15 per cent due to leaf rot. In West Bengal Dasgupta *et al.* (2000) reported the losses of betelvine due to foot rot and leaf rot by *Phytophthora parasitica*.

Etiology and Nomenclature of *Phytophthora* sp.

There is a considerable confusion regarding the nomenclature of the species of *Phytophthora* causing diseases under consideration. McRae (1928) established the parasitism of *Phytophthora* species. Later McRae (1934) by morphological study identified as *Phytophthora nicotianae* var *parasitica*. The *Phytophthora* species reported to attack betelvine includes *P. nicotianae* var. *parasitica, P. palmivora, P. parasitica*. Turner (1969) referred all isolates of *Phytophthora* from South-East Asia as *palmivora* type and this was stated to be true for the Indian isolates as reported by (Maiti and Sen, 1977). The isolates of *Phytophthora* collected from different locations of West Bengal were found to be the isolates of *Phytophthora parasitica* (Mahanty, 2004).In spite of the efforts of Mehrotra (1984), the *Phytophthora* sp. of betelvine has not been well investigated. Techniques have been given detailed by Ribeiro (1978) and some basic observations made by him are note worthy.It was found difficult by Ribeiro (1978) to apply a natural system of classification to this genus and of the 49 species suggested probably only a dozen are valid. The biggest problem facing the taxonomist is whether particular pathotype posses sufficiently stable new characters to a species novo or belongs to a subspecies classification. This dilemma will only be resolved when the extent of biological variation within each species is known. The problem of taxonomy is further compounded by the fact that *Phytophthora* species intercross readily, producing viable oospores.

Cultural Characteristics of *Phytophthora* sp.

Phytophthora produces profuse aerial mycelial growth on carrot agar medium (CAM). Soil extract and nitrate solution are best liquid media for sporulation. Abundant oospores are formed by single isolates on host plant and agar media. Best mycelial growth of isolates of *Phytophthora* were recorded in carrot agar at 25 °C - 30 °C. Isolates of *P. capsici* produced long pedicels and ellipsoidal to ovoid sporangia. Chlamydospore formation was observed in *P. palmivora* and one isolate of *P. capsici* (Santosh *et al.*, 1995). Tiwari (1973) reported that natural media like at Oat meal agar (OMA) and PDA supported best growth and sporulation of betelvine *Phytophthora* spp. Colony characteristics of different species of *Phytophthora* are best differentiated on cultures grown on corn meal agar (CMA) medium (Ribeiro, 1978).

Tiwari (1973) reported that natural media like OMA and PDA supported best growth and sporulation of betelvine *Phytophthora*. Soil extract and nitrate solutions were best liquid media for sporulation. Optimum temperature for radial growth and sporulation was 30 °C although sporangia germinated by zoospores production at a temperature slightly below the temperature at which it had actually developed.

Chaurasia and Thind (1975) tested the effect of nine vitamins on mycelia growth of *P. parasitica* var. *piperina*. Amongst nine vitamins tested, Folic Acid and Nicotinic acid ware found to be most favourable vitamins for the fungal growth. A limited amount of growth stimulation at 10, 25 and 50 ppm. concentrations were also found in case of ascorbic acid, pyridoxine, biotin, thiamine, riboflavin caused a significant inhibitory effects on growth. Contradictory report was observed by Ridings *et al.* (1969) who concluded thiamin requirement as a means for distinguishing *Pythium* from *Phytophthora*. Isolates growing in a glucose asparasgine medium without thiamine was likely to belong to the genus *Pythium*. Vyas and Chourasia (1978) worked on the formation of zoospores *in vivo* by *P. parasitica* var. *piperina*. The authors for the first time unequivocally demonstrated the formation of organics and oospores by betelvine *Phytophthora* infected host leaves. Ribeiro (1978) made a detail investigation of colony characteristics of different species of *Phytophthora* sp. based on cultures grown on corn meal agar (CMA) medium.

Morphological Characteristics of *Phytophthora* sp.

McRae (1928) established the parasitism of *Phytophthora* sp. on betelvine. Later McRae (1934) by morphological study identified that as *Phytophthora nicotianae* var *parasitica*. McRae (1934) reported that the length × breadth of sporangia of *P. parasitica*, ranged in 19 – 58 × 15 – 35 µm. Chlamydospore ranged between 20 – 35 µm and oospores between 14 – 40 µm and brown in colour. Dastur (1935) observed that the sporangia of *P. parasitica* var *piperenina* length × breadth ranged in 30 – 65 × 20 – 40 µm, oospores ranged between 18 – 30 µm, yellow in colour and chlamydospore not observed.

Thomas *et al.* (1948) studied in detail the morphology and stated that oospores production is the major criterion for the identification of *Phytophthora* sp.

According to Maiti and Sen (1977) the sporangial length × breadth of *Phytophthora palmivora* are 23 – 66 × 15 – 40 µm, chlamydospore ranged between 20 – 40 µm and oospores between 17 – 40 µm and light brown in colour.

In general characteristics of the genus *Phytophthora* as recorded by Ribeiro (1978) consist of mycelium which are hyaline, branched, coenocytic becoming septate with age, variable in diameter (5 – 8 µm), sometimes swollen, nodose or tuberculate. Hyphal diameter depends on physicochemical nature of the medium and on the nature of formation *i.e.* wheat on the surface, aerial, sub-merged or within the host cell. Hyphae branched at right angles (not in all case) and are often constricted at the base. Asexual reproduction is done by producing sporangia and/or chlamydospores. Sporangia, produced on sporangiophores, are variable in size and shape, usually subspherical, ovoid, limoniform, pyriform or obpyriform, hyaline to light yellow, may contain papilla. Chlamydospores are spherical to ovoid non-papillate, hyaline to deep brown, thin to thick walled, terminal or intercalary. Sexual reproduction is by antheridial and oogonial interaction resulting in the formation of oospores.

However, according to Balasubrahmaniyam *et al.* (1994) the sporangia one hyaline, papillate, range between 18 × 40 µm and 28 × 62 µm and do not produce any chlamydospore.

Pathogenecity Test

Pathogenecity of pathogens has been worked out severally and are briefly summarized below:

Pathogenecity of *P. capsici* and *P. palmovora* was studied on betelvine by inoculating plant material using either zoospore or mycelial suspension and the pathogen were reisolated. Typical lesions developed within 48 – 72 hrs of inoculation (Maiti, 1977). Pathogenecity of *Phytophthora parasitica* was studied on betelvine by inoculating excised leaves and stems using sporangial suspension and the pathogen was reisolated (Mahanty, 2004; Datta, 2012).

Survival of *Phytophthora* spp.

Mahmud (1941) and Chowdhury (1944) claimed to isolate the pathogen from the infected orchard soil. However, Mehrotra and Tiwari (1967) could not isolate the pathogen even from the artificially infected soil and they stated that the pathogen may survive in the soil as chlamydospores. *Phytophthora* species are not considered to be strong saprophytes and rarely survive in soil as mycelium if hosts or plant debris are removed from the field. Mycelium often survives by growing from year to year in living host tissues or in crop residues (Ribero, 1978). However, Newhook (1959) reported that *Phytophthora* has been found to persist in soil for at least 2 years after the death of trees in conifer forest. Other investigators have found that *Phytophthora* mycelium cannot survive over winter as a saprophyte.

Vyas and Chaurasia (1978) isolated the fungal oospores from infected soil and stated that the pathogen survives as oospores in the soil. On the other hand, Hickman (1958) reported that oospores of *P. fragariae* survived one and half year in non sterile soil. According to Bowers *et al.* (1990) oospores may be the over wintering propagules of *P. capsici.*

There are conflicting reports on the value of chlamydospores as survival structures in the life cycle of fungi. Ribeiro (1978) claimed that the chlamydospore is a primary unit of survival for *Phytophthora*. Wang *et al.* (1998) found that chlamydospores of *P. nicotiane* survived in soil for eight month. The infected host plant also plays a very important role in the survival of *Phytophthora* sp. (Liyanage and Wheeler, 1991).Agarwal *et al.* (1999) found that the survival period of about 10 months in the presence of host tissue is sufficient for carrying over the fungus from one year to next. However, the pathogen is predicted to survive less than one year due to its lack of saprophytic ability to colonize non-host plant (Quitugna and Trujillo, 1998).Temperature, pH and irrigation also play the very important role in the survival of *Phytophthora* sp. Thareja *et al.* (1989) reported that the maximum disease incidence occurred at 20 – 25 °C, average relative humidity of > 60 per cent under high rainfall and high soil moisture condition. High temperature (> 40 °C) is harmful for survival of *Phytophthora* (Hansen and Hamm, 1996) on the other hand low temperature is beneficial for the survival of *Phytophthora* (Wang *et al.*, 1998). Regarding pH, it is very critical for sporangia production of some species of *Phytophthora*. In general, high pH is toxic to sporangia. However, conflicting reports exist on the importance of pH on chlamydospore formation. Huguenin (1974) found no significant difference between pH 4.5 and 7.0 for chlamydospore formation and survival of it, while Chee (1973) reported that a pH of 5.8 is optimum. On the other hand Andrivon (1994) found that pH 5.4 were not suitable for survival *P. parasitica* in potato.

Soil moisture also plays a vital role in the survival of *Phytophthora* sp. It has been found that disease severely in infested plots increased as flooding frequency increased (Fallon *et al.*, 1991). Mahanty *et al.* (2011) worked on the effect of Irrigation Regime and IW:CPE Ratio on survival of *Phytophthora parasitica* (Dastur) causing foot rot and leaf rot of betelvine concluded that when interval of application of irrigation was increased, the survival of *P. parasitica* was also increased and depth of irrigation had no significant effect on survival of the pathogen. Ristaino *et al.* (1989) found that less frequency irrigation of infested plots of tomato caused a delay in disease on set *i.e.* the population of *Phytophthora* sp. remains low. So, it may be concluded that less frequent furrow irrigation is affective way to reduce the loss due to *Phytophthora* sp. (Cafe-Filho *et al.*, 1995).

The survival of *Phytophthora* spp. was studied under control condition with soil pH (5.4, 7.0 and 8.5), soil moisture (air dry, 25, 50 and 100 per cent moisture level) and temperature (10, 20, 30 and 40 0C) on survival ability of three isolates of *Phytophthora parasitica* (P21, P13, P8) causing foot rot and leaf rot of betelvine. The results showed that population of the pathogen increased up to first 2 weeks and then decreased gradually and disappeared within 13 weeks. This may be due to the germination of different spore forms and subsequent proliferation under natural condition with or without food base. *Phytophthora* could survive more than 10 weeks at moderate soil moisture level (50 per cent) and at a temperature range of 10-30 0C. High temperature of >40 0C is unfavourable for the survival of *P. parasitica*. At 50 per cent moisture, isolates 21 and 13 survived up to 13 weeks at 30 0C, while at 10 0C, all these isolates survived up to 14 weeks and at pH 5.4 all isolates had higher growth (Dasgupta *et al.*, 2012).

Epidemiology of *Phytophthora* sp.

Epidemiological studies under field condition were carried out by Stepwise multiple regression analysis of Per cent disease incidence of leaf and foot rot caused by *Phytophthora* sp on different varieties of betelvine (Bidhan Pan – I, Simurali Deshi and Simurali Gol Bhavna). The results of disease incidence during 2006-07, 2007-08 and 2008-09 were compiled to find out the important meteorological parameters which were responsible for incidence of both the diseases. The results revealed that among the five considered meteorological factors, minimum temperature was more responsible for increase the disease incidence or spread of the disease of foot rot and leaf rot disease in all the three varieties and in all the transformation models and in original data. In variety Simurali Deshi, in addition to min. temp. and max. RH was also responsible for increase the disease incidence or spread of the disease of foot rot disease in all the transformation models. It was also revealed that among the two transformation models for disease prediction Logit transformation fit best for disease prediction and it was confirmed high R^2 value. So, it was concluded that with increase in minimum temperature (› 26.5°C) foot rot and leaf rot disease incidence increase significantly in case of Ghanagette and Simurali Gol Bhavna varieties whereas increase in minimum temperature and maximum relative humidity (96.5 per cent) there was a significant increase in disease incidence of leaf rot of betelvine in Simurali Deshi variety (Datta *et al.*, 2009). The validation of the above results were made by further fitting the data against the disease incidence of foot rot and leaf rot of betelvine in 2009-10 using the effective meteorological parameters of the same year for validation of these equations and models. The results showed poor estimated value during initial disease incidence. Among the two diseases (foot rot and leaf rot) the estimated value of leaf rot incidence was nearer to original value. It was observed that in all prediction equation like Logit, Gompertz and Per cent transformation, after 10th week of observations, the estimated value was similar to that of original value. It was more validated on leaf rot disease but not in foot rot disease. Among the three varieties tested, in Ghanagette and Simurali Deshi varieties, the prediction equations which were prepared were also showed some validation in predicting the disease incidence of betelvine (Datta and Dasgupta, 2013). Datta (2012) compiled the results of 2006-07, 2007-08 and 2008-09 and validated with the year 2009-10 revealed that among the two diseases (foot rot and leaf rot) the estimated value of leaf rot incidence was nearer to original value. It was observed that in all prediction equation like Logit, Gompertz and Per cent transformation, after 8th week of observations, the estimated value was similar to that of original value. It was more validated on leaf rot disease than foot rot disease. Further, results of validation during the year 2007-08, 2008-09 and 2009-10 with 2010-11 showed poor estimated value during initial disease incidence. Among the foot rot and leaf rot diseases the estimated value of foot rot incidence was nearer to original value. It was observed that in all prediction equation like Logit, Gompertz and Per cent transformation, the estimated value was similar to that of original value in most of the observations (Datta, 2012; Dasgupta, 2013).

Management of *Phytophthora* Diseases of Betelvine

Biological Control

Of the two approaches of biological control *viz.* indirect through manipulation of associated microbiota by amendments and direct use of antagonist (Baker and Cook, 1974), the first that is the use of oil cake in betel vine plantation is an age old practice. Earlier workers (Tiwari and Mehrotra, 1968, Mehrotra and Tiwari, 1976) showed that dipping of cuttings in *T. viride* cell suspension effectively reduced the disease. Similarly use of corn straws and til oil cake also reduced disease (Mehrotra and Tiwari, 1976).

The effect of *Trichoderma viride* and plant extracts was tested against foot and leaf rot of *Piper betle* caused by *Phytophthora parasitica* var. *piperina* [*P. nicotianae* var. *parasitica*] by Yadav *et al.* (1998). The spore suspension of *T. viride* (10^2 spores/ml), leaf extracts of *Azadiracta indica* (2 per cent), *Pongamia pinnata* (3 per cent) completely inhibited sporulation of pathogen whereas spore suspension of *T. viride* (10^4 spores/ml) and 4 per cent concentration of leaf extracts of *A. indica, P. pinnats* completely stopped the mycelia growth of the fungus.

Anonymous (1997-98 and 1998-99 and 1999-2000) reported significant reduction of foot rots of betelvine with an isolates of *Trichoderma harzianum*. Leandro *et al.* (2004) went through the experiment where *Trichoderma* sp. populations monitored in field soil, it did not affect on root health but improved leaf area. Root rot pathogens (*viz. Phytophthora* sp. and *Pythium* sp.) were less frequently than untreated controls. Mio *et al.* (2002) showed the effect of *Trichoderma* spp. and solarization is effective in controlling *Phytophthora parasitica* during winter and summer. Evaluation was conducted by bait test to recover the pathogens. The results showed that the application of *Trichoderma* spp. along with the solarization of substratum for seedling production eliminates *Phytophthora parasitica* in solar collectors, solarized for 24 hours in winter and 48 hours in summer. Dasgupta *et al.* (2003) conducted a multilocational field trial on biological control of *Phytophthora* foot rot revealed that four soil applications of *Trichoderma* sp. through in oil cake at quarterly intervals significantly reduced the disease intensity (*viz.* leaf rot and foot rot of betelvine) and also increased the leaf yield.

Integrated Disease Management

In this way of management programme a promising combination of the fungicides/pesticides along with the biocontrol was determined. In few rows only chemical fungicides were applied and in other rows bioagents and fungicides were combinely applied. *Piper betle* plants were sprayed (at 4 ml/lit) with the potassium salt of phosphoric acid (potassium phosphate) at monthly intervals. It is concluded that potassium phosphonate should be used for the control of *Phytophthora palmivora*, as it had no phytotoxic effects (Johri and Chaurasia, 1998).

Dasgupta *et al.*(1999) tested three promising fungicides (fosetyl, chlorothalonil, Bordeaux Mixture) and phosphorus acid against 2 types of rots (root and foliage) caused by *Phytophthora* sp. The result showed that application of 1 per cent Bordeaux Mixture at monthly intervals reduced disease intensity, application of 0.3 per cent fosetyl-Al gave significantly higher leaf yield at bimonthly intervals and 0.4 per cent monthly application of phosphorus acid increased the fresh weight most significantly. Mahanty *et al.* (2008) conducted a field trial for two consecutive years for the management of two different rots of betelvine caused by *Phytophthora parasitica* using four different fungicides (*viz.* Mancozeb, Fosetyl-Al, Chlorothalonil, Bordeaux Mixture) and phosphorus acid (Akomin). The best treatment was Bordeaux Mixture at monthly interval led to the lowest foot rot (8.19 per cent) and leaf rot (10.74 per cent) disease incidence, the 2nd best treatment was application Fosetyl-al at monthly interval. The highest leaf yield and fresh weight of 100 leaves were obtained with Bordeaux Mixture application and also recorded highly remunerative and differed significantly from other treatments. In field experiment spray of mancozeb + carbendazim (0.2 per cent) was found effective with 33.38 per cent disease control (Patil *et al.*, 2009).

A salient literature on field trials using chemical fungicides in the control of *Phytophthora* rots on betelvine is listed below:

Sl.No.	*Chemical*	*Mode*	*References*
1.	Burgundy Mixture	Fungicide	Thompson, 1933
2.	Bordeaux Mixture	Preaching and Springing	McRae, 1934; Patil *et al.*, 1986; Dasgupta and Sen, 1996; Annonymous, 1998-2004
3.	Cuprous Oxide	Drenching	Chattopadhyay, 1967
4.	Blitox, Dexon, DM 45	Drenching	Maiti *et al.*, 1978
5.	Ridomil and Daconil	Drenching	Ramraj and Vidyasekharan, 1983
6.	Chlorothalonil and Fosetyl –Al	Drenching	Mahanty, 2004

Reports on IPM approach for *Phytophthora* diseases of some other crops are available (Anandaraj and Sarma, 1994; Chavez *et al.*, 1995) but these are not necessarily applicable to betelvine.

Maiti (1999) has meticulously identified the component of IDM that are now available for the *Phytophthora* diseases of betelvine, including disease resistance, plant nutrition, sanitation, chemical and biological control. The constraints to IPM and IDM development are identified as lack of forecasting system, non-availability of tools for breeding either conventionally or biotechnologically and a sketchy understanding of biological control system.Efforts on IDM of betelvine were initiated under AICRP on Betelvine project at several centers and some prima facie improvement in disease management and yield increase were reported (Anonymous, 1997-98, 1998-99 and 1999-2000). Mahanty *et al.* (2000) showed that the effectiveness of fungicides Akomin (phosphorus acid at 4 ml/lit), kavach (Chlorothalonil at 0.18 per cent), Aliette (Foretyl-al at 0.24 per cent), Dithane M-45 (Mancozeb at 2.5 g/lit) and Bordeaux Mixture (1 per cent $CuSO_4$) and 3 isolates of *T. harzianum* (T_3, T_{15}, T_{18}) against *Phytophthora parasitica* causing foot rot in betelvine. T_3 isolate mixed with Bordeaux mixture at monthly soil drenching was best combination of fungicide along with the biological agent.

Roy (2001) found that when Bordeaux Mixture was used in combination with earthing-up, sanitation and biocontrol agent (*T. harzianum*) the leaf and foot rot incidence of betelvine were reduced. Sanyal *et al.* (2002) reported the results of field trials in West Bengal, India, in 1990-2000 and 2000-2001 on biological (using *T. harzianum*) chemical (Bordeaux Mixture) and integrated control of *Phytophthora parasitica* and *Collectotrichum capsici* on betelvine.

Magdum *et al.* (2009) showed the minimum per cent disease incidence (PDI) of foot rot and leaf rot was 6.84 and 1.85 respectively due to the treatment with sanitation + one soil application of 1.0 per cent Bordeaux Mixture. The treatment also resulted highest (58.93 lac/hac leaves) yield and C : B ratio (1 : 2.60). Mohanty *et al.* (2011) reported that integrated use of sanitation, Bordeaux mixture and *T. harzianum* gave the best control of foot rot and leaf rot of betelvine. The treatment yielded maximum leaf production (3.448 million ha-1year-1) and the highest fresh weight of 100 leaves (351.85 g). The economic analysis showed that application of Bordeaux mixture along with *T. harzianum* and sanitation may be useful to obtain satisfactory control of both the diseases and economic benefit showing BCR of 1.48:1.

Conclusion

Although lot of information regarding management of *Phytophthora* diseases of betel vine have been generated by the AICRP on Betelvine and different workers but till date all these are not enough to combat this dreaded disease of betelvine, so it requires much more detailed study to make them fruitful to the farmers. There are ample scopes of research in the field of management of the *Phytophthora* diseases of betel vine. In IDM programme, the sources of initial inoculums need to be identified along with possibilities of their elimination. There are several reports that some co-lateral hosts harbour the pathogens. The irrigation water from the adjoining ponds may be the source of inoculum. The growers usually wash the infected material in nearby reservoirs and irrigate their crops with the same water, thus enabling the recurrence of the disease. So, the actual means of survival within and in the immediate vicinity of betelvine conservatory should be identified first and then factors affecting inoculum build-up and their dispersal need to be searched. Investigation on the presence of different mating types of *Phytophthora* sp. in the infected betelvine gardens is very much essential because it may lead to the production of vigorous strains of *Phytophthora* sp. causing epidemic.*Trichoderma* sp. was found to be an effective antagonist under *in-vitro* conditions. In the field of bio-control agent, the use of genetically engineered strains (Mutant) of *Trichoderma* sp. should be tried to over come the problems faced for the rhizosphere competence and less secretion of antifungal enzymes. The application of bio-protectant may be used on the foliage and also near the collar region which is very much sensitive to foot rot pathogens. The management strategy thus obtained should be confirmed following *in situ* farmer's plot demonstrations.Lastly it can be said that in the near future with the help of bio-technology disease free planting material and the varieties of betelvine, resistant to draught, flood and diseases will be developed for the benefit of growers.

References

Agarwal, G. P., Nama, K. G. and Bdiram, R. (1959). Fungi Causing plant diseases at Jabalpur, Madhya Pradesh. *J. Proc. Nath. Acad. Sci., India*, **29**: 310-315.

Agarwal. A., Priya Rani; Gurinderjit Kaur and Gupta, A. K. (1999). Aggressiveness and competitive fitness of *Phytophthora infestans* isolates collected from Potato and tomato in France. *Phytopathology*, **89**(8): 679-686.

Anandaraj, M. and Sarma, Y. R. (1994). Biological control of black pepper diseases. *Indian Cocoa, Arecanut and Spices J.*, **18**(1): 22-23.

Andrivon, D. (1994). Dynamics of the survival and infectivity to potato tubers of sporangia of *P. infestans* in three different soils. *Soil Biology and Biochemistry*, **26**(8): 945-952.

Anonymous (1997 & 1998). Proceedings of the fifth group meeting of the All India Coordinated Research Project on Betelvine. *Indian Institute of Horticultural Research, Bangalore*, India, pp. 143.

Anonymous, (1998 and 1999). Annual Report. All India Coordinated Research Project on Betelvine. *Indian Institute of Horticultural Research, Bangalore*, India, pp. 159.

Anonymous, (1999 and 2000). Annual Report, All India Net Working Project on Betelvine. National Research Centre for Medicinal and Aromatic Plants, Boriavi, Anand, Gujarat, India, pp. 165.

Anonymous (2000 and 2001). Annual Report, All India Net Working Project on Betelvine. National Research Centre for Medicinal and Aromatic Plants, Boriavi, Anand, Gujarat, India, pp. 137.

Anonymous (2001 and 2002). Annual Report, All India Net Working Project on Betelvine. National Research Centre for Medicinal and Aromatic Plants, Boriavi, Anand, Gujarat, India, pp. 152.

Anonymous (2002–2003 and 2003-2004).Biennial Report, All India Net Working Project on Betelvine. National Research Centre for Medicinal and Aromatic Plants, Boriavi, Anand, Gujarat, India, pp. 157.

Baker, K. F. and Cook, R. J. (1974). Biological Control Plant Pathogens, APS: St Paul.

Balasubramaniyam, V. R., Johri, J. K., Tripathi, R. D., Rawat, A. K. S. and Chaurasia, R. S. (1994). Betelvine (*Piper betle* L.) National Botanical Research Institute (NBRI), Lucknow, pp. 173.

Bowers, J. H., Papavizas, G. C. and Johnston, S. A. (1990). Effect of soil temperature soil water matric potential on the survival of *Phytophthroa capsici in Natural Soil. Plant Diseases*, **74**(10): 771-777.

Cafe Filho, A. C., Duniway, J. M. and Davis, R. M. (1995). Effects of the frequency of furrow irrigation on root and fruit rots of Squash caused by *Phytophthora capsici*. *Plant Disease*, **79**(1): 44-48.

Chattopadhyay, S. B. (1967). Diseases of plant yielding Drugs, Dyes and Spices, ICAR, New Delhi.

Chattopadhyay, S. B. and Maity, S. (1990). Diseases of betelvine and spices, New Delhi, Indian Council of Agricultural Research (ICAR), pp. 160.

Chaurasia, S. C. and Thind, T. S. (1975). Activity of some antibiotics against *P. parasitica* var. *piperina* incitant of foot-rot and leaf rot of Pan. *Hindusthan Anti. Bull.*, **18**: 28-30.

Chavez Alfaro, J. J., Zavaleta Mejia, E. and Telizortiz, D. (1995). Integrated control of blight in pepper (*Capsicum annuum* L.) caused by the fungus *P. capsici* Leo., in the Vatesequillo region Puebla. Mexico. *Fitopathologia*, **30**: 47-55.

Chee, K.H. 1973. Production, germination and survival of chlamydospores of *Phytophthora* from *Hevea braziliensis, Traqus. Brit. Mycol. Soc.*, **61**: 21-26.

Chowdhury, S. (1944). Diseases of 'Pan' (*Piper betle* L.) in Seylet, Assam, Assam-II. *Phytophthora* foot rot and leaf rot. *Proc. Indian Acad. Sci. Section-B*. **19**: 152-164.

Dasgupta, B. (2013). Epidemiological Studies of Foot rot and Leaf rot of Betelvine (*Piper betle* L.) caused by *Phytophthora* sp. under closed conservatory condition in West Bengal. Proceedings of the International Conference on Climate Change Impact and Adaptation. DUET, Gazipur, Bangladesh, 14-16 November, 2013, pp.613-621.

Dasgupta, B. and Sen, C. (1996). Loss assessment and use of chemicals in the management of *Phytophthora* root rot of betelvine. *Indian J. Mycol. Plant Pathol.*, **26:** 37-39.

Dasgupta, B. and Sen, C. (1997). Betelvine diseases and their management – A retropretrospective in perspective, . In: (Ed. M. K. Dasgupta) Pest Management in Changing Agricultural Situation, Viswa Bharati, Sriniketan, pp. 43-50.

Dasgupta and Sen, C. (1999). Assessment of *Phytophthora* root rot of betelvine and its management using chemicals. *J. Mycol. Plant, Pathol.*, **29**: 91-95.

Dasgupta, B., Roy, J. K. and Sen, C. (2000). Two major fungal diseases of betelvine. In Diseases of Plantation Crops, Spices, Betelvine and Mulberry (Ed. M. K. Dasgupta), pp. 133-137.

Dasgupta, B., Dutta, P. K., Muthuswamy, S. and Maiti, S. (2003). Biological Control of *Phytophthora* foot rot of betelvine (*Piper betle* L.). *Journal of Biological Control*, **17**(1): 63-67.

Dasgupta, B., Mohanty, B. and Datta, P. (2012). Survival of *Phytophthora parasitica* causing Foot and Leaf rot of Betelvine under different Soil pH, Moisture and Temperature regime. *SAARC J. Agri.*, **10**(1): 29-43.

Dastur, J.F.(1926.) Report, Department of Agriculture, C.P. and Berar 1924-1925.pp. 23-46.

Dastur, J.F.(1927). A short note on the foot rot diseases of pan in central provinces. *Agric. J. India*, **22**: 105-108.

Dastur, J. R. (1935). Diseases of Pan (*Piper betle*) in the Central Provinces. *Proc. Indian Acad. Sci.*, **1**(11): 26-31.

Datta Partha. (2012). Epidemiology and Management of Rots of Betelvine (*Piper betle* L.) caused by *Phytophthora parasitica* Ph. D. thesis. Dept. of Botany, Kalyani University, Nadia.

Datta, P., Dasgupta, B. Sengupta, C. and Sengupta, D. K. (2009). Epidemiological studies of stem rot of betelvine caused by *Phytophthora parasitica* under closed conservatory condition in West Bengal. *J. of Crop and Weed Sci.,* **5**(1): 258-260.

Datta Partha and Dasgupta, B. (2013). Testing and validation of developed Disease prediction model of Foot rot and Leaf rot of betelvine (*Piper betle* L.) caused by *Phytophthora* sp. In three varieties of betelvine under closed conservatory condition in West Bengal. *J.Mycopathol.Res.* **51** (1): 29-37.

Fallon, P. G., Greathead, A. S.; Mullen, R. J.; Benson, B. L. and Grogan, R. G. (1991). Individual and combined effects of flooding, *Phytopathora* rot and metalaxyl on asparagus establishment. *Plant Diseases,* **75**(5): 514-518.

Hansen, E. M. and Hamm, P. B. (1996). Survival of *Phytophthora lateralis* in infected roots of Port Oxford Cedar. *Plant Diseases.* **80**(9): 1075-1078.

Hickman, C.J. (1958). *Phytophthora*-plant destroyer. *Traqus. Brit. Mycol. Soc.,* **41**: 1-13.

Hugenin, B. (1974). Annales Phytopathol. Paris, 6: 425-440. (Cited in "A Ssource Book of Genus *Phytophthora*", by Ribeiro, O.K. 1978. Vaduz, J. Cramer, Germany, pp. 137.

Johri, J. K. and Chaurasia, R. S. (1998). A phosphonate for management of betelvine *Phytophthora. National Academy Science Letters,* **21**(7/8): 237-239.

Johri, J. K., Chaurasia, R. S. and Balasubramanyam, V. R. (1981). Fungal diseases in mahoba area – A Report. Proceedings Improvement of Betelvine cultivation. February, 27-28, 1981. NBRI, Lucknow (Edit S. D. Khanduja and V. R. Balasubramayam) pp. 78-83.

Johri, J. K., Chaurasia, R. S. and Balasubramanyam, V. R. (1984). Fungal disease in Mohoba area – A report proceeding improvement of betelvine cultivation. NBRI, Lucknow, pp. 78-83.

Leandro, L., Fergunson, L., Fernandez, G. and Louws, F. (2004). Evaluation of *Trichoderma* strains as biocontrol tools for integrated management of strawberry root rot. *Bulletin – OILB/SROP. 2004;* **27**(8): 229-233.

Liyanage, N. I. S. and Wheeler, B.*E.G.* (1991). Survival of *Phytophthora meadii* in Srilankan soil. *Plant Pathology,* **40**(3): 436-444.

Magdum, S. G., Shirke, M. S., Kamble, B. M., Salunkhe, S. M. and Tambe, B. N(2009). Integrated management of major diseases of betelvine. *Advances in Plant Sciences.* **22**(1): 35-36.

Mahanty, B. (2004). Some Biological aspects of *Phytophthora* sp. causing rots of Betelvine (*Piper betele* L.) and its management. Ph. D. thesis. Dept. of Plant Pathology, BCKV, Nadia, p. 48.

Mahanty, B. and Dasgupta, B. (2008). Management of foot rot and leaf rot of betelvine (*Piper betle*) caused by *Phytophthora parasitica* by using safer fungicides. *Journal of Mycopathological Research,* **46**(1): 81-84.

Mohanty Bibekananda, Datta Partha, Dasgupta B. and Sengupta Dalim Kumar. (2011). Integrated Management of Foot rot and Leaf rot of Betelvine. *SAARC J. Agric.* **9**(2): 83-91.

Mohanty Bibekananda, Dasgupta Basudeb and Datta Pranab Kumar.(2011). Effect of irrigation regime and IW: CPE ratio on survival of *Phytophthora parasitica* Dastur causing foot and leaf rot of betelvine (*Piper betle* L.). *J.Mycopathol.Res.* **49** (2): 353-355.

Mahanty, B., Roy, J. K., Dasgupta, B. and Sen, C. (2000). Relative efficacy of promising fungicides and biocontrol agent *Trichoderma* in the management of foot rot of betelvine. *Journal of Plantation Crops,* **28**(3): 179-184.

Mahmud, K.A. (1941). Report of Pan cultivation scheme, central provinces 1940-41. Govt. Printing Press, Nagpur, pp. 15.

Maiti, S. (1977). Diseases of *Piper betle* L. with emphasis on the foot rot caused by *Sclerotium rolfsii* Sacc. and its control. Ph. D. Thesis. B. C. K. V., Kalyani, India, pp. 181.

Maiti, S. (1999). Integrated disease management system in Betelvine. In: IPM system of Agriculture (Ed. R.K. Upadhyay; K.G. Mukherji and O.P. Dubey). Aditya Books Pvt. Ltd., New Delhi, India, vol. 7, Cash Crops, pp.253-261.

Maiti, S. and Sen, C. (1977). Leaf and foot rot of *Piper betle* caused by *Phytophthora palmivora*. *Indian Phytopath.,* **30**: 438-439.

Maiti, S. and Sen, C. (1979). Fungal diseases of betelvine. *PANS,* **25**: 150-157.

Maiti, S. and Sen, C. (1982). Incidence of major diseases of betelvine in relation to weather. *Indian Phytopathology,* **35**: 14-17.

Maiti, S., Khatua, D. C. and Sen, C. (1978). New leaf spot disease of betelvine. *Indian J. Mycol. Pl. Pathol.,* **8**: 217.

May De Mio., L. L., Ghini, R. and Kimati, H. (2002). Solarization to control *Phytophthora parasitica* on citrus seedling. *Fitopathologia-Brasileria.* **27**(3): 254-258.

McRac, W. (1928). Report of the Imperical Mycologist Scientist. Rept. Res. Inst., Pusa, 1927-28, pp. 56-57.

McRae, W. (1934). Foot rot diseases of *Piper betle* L. in Bengal. *Indian J. Agric. Sci.,* **4**: 585-617.

Mehrotra, R. S. (1981). Fungal diseases of betelvine and their control. In Proc. Group Discussion Improvement of Betelvine Cultivation. NIBRI, Lucknow, pp.3-12.

Mehrotra, R. S. (1984). Fungal diseases of betelvine and their control. In Proc. Improvement of Betelvine cultivation (Eds. S. D. Khanduja, V. R. Balasubrahmanyam) NBRI, Lucknow, pp. 3-12.

Mehrotra, R. S. and Tiwari, D. P. (1967). Studies on the foot rot and leaf rot of *Piper betle* L. behavior of *Phytophthora parasitica* var. *piperina* in soil. *Indian Phytopath.,* **20:** 161-167.

Mehrotra, R. S. and Tiwari, D. P. (1976). Organic amendments and control of foot rot of *Piper betle* caused by *Phytophthora parasitica* var. *piperina*. *Annals, Microbial.*, **27**: 415-421.

Newhook, F.J. 1959. The association of *Phytophthora* spp. with mortality of *Pinus radiate* and other conifers. *New Zealand Journal of Agricultural Research*, **2** (4): 808-843.

Patil, C. U., Zape, A. S. and Wathore, S. D. (2009). Efficacy of fungicides and bioagents against *Collectotrichum glocosporioides* causing blight in *Piper songum*. *International Journal of Plant Protection*, **2**(1): 63-66.

Patil, M. R., Waghe, S. V., Wagnikar, P.D. and Khune, N. N. (1986). Chemical control of wilt of betelvine. *Pesticides*, **20:** 28-29.

Paul, W.R.C. (1939). A leaf spot disease of betelvine, *Div. Plant Pathol. Adm*. Reptr. Div. Agric., Ceylon, pp. 41-45.

Quitugna, R. J. and Trujillo, E. E. (1998). Survival of *Phytopathora colocasiae* in field soil at various temperatures and water matric potentials, *Plant Disease*, **82**(2): 203-207.

Ramraj, B. P. and Vidyasekharan, P. (1983). Control of betelvine wilt. *Pesticides*, **17**(1): 8.

Riberio, O.K. 1978. A source of the genus *Phytophthora*, J. Cramer, Vaduz Liechtenstein, p. 417.

Ridings, W.H., Gallegly, M. E. and Lilly, V. G. (1969). Thiamine requirement as a means of distinguishing *Pythium* from *Phytophthora*. *Phytopathology*, **59**: 737-742.

Ristaino, J. B.; Duniway, J. M. and Marois, J. J. (1989). *Phytophthora* root rot and irrigation schedule influence growth and technology of processing tomato. *Journal of the American Society for Horticultural Sciences*, **114**(4): 556-561.

Roy, J. (2001). Management of two important diseases of betelvine (*P. betle* L.) with special emphasis on biological control. Ph. D. Dissertation, BCKV, Nadia.

Roy, T. C. (1948). Anthracnose disease of betelvine (*Piper betle* L.) caused by *Colletotrichum dastur*, Roy in Bengal. *J. Indian Bot. Sco.*, **27**: 96-102.

Saksena, S. B. (1977). *Phytophthora parasitica*, the scourge of 'Pan'. *Indian Phytopathology*, **30**: 1-16.

Santosh, A.F. D., Matsuoka, K., Alfenas, A.C., Maffia, L.A., Dos-Santos, A.F. (1995). Identification of *Phytophthora* spp. That infect *Hevea* sp. Fitopathologia – *Brasilier*, **20**: 151-159.

Sanyal, B., Sengupta, C., Poi, R., Dasgupta, B. and Sen, C. (2002). Management of major betelvine diseases using combination of pesticide and biocontrol agent. *Journal of Plantation Crops*, **30**(3): 48-52.

Sen, C., Maiti, S. and Khatua, D. C. (1984). Control of root rot and foot rot diseases of betelvine caused by fungal pathogens. Proceedings Improvement of Betelvine Cultivation NBRI. Lucknow, Feb. 27-28, pp. 33-46.

Su, M. T. (1937). Report of Mycologist, Mandalay for the year ended the 31 st March, 1973, p. 9.

Thareja, R. K.; Srivastava, M. P. and Singh, S. (1989). Epidemiology of buckeye rot of tomato caused by *Phytophthora nicotianae* var. *parasitica. Indian Phytopathology,* **42**(4): 582-584.

Thomas, K. M., Krishna Menon, K. and Ramakrishnan, T.S. (1948). Studies in the genus *Phytophthora, 11 Proc. Indian Acad. Sci.* B., **27:** 55-73.

Thompson, A. (1933). Division of Mycology. Annual Reports for 1931. Dept. Agric. Straits Settlements and Feb. Malaya States (Technical Report of the year 1931). Bull., 12 Gen., Ser., pp.48-52.

Tiwari, D. P. and Mehrotra, R. S. (1968). Rhizosphere and Rhizoplane studies of *Piper betle* L. in Sarwak. Trans. *British Mycol. Soc.*, **52**: 411-418.

Tiwari, D.P. (1973). Studies of foot rot and leaf rot of *Piper betle,* IV Factors influencing growth and sporulation in betelvine *Phytophthora* (*P. nicotiana* var. *parasitica. Indian Phytopathology,* **26**(3): 456-468.

Turner, G. J. (1969). *Phytophthora pulmivora* from *Piper betle* L. in Sarawak. *Trequs. Br. Mycol. Soc.,* **52**: 411-418.

Vyas, K. M. and Chaurasia, S. C. (1978). Formation of oospores *in vivo* by *Phytophthora parasitica* var. *piperina* Part. *Phytophthora News* 1. USA, **6**: 42-43.

Wang, G.; Zheng, Xd. B.; Lu, J.Y. and Li, T.F. (1998). Survival of chlamydospores and mycelia of *P. nicotianae* in soil. *J. Nanjing Agril. Univ.,* **21**(1): 41-45.

Yadav, B. P., Yadav, M. K. and Singh, S. N. (1998). Effect of plant extracts and *Trichoderma viride* on *Phytophthora parasitica* var. *piperina,* the stem and leaf rot pathogen of betelvine. *Journal of Applied Biology,* **8**(1): 95-99.

2015, Major Constrains and Verdict of Crop Productivity *Pages 198–214*
Editor: Dr. U.N. Bhale
Published by: ASTRAL INTERNATIONAL (P) LIMITED, NEW DELHI

Chapter 13

Problems and Prospects of Lentil (*Lens culinaris* Medik.) Cultivation in the Gangetic Plains of West Bengal

V. Yeka Zhimo*, Jayanta Saha, Birendranath Panja and Narayan Roy

Department of Plant Pathology, Bidhan Chandra Krishi Viswavidyalaya, Mohanpur – 741 252, West Bengal, India

Introduction

Lentil (*Lens culinaris* Medik.), one of the oldest and the only cultivated species of *Lens* genus, of the family *Leguminosae,* is a short, slender, bushy annual plant with a fairly erect growth habit (Webb and Hawtin, 1981). It is popular for its lens shaped seeds, which are consumed as food all over the world in stew or other forms for its protein, minerals (K, P, Fe, Zn) and vitamins, and because of its high lysine and tryptophane content, its consumption with wheat or rice provides a balance in essential amino acids for human nutrition (Bhatty, 1988). Lentil straw is also a valued animal feed (Erskine *et al.,* 1990). The crop originated from Turkey (Ladizinsky, 1979) and is well adapted to the cool seasons there and spread to other regions of the world such as central Asia, where it is grown in temperate summers and to south Asia, where it is grown in winter during the dry season after the monsoon rise in the northern part of the subcontinent.

* *Corresponding Author.* E-mail: vyzhimo@gmail.com

The major geographical regions of lentil production are South Asia and China (44.3 per cent), northern Great Plains in North America (41 per cent), west Asia and north Africa (6.7 per cent), sub-Saharan Africa (3.5 per cent) and Australia (2.5 per cent) (FAOSTAT, 2010). Lentil was introduced into the Indo-Gangetic plain around 2000 BC and now half of the world's current lentil production is in South Asia. In subtropical regions of Pakistan, India, Nepal and Bangladesh, lentil is grown as a winter ('rabi') crop primarily on residual soil moisture from monsoon rains.

Indian economy is largely affected from the fluctuations in world pulse demand and supply, as the country is one of the largest consumers of pulses due to a large population India accounts for the largest global area under lentil with 1.48 million ha area and 1.01 million ton production (AICRP 2010–2011). Indian subcontinent has a reputation of being an important player in the world's pulse scenario. In case of lentil, India had been quite dominant country in the world production schedule as it was the largest producer till 2004 and had been leading since a very long time. Canada took over India's lead in 2005 with the help of a sharp rise in its production level. Around 90 per cent of the production comes from the states of Uttar Pradesh, Madhya Pradesh, Bihar and West Bengal pertaining to the eastern and the northern part of the country (Rao *et al.*, 2011). The Gangetic plains of West Bengal includes Bardhaman, Birbhum, Medinipur, north and south Dinajpur, Hoogly, Howrah, non-Tarai parts of Jalpaiguri, Cochbihar, Maldah, Murshidabad, Nadia and northern 24-Parganas districts falls under the hot and sub-humid to humid eco-region with alluvium derived soils. The climate of the area is characterized by hot summers and mild to moderately cool winters. The moisture availability during November and December is moderately dry and is generally dry from January to April and lentil is cultivated during this period in residual moisture (Sehgal *et al.*, 1992).

The production and productivity in this region suffers from a wide range of socio economic problems and abiotic and biotic constraints. The major abiotic limiting factors to production are low moisture availability and high temperature stress in spring, and, at high elevations, cold temperatures in winter. Mineral imbalances like boron, and salinity and sodicity though localised do cause substantial yield loss. Among biotic stresses, foliar blight, rust, vascular rot and wilt are the most important fungal diseases in the Indian sub-continent. Leading to low productivity and reduced marketability by discolouring the seed. This chapter attempts to review the various problems and prospects in the production and productivity of lentil and strategies to improve the situation in the Gangetic alluvilal zone of West Bengal.

Constraints to Lentil Production in the Gangetic Plains of West Bengal

Socio-Economic Constraints

Even though India has the largest area under cultivation and the second largest lentil producer in the world, it imports large amount from the rest of the world. Farmers in India treat lentil as secondary crop. In general farmers' access to inputs is limited, both because of low purchasing power and accessibility to markets to sell the excess produce. Because of this situation, the farmers give first priority to staple cereals and cash crops for allocating inputs and the second priority to pulses as a whole. As a result, lentil continues to be grown on poor soils with low inputs. In addition, there is lack of policy support and post-harvest innovations related to pulse crops including lentil. Availability of quality seed of improved varieties and other inputs is one of the major constraints in increasing the production of grain legumes (David *et al.*, 2002).

Productivity Constraints

Abiotic Constraints

The average productivity of lentil in south Asian countries including India is less than the world average mainly due to prevalence of different biotic and abiotic stresses. Yield gaps of 30-105 per cent with an average of 42 per cent have been reported from different production zones in India (Reddy and Reddy, 2010; Ali and Gupta, 2012). Most of the lentil in India is grown in low fertility, problematic soils and unpredictable environmental conditions. Drought is the major abiotic stress for lentil in many countries of the world (Johansen *et al.,* 1994) and it often affects the crop concurrently with heat stress. Water stress can cause heavy yield losses depending on the crop stage, drought severity, evaporative demand of the atmosphere and moisture holding capacity of the soil (Erskine *et al.,* 1994; Subbarao *et al.,* 1995). Lentil in West Bengal as also in Bangladesh is grown as a rabi crop and the prevailing conditions of short winter and atmospheric heat stress, especially when linked to moisture stress during flowering and pod filling drastically reduces seed yield in lentil because of damage to reproductive organs, accelerated development and shortened reproductive period (Siddique *et al.,* 2002; Boote *et al.,* 2005). Several researchers have reported useful genetic variation in response to drought stress under different conditions within the cultivated germplasm (Erskine *et al.,* 1994; Sarker *et al.,* 2005; Shrestha *et al.,* 2005, 2006; Kumar *et al.,* 2009). There is clearly scope to select for improved heat and drought stress in lentil. With recent changes in the global temperatures the grain yield is likely to be drastically affected by temperature extremities.

Lentils are often grown in acidic soil with imbalance in key minerals such as iron and boron constraining yield in West Bengal. However, these stresses remain of local importance. In acidic soils, it leads to deficiency of P, K, Ca, Mo and B and can also lead to toxicity of Al, Fe and Mn. Molybdenum deficiency leads to hampered nodulation and symbiotic fixation of legumes. Owing to the above effects, lentil productivity was reduced by 71 per cent in acidic soil as compared to normal soil (Dwivedi, 1996). Boron toxicity has been reported from South Australia, India, Iraq, Pakistan, USA and the Mediterranean environment (Yau and Erskine, 2000).

All of the above locally-important environmental stresses and soil mineral imbalances in lentil can be managed through selection within the cultivated germplasm, and more importantly by assisting farmers to adopt better crop and soil management practices that reduce drought by both enhancing rainfall infiltration and retention in the soil.

Biotic Constraints

The richness of legumes in nitrogen and phosphorus makes lentil attractive for insect pests and diseases (Sinclair and Vadez, 2012). The crop is relatively free from any serious infestation of insect pests except occasionally pod borer (*Etiella zinkenella*) and aphids (*Aphis craccivora*). The present lentil productivity is very low in India due to many factors and fungal diseases are the major biotic constraints to lentil production. It is infected throughout its geographic distribution by *Alternaria* blight (*Alternaria* spp.), *Ascochyta* blight (*Ascochyta lentis*), anthracnose (*Colletotrichum truncatum*), Botrytis stem and pod rot (*Botrytis cinerea*), rust (*Uromyces fabae*), *Sclerotinia* stem and pod rot (*Sclerotinia sclerotiorum*), *Stemphylium* blight (*Stemphylium botryosum*), wilts (*Pythium* spp., *Rhizoctonia* spp., *Fusarium oxysporum*), and downy mildew (*Peronospora lentis*) (Khare, 1981). The crop is known to be naturally infected by more than 30 viruses, among which the most important are Bean leafroll virus (BLRV), Bean yellow mosaic virus (BYMV), Beet western yellows virus (BWYV) etc. Some of these diseases are common in most lentil-growing regions worldwide, whereas others are limited to certain production areas. The economic importance of a particular disease is not necessarily related to its geographic distribution. A disease may have limited occurrence, but still it may cause significant economic losses. Under conducive conditions, these diseases, singly or collectively can cause significant reductions in both grain yield and quality.

Among the fungal diseases, foliar blights caused by *Stemphylium botryosum* Wallr., rust caused by *Uromyces vicia-fabae* Pers. J. Schroet and collar rot caused by *Sclerotium rolfsii* and are the prevalent and most destructive ones causing losses to the tune of 50-60 per cent (Bakr, 1993) in the eastern Indian sub-continent including West Bengal.

Stemphylium Blight

Stemphylium blight caused by *Stemphylium botryosum* Wallr has emerged as an important disease in the Gangetic plains of West Bengal and Bangladesh and cause up to the extent of 62 per cent yield reduction (Bakr, 1993). The disease has also been reported from northeast India (Sinha and Singh, 1991), where (in Bihar) disease intensity as high as 83 per cent was observed on an unsprayed local cultivar, causing >90 per cent yield loss (Sinha and Singh, 1993). The disease is caused by a ubiquitous, dematiaceous filamentous fungus *Stemphylium botryosum*, the anamorph of *Pleospora herbarum*, (Kiffer and Morelet, 2000). Sinha and Singh (1993) reported that an average mean temperature of 18±2°C and RH of 85 to 90 per cent in the morning were favorable for the appearance, development and spread of *Stemphylium* blight of lentil in India, whereas RH of >50 per cent in the afternoon was essential. Genetics and inheritance of resistance to *Stemphylium* blight of lentil have not been completely elucidated. Based on the frequency distribution of a recombinant inbred line (RIL) population developed from the cross Barimasur-4 X CDC Milestone, Kumar (2007) reported that resistance was quantitatively inherited. Limited efforts have been made to screen lentil germplasm against the pathogen. *Stemphylium* blight was observed at 55-58 days of sowing in the experimental fields in West Bengal (Zhimo, 2012). The initiation of disease symptoms coincided with the initiation of flowering of the plants. The disease first appeared as small brown colored lesions on the leaflets which later enlarged, became dark brown to blackish and covered the entire leaf surface. The symptoms differ from other foliar lesions by being larger and spreading across or along the entire leaflet. Defoliation occurs rapidly, leaving the branches with terminal leaves. The stems and branches also bend down, dry up and gradually turn ashy white, but the pods remain green. Infected foliage and branches were dull yellow giving a blighted appearance.

Drooping and drying of the stems and branches were common which gradually turned ashy white (Plates 13.1a and b). More than 50 per cent severity was observed in susceptible genotypes like L 4076, L 4147, HPLC 648 and IPL 215 for successive years under natural conditions in West Bengal while many genotypes like RLG 109, K 75, KLB 104, HUL 57, KLB 09-2, PLO 98, L 4702, RLG 112, LL 1114, L 4589 showed moderate resistance to natural infections of *Stemphylium* blight (Zhimo, 2012). Beare (2002) indicated that Crimson and Eston had moderate resistance to *Stemphylium* blight whilst CDC Glamis and CDC Milestone were susceptible. Erskine and Sarker (1997) had also screened out Precoz (ILL - 4605) and Barimasur - 4 as two resistant genotypes against *Stemphylium* blight of lentil from Argentina and Bangladesh respectively. Despite considerable progress in breeding for resistance to individual diseases, there is a need for more cultivars with multiple disease resistance that is durable.

Rust

Among foliar diseases, rust caused by *Uromyces vicia-fabae* (Pers.) Schroet is a serious yield constraint in Bangladesh, Ethiopia, India, Morocco, Nepal, Pakistan and South America. The extent of damage depends upon the crop stage, host genotype, inoculum load and severity of the infection. Yield losses to the extent of 60–69 per cent have been reported in India (Singh *et al.*, 1986) and up to 100 per cent in Ethiopia (Negussie *et al.*, 2005). It infects all the aerial plant parts under severe infection plants dry even before they form seeds. Circular brown patches of dry plants could be seen in the field (Khare and Agrawal, 1978). Rust symptoms in lentil were first observed 65-70 days after sowing as pale white aecial cups on the leaves which then gradually spread to pods and stems in experimental fields in West Bengal. The uredial and telial stages were observed as dark brown to black lesions on the leaves in the later stage. Circular brown patches of dry plants could be seen in the field. Rust infected plants showed excessive defoliation in case of severe infections on susceptible varieties (Plates 13.2a and b).

Joshi and Tripathi (2012) studied on the epidemiology of lentil rust and observed that the optimum germination of aeciospores, urediospores and teliospores was recorded at 20°C. Viability of aeciospores and urediospores decreased with increase in time, whereas, germination of teliospore after 8 months of storage gave positive results while, 24 hour leaf wetness after inoculation was found to be optimum for rust development. Most studies on genetics of rust resistance in lentil have revealed that resistance is monogenic and dominant (Kumar *et al.*, 1997). They reported that resistance to *Uromyces fabae* in five lentil genotypes (L 2991, L 2981, L 1534, L 178 and HPLC 8868) was governed by single dominant genes; whereas in one genotype,

Plate 13.1a: Lentil Plots Infected with *Stemphylium* Blight.

Plate 13.1b: Drooping and Drying of Leaves. Inset: Microphotograph of *Stemphylium botryosum* spores.

Plate 13.2a: Aceial Cups on Lentil Leaves. Inset: Microphotograph of cross section of aecial cups.

Plate 13.2b: Telial Stage on the Leaves and Stem. Inset: Microphotograph of cross section of lentil stem with teliospores.

Precoz, it was conditioned by two dominant genes. Chahota *et al.* (2002) reported that resistance to rust is controlled by duplicate, non-allelic and non linked dominant genes. Although a breakdown of rust resistance in lentil has not been reported, the likelihood of such an event in varieties with monogenic resistance cannot be ruled out (Negussiee *et al.,* 1998). When attempts were made to screen rust resistant genotypes under field conditions for successive years in West Bengal, it was observed that there were significant variations in the level of severities and genotypes like VL 140, LH 84-8, HPLC 648, L 4147, LL 1122, RLG 109, IPL 316, K 75, KLB 104, KLB 09-2, PLO 98, L 4702, RLG 112, LL 1114 and IPL 319 showed promising resistance to rust (Zhimo, 2012). Similar type of observation was earlier recorded by Singh and Sirohi (2000) when they conducted a screening experiment against rust with 96 lentil genotypes and they found only two genotypes *viz.* L 261 and EC 1 were immune. Hussain *et al.* (2008) also conducted experiments under field conditions with 590 local accessions and found only two lines *viz.* 66016-3 and 66013-4 were resistant to rust. In a separate screening experiment, Kumar *et al.* (1997) also reported five lentil genotypes *viz.* L- 2991, L- 2981, L -1534, L- 178 and HPLC- 8868 as resistant to this disease.

Collar Rot

Collar rot of lentil is caused by *Sclerotium rolfsii* Sacc. is a major disease in all the lentil growing areas of the country and causes 50 per cent loss at farmer's field. The disease is also widely prevalent in Utter Pradesh and causing 25 to 35 per cent loss (Khare *et al.*, 1979). High soil moisture, dense planting, and frequent irrigation promote infection (Aycock, 1966). High temperatures and moist conditions are associated with germination of sclerotia (Punja, 1985). The collar rot disease was first observed in the fields on 8 day old seedlings which continued until the seedlings were 4 weeks old. Affected seedlings turned yellow though the leaves did not droop. Young seedlings collapsed but older seedlings remained dry without collapsing. When uprooted, the seedlings showed rotting at the collar region and downwards. The rotten portions were seen covered with whitish mycelial strands (Plates 12.3a and b). A white mycelial coating could be seen on the taproot of completely dried seedlings, even after several days of death.

When germplasms were screened to assess the per cent mortality due to seedling diseases, it was observed that there were variations in seasons within the same genotypes studied. Incidences of seedling diseases of any plant depend on the inoculum density in the soil or the level of seed infections. Variations observed in the two different years within the same genotypes with respect to disease incidence may be due to uneven distribution of causal pathogens in the soil. These variations in reaction to seedling disease may be due to the inherent genetic variations within the genotypes as influenced by especially soil environmental factors. Genotypes SL 2-28, RLG 112, L 4588, HPLC 648, LH 84-8, PL 104, VL 141, PLO 097, L 4589, HPCL 617 and ASHA exhibited very low mortality rates due to collar rot disease in the fields of West Bengal (Zhimo, 2012).

Plate 12.3a: Seedlings of Lentil with Collar Rot Symptoms.

Plate 12.3b: White Mycelial Coating on the Taproot of Dried Infected Seedlings.

Other Strategies to Improve Lentil Production and Productivity in West Bengal

The Gangetic Plains of India offers a vast scope for enhancing legumes production both under irrigated and rainfed agro-ecosystems. Some prospects for improving their production and productivity are discussed below-

Biological Intervention in Soil and Plant Health Improvement

Pulse crops have always been a critical component of agroecosystems throughout the world because of their ability to convert atmospheric nitrogen (N) into usable plant protein, their ability to grow in N-poor soils, and their contribution to the pool of soil N that can be used by succeeding crops. The approximately 250 million ha of legumes grown in the world fix about 90 trillion g of N each year (Graham and Vance, 2000). The fixed N is used directly for plant growth and maintenance, and provides an excellent source of protein for humans and livestock. Six main genera of bacteria are involved in the symbiotic relationship with legumes: *Rhizobium, Bradyrhizobium, Sinorhizobium, Azorhizobium, Mesorhizobium* and *Allorhizobium,* and are collectively known as rhizobia. The contribution of the rhizobia-lentil symbiosis to soil health is affected by intrinsic biotic factors, such as plant genotype, strain of *Rhizobium leguminosarum* bv. *viciae,* and interactions between the two. Indigenous populations of rhizobia are often not adequate to support nitrogen fixation in lentils. Soils that have not been 'conditioned' to grow lentils typically are devoid of effective rhizobial strains and must be inoculated. Because nodulation, nitrogen fixation and photosynthesis are linked, factors that reduce photosynthate production and nodulation including many plant pathogens can adversely affect symbiosis and nitrogen fixation in lentil. It has been estimated that application of inoculation plays a vital role for successful lentil crop cultivation to enhance number of nodules/plant. Lentils inoculated with phosphate-solubilizing bacteria can increase above- and below-ground biomass, nodulation, and P-use efficiency (Singh *et al.,* 2005). In recent times, there has been a worldwide swing to the use of eco-friendly methods for protecting the crops from pests and diseases. In the present context, biological control of lentil diseases with bioagents offers a great promise. A biological control agent colonizes the rhizosphere, the site requiring protection and leaves no toxic residues as opposed to chemicals (Dubey *et al.,* 2007). *Trichoderma* species are typically known to be soil borne, green-spored ascomycetes that can be associated with the roots of plants as well as in the rhizosphere. The *Trichoderma* species are commonly considered a key genus in agricultural soils. These species are known for their potential to control plant disease in what can be a close association with many aspects typical of endophytic associations for plant health and growth (Bailey *et al.,* 2008; Bennett and Whipps, 2008). The suppression of disease by *Trichoderma* is based on hyperparasitism, antibiosis, induced resistance in the host plant and competition for nutrients and space (Harman *et al.,* 2004). Several strains of *Trichoderma* spp. have been found to be effective as biocontrol agents of various soil-borne plant pathogenic fungi such as *Sclerotium rolfsii* (Papavizas, 1985).

Utilization of Rice Fallows

Vast areas in the Indo-Gangetic Plains are mono-cropped under medium- and long duration rice. The non-availability of irrigation water and delay in vacating the field after rice does not normally permit early sowing of lentil when moisture is still available. The top soil layer generally dries out at the time of harvest of rice and thus planting of a post rainy season crop is not feasible. Under such conditions relay cropping of small-seeded lentil in the standing rice crop field, just before the harvest to ensure germination using the residual moisture and to avoid tillage operations during pulse growing is a common practice. Such a relay cropping operation (known by the terms utera or paira) could increase lentil production and sustain productivity of the rice-based system. Lentil seeds are broadcast in the standing crop of rice 7-15 days before harvest when there is adequate moisture for germination in the top layer of soil (Ali *et al.*, 2000). Expansion of this system will depend upon development of genotypes specially suited for relay cropping and of matching agro-technology, which has not received adequate attention so far. Relay cropping, commonly practiced in South Asia (Ali *et al.*, 1999) provides the plants more time for vegetative growth and reduces production cost due to zero tillage compared to sole cropping. Research on relay cropping by the national programs of Bangladesh, Nepal and India has led to yield increases by optimizing seed rate (40–45 kg/ha) and time of seeding (Ali *et al.*, 2009).

Improved Agronomic Practices

One of the key factors to achieve higher yield is the use of appropriate crop management practice. Improved production packages have to be developed considering factors like soil type, sowing date, seed rate, seeding depth, crop harvest time and other crop management options including weed control.

Matching crop maturity duration to available cropping window, including soil moisture availability, is a major strategy to avoid drought stress. Hence, emphasis should be given in crop improvement programs to develop high-yielding, short-duration cultivars which escape terminal drought. Moreover, these short duration varieties provide opportunities for inclusion of a given crop/variety in the cropping systems with a narrow cropping window or new production niches.

Seed priming where seed is soaked in water overnight, surface dried and then sown, is another technology which has been shown to improve plant stand, enhance early vigour and provide benefits in terms of earlier maturity, reduced disease and increased yield. Seed priming has shown that a 29– 38 per cent yield increase can be achieved through priming for 8 h in improved varieties (Ali *et al.*, 2009).

Lentil is a very poor competitor against weeds. Early sowing with adequate provision for weed control generally results in major yield gains. Lentil is normally planted at the beginning of the winter season except in cold-prone highland areas where the crop is spring-sown. Notwithstanding the development of herbicide tolerance in lentil, growers must learn to control weeds effectively using integrated weed management. Integrating existing practices is the likely answer to sustained lentil production within a cropping system (Yenish *et al.*, 2009). The avenue of improved crop management to manage water stress must also be explored–especially the role of conservation agriculture involving minimum tillage or zero tillage system as mentioned earlier paragraph, legume rotations, improve soil characteristics and increase nutrient and water use efficiency (Siddique *et al.*, 2010).

Adoption of Improved Production Systems

There is wide gap in the yield data for the last decades indicating differences in countries. Among the poorer lentil producers Bangladesh and Nepal stand out positively. Cooperation between ICARDA and the national program has resulted in the release of a cultivar, BARI Masur 4, with a high yield potential and resistance to the major diseases– rust and *Stemphylium* blight. An extensive campaign of field demonstrations coupled with seed multiplication and its widespread availability has facilitated the rapid spread of the new cultivar which is highly profitable for small holders to grow. This study has highlighted the fact that a profitable new technology can spread quickly among poor small holders and contribute to alleviation of their poverty providing it is coupled with an outreach program. Another household survey in Ethiopia and the gross national data of Nepal provide additional examples of how local and national economic impacts of new lentil technologies can be substantial (Aw Hassan *et al.*, 2009; Neupane *et al.*, 2008); and emphasize the need to connect research outputs with outreach campaigns. Unfortunately, putatively profitable technologies have not yet reached farmers in many other countries like India.

Conclusion

Lentil will undoubtedly continue to play a key role in the food security of millions of Asians especially in the Indian sub continent particularly within the poorer sectors of society who rely on vegetable sources for their protein and energy requirements. The crop will also retain an important position within rainfed cropping systems as a source of nitrogen and hence soil fertility, as rising oil prices increase the cost of inorganic fertilizer alternatives. Considering first technical productivity issues, the major challenge is to increase the competitiveness of lentil against cereals and other more remunerative alternative crops like oilseeds. Considering research needs, lentil yields are low because of the crop's limited yield potential like lack of seedling vigour, slow leaf area development, low harvest index, lack of lodging resistance, low or no response to inputs and vulnerability to an array of abiotic stresses and devastating diseases. Under conducive conditions, the diseases, singly or collectively can cause significant reductions in both grain yield and quality. Diseases also interact with agronomic practices that change the environmental conditions in which the crop grows and crop growth, thus impacting on the adaptation of disease and host. Management techniques and lentil genotypes grown have evolved over a long period of time and are relatively stable in traditional lentil-growing areas. Resistance or tolerance to these diseases is a key to the adaptation of lentil in most areas, especially where cultural controls such as fungicides are unavailable or uneconomic and stress must be given in screening and evaluation of genotypes against these diseases as an attractive method of disease control to ensure sustainable productivity and profitability of lentil. With globalization, lentil germplasm has been distributed around the world and new methods and tools have become available to farmers. These can be used to improve yield and profitability if optimal management practices are used with genotypes adapted to the practices.

Adequate variability for many of the crop's genetic constraints exists within the crop gene pool allowing manipulation through plant breeding. However, several other important traits, such as biomass yield, pod shedding and nitrogen fixation are not currently addressable by breeding because of insufficient genetic variation. There must be a major policy shift to increase overall investment in the sustainable intensification of production systems which have been especially neglected in crops like lentil. Looking ahead, escalating costs of producing inorganic nitrogen fertilizer, reducing availability of water for agriculture, climate change, food insecurity and an increasingly nutrition-conscious consumer society collectively give a bright future for a highly nutritious food produced by a nitrogen-fixing crop such as lentil adapted to the cereal-based farming systems of marginal lands.

References

Ali, A., Johnson, D. L., and Stushnoff, C. (1999). Screening of lentil (*Lens culinaris*) for cold hardiness under controlled conditions. *Journal of Agricultural Science*, **133**: 313–319.

Ali, M. and Gupta, S. (2012). Carrying capacity of Indian agriculture: Pulse Crops. *Current Science* **102**: 874-881.

Ali, M., Joshi, P.K., Pande, S., Asokan, M., Virmani, S., Kumar, M. and Kandpal, B.K. (2000). *Legumes in the Indo-Gangetic Plain of India.* In: Legumes in rice and wheat cropping systems of the Indo-Gangetic Plain - constraints and opportunities. International Crops Research Institute for the Semi-Arid Tropics, Patancheru, Andhra Pradesh, India, pp. 35-70.

Ali, M., Singh, K. K., Pramanik, S. C., and Ali, O. (2009). Cropping systems and production agromony. In: *The lentil–Botany, production and uses* (W. Erskine, F. J. Muehlbauer, A. Sarker, and B. Sharma Eds.) *Wallingford: Comm. Agric. Bureau. Int.*, pp. 213–228.

All India Coordinated Research Project, (2010–11). Project coordinator's report on MULLaRP, rabi 2009–2010. *Indian Institute of Pulses Research, Kanpur, India*.

Aw Hassan, A. A., Regassa, S., Shafiqul Islam, Q. M., and Sarker, A. (2009). The impact of improvement research: the case of Bangladesh and Ethiopia. In: *The lentil–Botany, production and uses* (W. Erskine, F. J. Muehlbauer, A. Sarker, and B. Sharma Eds.) Wallingford: Comm. Agric. Bureau. Int., pp. 425–446.

Aycock, R. (1966). Stem rot and other diseases caused by *Sclerotium rolfsii. N.C. Agr. Expt. St. Tech. Bul*. No. 174.

Bailey, B.A., Bae, H., Strem, M.D., Crozier, J., Thomas, S.E., Samuels, G.J., Vinyard, B.T. and Holmes, K.A. (2008). Antibiosis, mycoparatism and colonization success for endophytic *Trichoderma* isolates with biological control potential in *Theobroma cacao*. *Biological Control*, **46** (1): 24–35.

Bakr, M. A. (1993). Plant protection of lentil in Bangladesh. In (), Lentil in South Asia (Erskine, W. and Saxena, M. C. Eds.). Aleppo: ICARDA, pp. 177– 186.

Beare, M. (2002). Investigation into *Stemphylium botryosum* resistance in lentil. *Undergraduate thesis*. University of Saskatchewan, Saskatoon, Saskatchewan.

Bennett, A.J. and Whipps, J.M. 2008. Beneficial microorganism survival on seed, roots and in rhizosphere soil following application to seed during drum priming. *Biological Control*, **44** (3): 349–361.

Bhatty, R. S. (1988). Composition and quality of lentil (*Lens culinaris* Medik.): a review. *Canadian Institute of Food Science and Technology*, **21**: 144–160.

Boote, K. J., Allen, L. H., Prasad, P. V. V., Baker, J. T., Gesch, R. W., Snyder, A. M., *et al.* (2005). Elevated temperature and CO2 impacts on pollination, reproductive growth and yield of several globally important crops. *Journal of Agricultural Meteorology Japan*, **60**: 469–474.

Chahota, P.K., Gupta, V.P. and Sharma, S.K. (2002). Inheritance of rust resistance in lentil. *Indian Journal of Genetics and Plant Breed*ing, **62**: 226-227.

David, S, Mukandala, L. and Mafuru, J. (2002). Seed availability, an ignored factor in crop varietal adoption studies: a case study of beans in Tanzania. *Journal of Sustainable Agriculture*, **21**: 5-20.

Dubey, S.C., Suresh, M.S. (2007). Evaluation of Trichoderma species against *Fusarium oxysporum* f. sp. *ciceris* for integrated management of chickpea wilt. *Biol. Contamin.* **40**: 118-127.

Dwivedi, G.K. (1 996) Tolerance of some crops to soil acidity and response to liming. *Journal of the Indian Society of Soil Science*, **44**: 736-741.

Erskine, W. and Sarker, A. (1997). Bangladesh in a big way–and the results have been satisfying. ICARDA has been helping breed the varieties of the future. *ICARDA Caravan*, **6**: 8-10.

Erskine, W., Rihawe, S., and Capper, B. S. (1990). Variation in lentil straw quality. *Animal Feed Science and Technology*, **28**: 61–69.

Erskine, W., Saxena, N. P., and Saxena, M. C. (1993). Iron deficiency in lentil: Yield loss and geographic distribution in a germplasm collection. *Plant and Soil*, **151**: 249–254.

Erskine, W., Tufail, M., Russell, A., Tyagi, M. C., Rahman, M. M., and Saxena, M. C. (1994). Current and future strategies in breeding lentil for resistance to abiotic and biotic stresses. *Euphytica*, **73**: 127–135.

FAOSTAT, (2010). FAOSTAT statistical database of the United Nations Food and Agriculture Organization (FAO). Italy: Rome.

Graham, P.H. and Vance, C.P. (2000). Nitrogen fixation in perspective: an overview of research and extension needs. *Field Crops Research*, **65**: 93-1 06.

Harman, G.E., Howell, C.R., Viterbo, A., Chet, I., Lorito, M. (2004). *Trichoderma* species opportunistic, avirulent plant symbionts. *National Review Microbiology*, **2**: 43-56.

Hussain, S., Ali, A., Zahid, M. A. and Iqbal, S. M. (2008). Multiple disease resistance in lentil germplasm from Pakistan. *Pakistan Journal of Botany*, **40**: 897–903.

Johansen, C., Baldev, B., Brouwer, J. B., Erskine, W., Jermyn, W. A., Li Juan, L., *et al.* (1994). Biotic and abiotic stresses constraining productivity of cool season food legumes in Asia, Africa and Oceania. In: Expanding the production and use of cool season food legumes (Muehlbauer, F. J. and Kaiser, W. Eds.), The Netherlands: Kluwer. pp. 175–194.

Joshi, A. and Tripathi, H.S. (2012). Studies on epidemiology of lentil rust (*Uromyces viciae fabae*). *Indian Phytopathology,* **65**: 67-70.

Khare, M. N., Agrawal, S. C. and Jain, A. C. (1979). Lentil diseases and their control. *Technical Bulletin*, Jawaharlal Nehru Krishi Vishwa Vidyalay, Jabalpur, Madhya Pradesh, India, pp. 1-29.

Khare, M. N. (1981). Diseases of lentils. In Lentils (Webb, C. Hawtin, G. C. Eds.), *Wallingford: Comm Agric Bureau*. pp. 163–172.

Khare, M.N. and Agarwal, S.C. (1978). Lentil rust survey in Madhya Pradesh. *Proc. of All India Pulse Workshop*, Baroda, pp. 3.

Kiffer, E. and Morelet, M. (2000). The Deuteromycetes: mitosporic fungi, classification and generic keys. Science Publishers, Enfield, NH.

Kumar, J., Solanki, R. K., Singh, S. K., Basu, P. S., and Kumar, S. (2009). Evaluation of lentil germplasm for traits imparting drought tolerance. In: *The 3rd International Conference on Integrated Approaches to Improve Crop Production under Drought Prone Environments, Abstract (P6.20),* Shanghai Academy of Agricultural Sciences, Shanghai (China) Oct 11–16, 2009, pp. 195–196.

Kumar, P. (2007). Genetics of resistance to *Stemphylium* leaf blight of lentil (*Lens culinaris*) in the cross Barimasur-4 × CDC Milestone. M.S. Thesis, University of Saskatchewan, Canada.

Kumar. P. and Agarwal, J.P. (1993). Response of Lentil (*Lens esculenta*) to Rhizobium inoculation, nitrogen and phosphorus fertilization. *Indian Journal of Agron*omy, **38**(2): 318-320.

Kumar, V., Singh, B.M. and Singh, S. (1997). Genetics of lentil resistance to rust. *Lens,* **24**: 23-25.

Ladizinsky, G. (1979). The origin of lentil and wild gene pool. *Euphytica*, **28**: 179-187.

Negussie, T., Pretorius, Z.A. and Bender, C.M. (1998). Components of rust resistance in lentil. *Euphytica*, **142**: 55–64.

Negussie, T., Pretorius, Z.A. and Bender, C.M. (2005). Effect of some environmental factors on in vitro germination of urediniospores and infection of lentils by rust. *J. Phytopath*., **153:** 43-47.

Neupane, R., Sarker, A., Shrestha, R., Yadav, N. K., Erskine, W., and Francis, C. (2008). Lentil production in Nepal. A success story of fruitful partnership between the Nepal Agricultural Research Council, the Centre for Legumes in Mediterranean

Agriculture and the International Center for Agricultural Research in the Dry Areas. APAARI Publication: 2008/1, APAARI, Thailand, pp. 1–42.

Papavizas, G.C. (1985). *Trichoderma* and *Gliocladium* biology, ecology and the potential for biocontrol. *Annual Review of Phytopathology*, **23**: 23-77. Punja, Z.K. (1985). The biology, ecology, and control of *Sclerotium rolfsii*. *Annual Review of Phytopathology*, **23**: 97-127.

Rao, S. D., Dhuppar, P., Subhash, C. and Chintapalli, B. (2011). Emerging Trends in Lentil Crop Production System, Profitability and Global Concern. In: *Proceedings of the International Conference on Economics and Business: Analysis and Application*, EBA DEI, Agra.

Reddy, A. A. and Reddy, G. P. (2010). Supply side constrains in production of pulses in India: A case study lentil. *Agricultural Economics Research Review*, **23**: 129-136.

Sarkar, A., W. Erskine and M. Singh. (2005). Variation in shoot and root characteristics and their association with drought tolerance in lentil. *Genet. Resour.Crop Evol.*, **52**: 89-97.

Sehgal, J. L., Mandal, D. K., Mandal, C. and Vadivelu, S. (1992). Agro-Ecological Regions of India. In: National Bureau of Soil Survey and Land Use Planning (ICAR), NBSS Publ. 24, *Technical Bulletin*, pp. 82-84.

Shrestha, R., Siddique, K. H.M., Turner, N. C., Turner, D.W., and Berger, J. D. (2005). Growth and seed yield of lentil (*Lens culinaris* Medik.) genotypes of West Asian and South Asian origin and crossbreds between the two under rainfed conditions in Nepal. *Australian Journal of Agricultural Research*, **56**: 971–981.

Shrestha, R., Turner, N. C., Siddique, K. H. M., Turner, D. W., and Speijers, J. (2006). A water deficit during pod establishment in lentils reduces flower and pod numbers but not see size. *Australian Journal of Agricultural Research*, **57**: 427–438.

Siddique, K. H. M., Loss, S. P., and Thompson, B. D. (2002). Cool season grain legumes in dryland Mediterranean environments of Western Australia: Significance of early flowering. In: *Management of agricultural drought: Agronomic and genetic options* (Saxena, N. P. Ed.), Enfield: Scientific. pp. 151–162.

Siddique, K. H. M., Johansen, C., Turner, N. C., Jeuffroy, A. M. -H., Hashem, A., Sakar, D., *et al.* (2010). Innovations in agronomy of food legumes. In: *5th International Food Legume Research Conference and 7th European Conference on Grain Legumes*, Antalya, Turkey (In press).

Sinclair, T.R. and Vadez, V. (2012). The future of grain legumes in cropping systems. Crop and Pasture Science. http: //dx.doi.org/10.1071/ CP12128.

Singh, A. and *Sirohi*, A. *(2000)*. Identification of sources of resistance in *lentil* against rust. *Crop Improvement*, **27**(1): 103-104.

Singh, K., Jhooty, J. S., and Cheema, H. S. (1986). Assessment of losses in lentil yield due to rust caused by *Uromyces fabae*. *LENS Newsletter*, **13**: 28–29.

Singh, K.K., Srinivasarao, C. and Ali, M. (2005). Root growth, nodulation, grain yield, and phosphorus use efficiency of lentil as influenced by phosphorus, irrigation, and inoculation. *Communications in Soil Science and Plant Analysis*, **36**: 1919-1929.

Sinha, J. N., and Singh, A. P. (1991). *Stemphylium sarciniforme* on *Lens culinaris*. *Indian Phytopathology*, **44**: 421.

Sinha, J. N., and Singh, A. P. (1993). Effect of environment on the development and spread of stemphylium blight of lentil. *Indian Phytopathology*, **46**: 252–253.

Subbarao, G. V., Johansen, C., Slinkard, A. E., Nageswara Rao, R. C., Saxena, N. P., and Chauhan, Y. S. (1995). Strategies for improving drought resistance in grain legumes. *Critical Reviews in Plant Sciences*, **14**: 469–523.

Webb, C. and G. Hawtin. (1981). Lentils: (Webb, C. and Hawtin, G. Eds.). CAB/ ICARDA publication.

Yau, S. K., and Erskine, W. (2000). Diversity of boron toxicity tolerance in lentil growth and yield. *Genetic Resources and Crop Evolution*, **47**: 55–61.

Yenish, J. P., Pala, M., and Haddad, A. (2009). Weed management. In The Lentil– Botany, production and uses (Erskine, W., Muehlbauer, F. J., Sarker, A. and Sharma, B. Eds.), Wallingford: pp. 326–342.

Zhimo, V.Y. (2012). Studies on disease of important pulse crops in the Gangetic alluvial zone of West Bengal. M.Sc. Thesis. Bidhan Chandra Krishi Viswavidyalaya, Mohanpur, West Bengal, India. pp. 70-76.

2015, Major Constrains and Verdict of Crop Productivity *Pages 215–226*
Editor: **Dr. U.N. Bhale**
Published by: **ASTRAL INTERNATIONAL (P) LIMITED, NEW DELHI**

Chapter 14

Fungal Biodiversity: Major Constrains of Crop Productivity in India

P.N. Chowdhry*, Nidhi Anand and Suman Lata

National Centre of Fungal Taxonomy (NCFT), Inderpuri, New Delhi – 110 012, India

Introduction

India is an agrarian economy and vast majority of land is used for farming and a wide range of crops are cultivated in its different agro-ecological regions. With a production of 93.9 million tons (Mt) of wheat, 104.6 Mt of rice, 21.6 Mt of maize, 20.7 Mt of millets, 357.7 Mt of sugarcane, 8.1 Mt of fiber crops (Jute, Mesta, Cotton), 17.2 Mt of pulses and 30.0 Mt of oilseeds crops, in the year 2011-12 (MoA, 2012), During normal monsoon(2011) season, Indian agriculture accomplished an all-time record production of 85.9 million tons of wheat, a 6.4 per cent increase from a year earlier. Rice output in India also hit a new record at 95.3 million tones, a 7 per cent increase from the year earlier (Bloomberg, 2011) Lentils and many other food staples production also increased year over year. Indian farmers thus produced about 71 kilograms of wheat and 80 kilograms of rice for every member of Indian population in 2011. The per capita supply of rice every year in India is now higher than the per capita consumption of rice every year in Japan (Japan Review, 2006). Agricultural ecosystems have become an important and growing research field in recent years. Agricultural practices influence the environment; have changed overtime, the plant pathogens

* *Corresponding Author.* E-mail: pnchowdhry@gmail.com

and host plant species that inhabit the Agri-environment have adapted to each other (Warren *et al.,* 2008).

Fungi are one of the most important groups of organisms on the planet. They are destructive agents causing losses of agricultural commodities in many parts of the world. This is easy to overlook, largely hidden, with unseen actions and growth. Plant pathogenic fungi cause devastating damage to crop production worldwide. Some fungi are parasites of plants. Most of our common crop plants are susceptible to fungal attack of one kind or another. Spore production and dispersal is enormously efficient in fungi and plants of the same species crowded together in fields are ripe for attack. Fungal diseases can on occasion result in the loss of entire crops if they are not treated with antifungal agents. Worldwide Fungal diseases are a huge problem. The fungal stem rust (*Puccinia gravinis*) of wheat was effectively controlled through introgression, decades ago, of the Sr31 resistance gene by Norman Borlaug and colleagues and has been remarkably durable, but a resistant strain of the rust has recently emerged in Africa and, in this age of globalization, represents a potential worldwide threat if not addressed in a timely fashion. Powdery mildews affect a wide range of crops, including major cereals like wheat, sorghum, and millet; fungal anthracnose affects crops such as sorghum, beans, and cassava; leaf spot and root rots also plague beans and other important crops; turcicum and gray leaf spot diseases are serious pests of maize in Africa; blast disease is serious for rice; and Black Sigatoka limits banana production worldwide. Small farm environments seriously promote the development of fungi that lead to mycotoxin accumulations; better statistics are sadly needed, but the little we have suggests the health effects of mycotoxins on the poor are much more serious than recognized previously. India is a global agricultural powerhouse.It is the second largest producer of rice, wheat, cotton, sugarcane, farmed fish, sheep and goat meat, fruit, vegetables and tea. The country has some 195 m ha under cultivation of which some 63 per cent are rain fed (roughly 125m ha).The growing Indian population necessitates reduced crop losses to improve food security, and the control of fungal plant pathogens is vital to help maintain food production. Moreover, Indian farmers are under direct threat of diverse fungal pathovar population on important crops like cereals, Legumes, vegetables and fruit crops. Fungal diseases (incited by number of fungi) are important constrain of crop productivity in India and have been enumerated as follows:

Fungal Diversity on Cereals

Constrain of Wheat

Wheat is facing a serious threat from a fungal disease if, not quickly contained that could wipe out the world's crop (Christy 2014). Wheat leaf rust is fungal disease that affects wheat, barley and rye stems, leaves and grains. In temperate zones it is destructive on winter wheat because the pathogen overwinters. It causes serious epidemics in North America, Mexico and South America and is a devastating seasonal disease in India. All three types of *Puccinia* are heteroecious requiring two distinct and distantly related hosts (alternate hosts). There is mounting concern at the dangers posed to global food security.Wheat rusts (yellow, leaf and stem rust) are fungal diseases that affect wheat in almost every country in which it grows, as infectious spores are transmitted by wind and through soil. Seven to 15 days after infection, small red dots called uredinia begin to form on the leaf or stem surface

The rusts of wheat belong to genus *Puccinia,* family Pucciniaceae, order Uredinales and class Basidiomycotina. Wheat rust, a devastating disease known as the "polio of agriculture", has spread from Africa to South and Central Asia, the Middle East and Europe, with calamitous losses for the world's second most important grain crop, after rice. Wheat suffers from three rust diseases namely stem, leaf and stripe rust. *Puccinia. graminis* cause stem rust, *Puccinia recondita* cause leaf rust but currently *Puccinia triticina* was suggested to be the preferred name, *P. glumarum* cause of yellow rust of wheat is caused by but later changed as *P. striiformis*. Rust fungi which are morphologically identical but attack different host genes are known as special forms (*formae specialis*) for example, *P. graminis* on wheat is *Puccinia graminis f. sp. tritici, Puccinia. recondita* as *Puccinia recondita* f. sp. *graminis* and *Puccinia striiformis* as *Puccinia recondita* f. sp. *graminis*. In each special form of these rusts have several pathogenic (physiological) races which can be detected on different varieties of the same species by inoculation and these varieties are known as differential hosts. The stem rust fungus, *Puccinia graminis,* attaches to another organism to survive and killing the host in the process

Other important pathogens are *Ustilago tritici* causing black powdery mass or Loose Smut of Wheat, *Tilletia indica* or *Neovossia indica* cause of Karnal Bunt of Wheat. *Erysiphe graminis* f. sp. *tritici* causes Powdery Mildew of Wheat. *Alternaria triticina* causes leaf spots while, *Clavibacter tritici* causes of Yellow Ear Rot of Wheat, *Blumeria graminis* f. sp. *tritici* causing Powdery Mildew, *Pyrenophora tririci- repentis* causing Tan spots, *Septoria tritici* cause of leaf blotch, *Pheosphaeria nodorum* causing glume blotch, *Fusarium graminearum* causing head blotch, *Drechslera sorokiana* cause of leaf spot blotch while *Fusarium culmorum, F. pseudograminearum* and *F. graminearum* are the cause of root rot diseases.

Constrain of Rice

Rice is the second largest produced cereal and will be the main daily staple of about 4.6 billion people by 2025. Considering the fact that (FAO 2009) the forecasts estimate to food production must double by 2050, increasing rice production and crop yield appears crucial to human life. But rice is subjected to important fungal pathogens which dramatically reduce crop yields: i) *Magnaporthe oryzae* the rice blast, a devastating pathogen induced by the, affected cereal grain crops such as rice, wheat, rye, and barley; about 30 per cent of cereal yields are lost because of fungi annually. In India, blast disease occurs in rice crop of all the states where rice is commercially grown. The disease is quite common in areas of high rainfall and in irrigated areas where the crop is highly doses with nitrogen. Several epidemics due to blast have been experienced in different parts of the world resulting losses in yield from 50 to 90 per cent. In India, the yield loss during the year 1960-61 was estimated to 2, 66,340 tones. ii) *Xanthomonas oryzae* the bacterial blight is responsible for as much as over 70 per cent of yield losses in Asia. Furthermore, drought stress is an increasing, worldwide problem which affects about 50 per cent of the global rice production each year. iii) *Drechslera oryzae* causing Brown Spot Disease of Rice and

iv) *Rhizoctonia solani* develops Sheath Blight of Rice and iv) *Drechslera turcicum* causing leaf blight

Constrain of Maize

Stalk rots are caused by several fungi and bacteria which affects the plants near maturity. Losses from stalk rot vary region to region and are estimated 10- 20 per cent. Losses are caused either by poor filling of the cobs or due to lodging of affected plants. The following pathogenic are associated with stalk rot of maize caused *Colletotrichum graminicola., Diplodia maydis, Fusarium moniliforme* var. *subglutinans, Gibbrella zeae, Fusarium graminearum, Setosphaeria turcica, Exserohilum turcicum, Pythium aphanidermatum, Rhizoctonia solani* sub sp. *sasakii, Cochliobolus heterostrophus.* Anamorph: *Bipolaris maydis.*

Constrain of Barley

Cultivated over a wide range of geographic area (Paulitz and Steffenson, 2011). Root and foliar pathogenic fungi function - as necrotrophs, killing tissue in advance of growing hyphae; Foliage diseases of Barley, *Fusarium head blight* (FHB) - or scab, devastating and insidious disease of barley. *Puccinia graminis* cause of Stem rust, one of the most serious diseases - of small grain cereals; *Puccinia striiformis* f.sp. *hordei* cause of Stripe rust or yellow rust - disease, and *Puccinia coronate* cause of crown rust plaguing barley in southern Asia, eastern Africa, Western Europe, and Middle East. While Powdery mildew, caused by Ascomycetous fungus *Blumeria graminis* f. sp. *hordei.*

Fungal Diversity on Legumes and Oilseeds

The investigation revealed (Dhandhalya and Shiyani, 2009) a wide yield gap in major oilseed crops of Saurastra region. The average yield gap II was about 710 kg/ ha and 424 kg/ha in groundnut and sesame, respectively. Yield gap is also large in both the crops. A large production potential is still unexploited in both the crops. Among biotic constraints water stress is the single most limiting factor in Saurastra, which caused about 3.6 lakh tones and 0.33 lakh tones production loss amounting to Rs.507.37 crores and Rs. 48.97 crores in case of groundnut and sesame, respectively. Among biotic constraints, Tikka disease in groundnut and brown angular spot in sesame topped the list causing maximum value loss of Rs. 92.93 crores and Rs.10.97 crores, respectively. More attention towards watershed development and drip irrigation system, soil fertility management, regular power supply and strengthening extension services are the major suggestions emerged for crop protection.

Constrain of Soybean

Phakospora pachyrhizi is the most important fungal pathogens to attack soybean in the rice lands are the soil borne fungi and is the only economically important fungal foliar pathogen of soybean. Rust is the most widely spread and economically important fungal foliar disease of soybean grown at Africa, Asia, and Latin America,North American varieties have had good resistance until recently as the pathogen has merged in some areas. Asian soybean rust affects the soya crop in Brazil - the world's biggest producer. There are more than 244 confirmed cases this year. Industry spending on insecticides and fungicides is expected to reach \$11bn (£6.6bn). The fungus can reduce yields between 30 per cent and 90 per cent (Yang 1980) Light brown to reddish pustules on the underside of leaves are predominantly develops. The infected leaves may drop off. The Asian Vegetable Research and Development Center (AVRDC), Taiwan has a regional research program on rust of soybean in Asia. However, no satisfactory host plant resistance to rust is available. Seed treatment with fungicides is the recommended.

Other Phytopathogenic fungi associated are *Pythium ultimum* and *Pythium debaryanum* causing *Pythium* root and seedling rot. These fungi develop "wet" roots and seedlings turn brown. Planting good quality fresh seed can minimize seed with any common seed fungicidal dressing before planting. *Fusarium oxysporum* is another pathogen causing root rot. The fungal organism normally appears in wet weather followed by heavy rain or floods. Characteristic symptoms of the disease develops including rotting of seedling roots and dark brown patches on stems. Seed treatment with common fungicides can reduce the incidence of root rot. Rhizoctonia root rot. The disease is caused by the common soil borne fungus. *Rhizoctonia solani* develops brown or reddish brown patches on the lower stem and seedling hypocotyls are the most conspicuous symptoms of the disease. The fungus cannot be economically controlled by fungicides; however, the simple cultural method of ridging soil around the base of the plants reduces the damage. *Phytohthora sojae* is another serious pathogen developing roots and stem rot. The stem just above the soil surface turns dark brown. Plants wilt and die. These fungi predominate in low-lying, poorly drained areas, and heavy clay soils. Resistant cultivars and improvement in soil drainage help in minimizing the incidence disease. *Macrophomina phaseolina* (sclerotial state R*hizoctonia bataticola*) infects lower stem shows black patches like powdered charcoal causing Charcoal Rot. The fungus becomes active in hot, dry weather in dry soil. Crop rotations with a non-host crop can reduce charcoal rot incidence. Anthracnose is caused by the seed and soil borne fungi, *Colletotrichum truncatum* and *Colletotrichum truncatum* var. *destructivum* cause Anthracnose of soybean. These fungi attack both young seedlings and older plants and produce dark brown patches on the stem. Crop rotation and planting of disease-free, fungicide-treated seeds are the recommended as effective control measures. *Cercospora kikuchii* cause blight and leaf spot (purple seed stain) and infected seeds can produce diseased seedlings. Later the infection spreads to the stem and leaves and produces pale to dark purple stained seeds.

Constrain of Mung Bean and Black Gram

More than sixteen different plant pathogenic fungi have so far been recorded on Mung bean and twenty one on black gram from Bangladesh alone (Fakir, 1983; Ahmed, 1985).A similar disease situation has been observed wherever Mung bean and black gram are grown in rice fallow lands in India and Nepal (Pande and Joshi, 1995). Three fungal pathogens that attack both the pulses are considered economically important. These are yellow mosaic (Mung bean yellow mosaic virus), *Cercospora* leaf spot (*Pseudocercospora cruenta*), and powdery mildew (*Erysiphe polygoni* and *Oidium* sp).

Constrain of Lentil

Lentil is an important food legume which is commonly grown in rice fallow lands it is largely relay cropped in the rice lands in India, Nepal, and Pakistan and Bangladesh (Johansen *et al.*, 1994). Although the area under lentil is increasing in the Indian subcontinent, the crop, like other food legumes, suffers from number of diseases caused by fungi, bacteria, viruses, and nematodes. The important diseases of lentil, which are the potential constraints to lentil, are *Fusarium oxysporum* f. sp *lentis* the important root and stem pathogen other fungi that affect lentil are: *Rhizoctonia solani* cause of vascular wilt); (*Macrophomina phaseolina;* sclerotial state *Rhizoctonia bataticola*) wet root rot dry root rot, (*Sclerotium rolfsii* collar rot, *Sclerotinia sclerotiorum* stem rot, *Fusarium solani* black root rot *Aphanidermatum* damping-off, and *Pythium ultimum* root rot. The pathogens causing root rot and stem diseases are soil borne and attack the lentil crop at the seed, seedling, and adult stages resulting in seed rot, damping-off, wilt, and root and stem rots (Khare *et al.*, 1979; Khare, 1981).

The important foliar fungi in lentil growing regions are *Ascochyta fabae* f. sp *lentis* causing Blight, *Uromyces viciae-fabae* rust, *Erysiphe polygoni* and *Leveillula taurica* powdery mildew, *Peronospora lentis* Downy mildew, and *Botrytis cinerea* gray mold. *Colletotrichum truncatum* Anthracnose and *Alternaria alternata* leaves blight also affect lentil. *Stemphylium botryosum* is a new blight and most destructive disease of lentil which often causes greater losses in association with *Botrytis* gray mold.

Constrain on Chickpea

Chickpea is traditionally cultivated in the rice fallows of India, Nepal, Pakistan, and Bangladesh (Pande and Joshi, 1995). It is sensitive to excessive soil moisture, high humidity, cloudy, and foggy weather which limit crop establishment, flower production, and fruit set (Kay, 1979) and also increase severity of common soil borne (root rots and wilts) and foliar diseases. Some of the potentially serious diseases of chickpea pathogenic fungi are: *Ascochyta rabiei* causing *Ascochyta* blight, *Botrytis cinerea* cause *Botrytis* gray mold (BGM) and *Fusarium oxysporum* f. sp *ciceris* cause of *Fusarium* wilt, *Macrophomina bataticola*; sclerotial state *Rhizoctonia bataticola* cause of dry root rot *Phytohthora megasperma* cause of root rot and *Pythium ultimum* damping-off,

Constrain of Pigeonpea

Over 60 pathogens have been reported to attack pigeonpea (Nene *et al.*, 1996), but only a few of them are widely distributed and are of economic importance. *Fusarium udum* cause of Wilt disease, *Phytopththora drechsleri* f. sp *cajani* cause of Leaf blight, *Cercospora* spp for leaf spot, and *Alternaria* spp for *Alternaria* leaf blight are economically important, wherever pigeonpea is grown. Yield losses varied from country to country and season to season (Reddy *et al.*, 1990).

Constrain of Mustard

Sclerotinia rot and *Alternaria* blight were rated as the two primary constraints experienced by mustard growers. *Sclerotinia* rot was considered the most important threat to mustard crop in due to extent of yield loss to the crop. Application of aqueous garlic clove extract as seed treatment (1 per cent w/v) followed by a foliar spray at 50 days after sowing could be a user-friendly, effective and economic non-chemical means of managing *Sclerotinia* rot of mustard (Meena *et al.*, 2006).

Fungal Diversity of Vegetables

Constrain on Chilies

Colletotrichum capsici is one of the most important plant pathogens worldwide especially tropical and subtropical regions causing the economically important disease anthracnose in a wide range of hosts including cereals, legumes, vegetables, perennial crops and tree fruits (Bailey and Jeger, 1992). Anthracnose disease is a major problem in India and one of the more significant economic constraints to chilli production,. The disease causes both pre- and post-harvest fruit decay. Chilli anthracnose usually develops under high humid conditions when rain occurs after the fruits have started to ripen with reported losses of up to 84 per cent. Economic losses caused by the disease are mainly attributed to lower fruit quality. Accurate taxonomic information is necessary (Thane *et al.*, 2008) for effective disease control management In the *Colletotrichum* pathos-system, and different *Colletotrichum* species can be associated with anthracnose in of the same host. Chilli anthracnose disease caused by *Colletotrichum* species[§].

Constrain on Potato

Phytophthora infestans cause of Late Blight according to the International Potato Center, is the single most costly biotic constraint to global food production. Potato blight is a fungus-like organism that thrives in damp and humid conditions. The speeds with which blight infection occurs and the devastating impact on the crop make it the biggest threat to the six million tones of potatoes produced in the UK each year. *Alternaria solani* is the cause of another disease popularly known as Early Blight of Potato is the disease of world-wide occurrence and distribution wherever potato is cultivated. The fungus inflicts young crop and may cause yield loss up to 40 per cent if not contained. The disease appears both in tropical and temperate regions. Rhizoctonia solani is also an important fungal organism developing Black Scurf or Rhizoctonia Stem Canker symptoms.

Constrain on Tomato

Alternaria solani develops the symptoms of Early Blight of Tomato. This is common disease of tomato occurring all over India and other tomato growing countries. The fungal pathogen attacks the foliage causing characteristic leaf spots and blight. *Fusarium oxysporum* f. *lycopersici* makes the tomato plant wilted while *Phytophthora infestans* is also a serious pathogenic fungus developing of Late Blight of Tomato crops and fruit.

Fungal Diversity on Fruits

Constrain of Mango

Oidium mangifera is a serious fungi developing as Powdery mildew.It attacks the leaves, flower scales, buds of tender flower heads, axils, stalks and fruits. It manifests itself as wefts of white mycelium on the affected parts (Rawal and Saxena, 1992). The outbreaks have been noticed in Uttar Pradesh in the last 6-7 years. *Colletotrichum gloeosporioides* perfect state is *Glomerella cingulata* is known to be serious pathogen developed anthracnose disease. It spreads in and on all the tender parts of the plant and is serious on tender twigs, flowers, flower stems and fruits in storage. In India, it is widely distributed in Bihar, Punjab, Maharashtra, Gujarat, Kerala, Uttar Pradesh, Rajasthan and Karnataka. Prakash and Srivastava (1987) reported 30-90 per cent losses in Lucknow. *Fusarium moniliforme* var. subglutinans is a serious pathogen causing malformation and witches broom or bunchy top or vegetative malformation with a proliferation of infected tissue. The flowering panicles turn into a compact mass of flowers. *Botryodiplodia theobromae* develop symptoms, namely, floral Stem end rot as the dark epicarp around the base of the pedicel in the initial stage. Also manifested by discoloration and darkening of the bark some distance from the tip of the twigs. The dark area advances and young green leaves start withering first at the base and then extending outwards along the vein. *Cephaleuros mycoides* an algae developing red spot is a in the tarai and in the other humid regions of India. It's occurred in Bihar, Karnataka and Uttar Pradesh. It attacks the foliage, bark and twigs of the host plant. In serious infections, the bark becomes thickened, the twigs get enlarged and remain stunted and the foliage becomes sparse and finally dries up. Initially the spots are greenish-grey in colour and velvety in texture, but later on the surface turns reddish-brown (Pathak, 1980).

Constrain of Citrus

Citrus ranks the third place among the fruits grown in India and thus assumes considerable importance. It grows much better in tropical and subtropical conditions and there is an increase in area under citrus cultivation from 90.7 thousand hectares in 1961 to 279.4 thousand hectares in 1988.

Phytophthora nicotianae var. *parasitica* causing Gummosis, Leaf fall, Root rot, Foot rot or Fruit rot and occurs in all the citrus growing areas especially in the high rainfall affected regions. *Fusarium lateritium* is widespread in Assam and is often associated with (Chowdhuri, 1951) symptoms of decline through: (1) rotting of the rootlets and (2) girdling of the trunk and a dropping of blighted leaves. *Colletotrichum gloeosporioides* develops anthracnose and the disease has been reported from Punjab, Madhya Pradesh, Uttar Pradesh, Assam, Rajasthan and South India. The disease is also known as withering tip which in recent years is responsible for decline in production of lime in North India, particularly in Kanpur area. Wither tip, leaf-spot and fruit rot, (Pathak, 1980). Anthracnose attacks the young leaves, shoots, blossoms and fruits of small acid limes. Young foliage and blossoms are blighted and distinct lesions formed on leaves and fruits. Affected fruits frequently drop prematurely. Wither tip is characterized by shedding of leaves and die-back of twigs. Leaves show light green spots which turn brown. On dead twigs, black dot like acervuli appear in concentric rings. *Acrosporium tingitanium* infects all the aerial parts of the plant are attacked. The disease is characterized by a whitish, mealy powdery growth on young leaves and twigs, which covers the entire leaf and young growing shoots. Even the small fruits also get infected. Such fruits shrink and become mummified. Affected leaves become distorted. *Elsinoe faweetti* commonly known for Scab of sour orange,

Rough lemon and Tangelo, and has been reported from West Bengal, Assam, Uttar Pradesh, Andhra Pradesh and Karnataka and is mostly prevalent in the nursery. It is also recorded on sweet oranges in Karnataka (Rawal and Ullasa, 1988). About 25-55 per cent of the fruits are completely spoiled in Assam (Chowdhuri, 1955) by scab. The fungus rarely leads to the leaf drop but foliage distortion is by the buff colored warts may develop thinness of the top in nurseries. *Colletotrichum gloeosporioides* characterized as die-back or twigs blight by shedding of leaves and is present in citrus is growing farms. *Fusarium* and *Diplodia natalensis* is generally found associated this disease as well is also caused by the association of *Fusarium* spp. and *Diplodia natalensis* along with *Colletotrichum gloeosporioides*.

Constrain of Jackfruit

Jackfruit (*Artocarpus heterophyllus*) is consumed as fruit as well as vegetable. The trees are supposed to be hardy but even then attack by the following fungi and become a limiting factor in its cultivation.

Colletotrichum gloeosporioides causing Leaf spot is the most common occurring disease on Jackfruit. The fungus develops dark brown to brick red spot on both the leaf surfaces. These spots after few days turn into grayish white centers with dark brown boundaries. Dark colored fungus acervuli are observed on grayish region which, when torn, ooze out profuse spore masses. The fungus has been seen to infect the young tender shoots and petioles of the leaves also. *Botryodiplodia theobromae* is a destructive pathogen of Jackfruit. . On the onset of die-back it becomes evident by discolorations and darkening of the back some distance from the tip. The dark area advances and young green twigs start withering first at the base and then extending outwards along the veins of leaf edges. *Botryobasidium solmonicolor* is widespread in tropical and subtropical areas. The fungus appears as a pinkish powdery coating on the stem. Pink colour represents profuse conidial production of fungus. Young woody branches of the affected trees lose their leaves and show die-back.

Constrain of Litchi

Litchi chinensis as such does not have much of fungal infection problems (Galan 1989). However, *Pestalotiopsis paueiseta, Botryodiplodia theobromae* and *Colletotrichum gloeosporioides* are fungal pathogens develop tar leaf spots, older leaves show higher percentage of infections than the younger leaves. The diseased leaves serve as the source of primary infection. The leaf spots spread rapidly during humid and rainy months. *Microdiplodia litchi* is another serious fungus of leaf spot of litchi. The diseased leaves show yellow brown to brick red areas mostly towards margins. The colored areas gradually become light brown and show black dot like Pycnidia. The other fungus recorded on litchi is *Clitocybe tabescence* in association with *Fusarium wilt*. *Phomopsis* die-back and *Gloeosporium* foliar necrosis is recorded from litch growing places.

Constrain of Apple

The apple is the pomaceous fruit of the apple tree refer to species *Malus domestica* in the rose family (Rosaceae). It is one of the most widely cultivated tree fruits, and the most widely known of the many members of genus *Malus* that are used by humans.

Apples grow on small, deciduous trees. The tree originated in Central Asia, where its wild ancestor, *Malus sieversii,* is still found today. Apples have been grown for thousands of years in Asia and Europe, and were brought to North America by European colonists. In India the production reaches 2, 203,000 Mt. tons in year 2012

Podosphaera leucotricha, an obligate biotrophic ascomycete fungus cause of Powdery mildew, is one of the major pathogen of cultivated apple (*Malus* × *domestica*) throughout the world. Infected plants are characterized by reduced photosynthesis and transpiration, resulting in suboptimal carbohydrate assimilation and reduced growth which is characterized by light grey powdery patches appearing on the leaves, shoots and flowers, normally in spring. The flowers will turn a creamy yellow color and will not develop correctly. This can be treated in a manner not dissimilar from treating Botrytis; eliminating the conditions which caused the disease in the first place and burning the infected plants are among the recommended actions to take (Urbanietz and Dunemann, 2000) The yield losses are caused by primary infection of flower buds and a general loss of vitality of the infected plant due to reduced assimilation. An infection of the fruits, however, is very rare.The pathogen is supposed to spread almost exclusively asexually, although indicated that ascospores might be an underrated additional source of infection. Mycelium overwintering in dormant buds normally initiates primary infection on young leaves, which produce inoculum in the form of conidia for the secondary cycles.

Venturia inaequalis is a heterothallic, haploid ascomycete, reproducing sexually and asexually (Khajuaria *et al.,* 2012) pathogen of Apple scab is causing major losses to apple growers throughout the world. Apple scab causes leaves to develop olive-brown spots with a velvety texture that later turn brown and become cork-like in texture. The disease also affects the fruit, which also develops similar brown spots with velvety or cork-like textures. Apple scab is spread through fungus growing in old apple leaves on the ground and spreads during warm spring weather to infect the New Year's growth (Bradley *et al.,* 2009).

References

Ahmed, H.U. (1985). Disease problems of pulses and oilseed crops in Bangladesh. Presented at the First Biennial Conference of the Bangladesh Phytopathological Society, 13-14 Apr 1985, Bangladesh Agricultural Research Institute, Joydebpur, Gazipur, Bangladesh.

Bailey, J.A and Jeger, M.J. (1992) *Colletotrichum*: Biology, Pathology and Control CAB International, Wallingford (**1992**), pp. 88–120.

Bloomberg. (2011). India Allows Wheat Exports for the First Time in Four Years."

Bradley, Fern Marshall; Ellis, Barbara W.; Martin, Deborah L., ed. (2009). *The Organic Gardener's Handbook of Natural Pest and Disease Control*. Rodale, Inc. pp. 32–34. ISBN 978-1-60529-677-7.

Chowdhuri, S. (1951). Gummosis of Citrus in Assam. *Indian J. Agric. Sci.,* **16**: 570-571.

Chowdhuri, S. (1955). *Citrus melanose* in Assam and its control. *Sci. and Cult.,* **20**: 323-335.

Christy Chamy (2014) Wheat rust: The fungal disease that threatens to destroy the world crop.The Independent Apr 20, 2014, 05.15 PM IST.

Dhandhalya M.G. and Shiyani, R.L. (2009) Production potentials, yield gaps and research prioritization of production constrains in major oil seed crop of Saurastra region. *Indian J. Agric. Res.*, **43** (1): 18-25,

Fakir.A. (1983). Pulse diseases and their control. (In Bangla.) Mymensingh, Bangladesh: Bangladesh Agricultural University. 14 pp.

FAO (2009) "Rapid growth of select Asian economies". Food and Agriculture Organisation of the United Nations.

Galan, S. V. (1989). Lychee Cultivation. FAO Plant Protection and Production Paper 83, FAO, Rome. 1991.

Japan Review. (2006)."Fish and Rice in the Japanese Diet."

Johansen, C., Baldev, B., Brouwer, J.B., Erskine, W., Jermyn, W.A., Li-Juan, L., Malik, B.A., Ahad-Miah, A., and Silim, S.N. (1994). Biotic and a biotic stresses constraining productivity of cool season food legumes in Asia, Africa, and Oceania. Pages 175-194 *in* Expanding the production and use of cool season food legumes (Muehlbauer, F.J., and Kaiser, W.J., eds.). Dordrecht. The Netherlands: Kluwer Academic-Publishers.

Khajuaria, Y.P; Kaul, S and Dhar, M.K (2012). Molecular characterization *Venturia inaequalis* causing Apple Scab in Kashmir.1: 399. Doi.10.4172/scientific report 339.

Kay, D.E. (1979). Chickpea (*Cicer arietinum*). Pages 48-71 *in* Crop and Product Digest No. 3, Food legumes. London, UK: Tropical Products Institute.

Khare, M.N. Agarwal, S.C., and Jain, A.C. (1979). Diseases of lentil and their control. Technical Bulletin. Jabalpur, India: Jawaharlal Nehru Krishi Vishwa Vidhyalaya

Khare, M.N. (1981). Diseases of lentil. Pages 163-172 *in* Lentils (Webb, C., and Haw tin, G., Eds) Farnham Royal, England, UK: Commonwealth Agricultural Bureaux.

Khajuaria, Y.P;Kaul, S and Dhar, M.K. (2012) Molecular characterization *Venturia inaequalis* causing Apple Scab in Kashmir.1: 399. Doi.10.4172/ *Scientific report*, p. 339.

Meena P.D, Sharma A.K, Jha S.K and Chattopadhyay C (2006) Impact of Fungal Diseases on Mustard Yield - Farmers' Perception and Garlic Clove Extract in Management of Sclerotinia Rot. *Indian Journal of Plant Protection*, **34**(2): 229-232

MoA(2012) *Ministry of Agriculture,* Department of *Agriculture* and Cooperation Government of India. *Annual report*. 2011-2012.

Nene, Y.L., Sheila, V.K., and Sharma, S.B. (1996). A world list of chickpea and pigeonpea pathogens. Fifth edition. Patancheru 502 324, 152 Andhra Pradesh, India: International Crops Research Institute for the Semi-Arid Tropics, pp. 27. (Semi-formal publication.)

Pande, S., and Joshi, P.K. (1995). Constraints and prospects of legumes in the rice-wheat based cropping system in Terai region of Nepal, Trip Report 7 Dec-31 Dec 1995. Patancheru 502 324, Andhra Pradesh, India: International Crops Research Institute for the Semi-Arid Tropics, (Limited circulation), 5 pp.

Paulitz, T C. and Steffenson B. J. (2011) Biotic Stress in Barley: Disease Problems and Solutions DOI: 10.1002/9780470958636.

Pathak, V.N. (1980). Diseases of Tropical Crops. Oxford and IBM Publishing Co. pp. 309.

Prakash, Om and Srivastava, K.C (1987). Mango Diseases and their Management-A World Review. Today and Tomorrow Pub., New Delhi., p 165.

Rawal, R.D. and A.K. Saxena. (1992). Influence of weather parameters on development of mango powdery mildew (abst.). 4th Inter. Symp. Mango held during July 4-11, 1992 at Florida, U.S.A.

Reddy, M.V., Sharma, S.B., and Nene, Y.L. (1990). Pigeonpea: disease management. Pages 303-347 *in* The pigeonpea (Nene, Y.L., Hall, S.D., Sheila, V.K., eds.). Wallingford, Oxon, UK: CAB International.

Prihastuti, H , Phoulivong S , Paul W.J. T, and . Hyde, K. D (2008) chilli Anthracnose diseases caused by *Collectotrichum* species. *J Zhejiang Univ Sci B.,* **9**(10): 764–778.

Urbanietz, A. and Dunemann, F (2000) Isolation, identification and molecular characterization of physiological races of apple powdery mildew (*Podosphaera leucotricha*). *Plant Pathology,* **2**(4): 125–133.

Warren, J.; Lawson, C. and Belcher, K. (2008). *The Agri-Environment.* Cambridge University Press. p. 224.

Tang, C.Y. (1980). Developments in crop protection in *Glycine.* Pages 323-325 *in* advances in legume sciences (Summerfield, R.J., and Bunting, A.H., eds.). London, UK: HMSO.

2015, Major Constrains and Verdict of Crop Productivity Pages 227–233
Editor: Dr. U.N. Bhale
Published by: ASTRAL INTERNATIONAL (P) LIMITED, NEW DELHI

Chapter 15

Physiological Impact of Sapota Post-harvest Fruit Fungi on Toxin Production of *Vigna aconitifolia* Seeds

U.N. Bhale[1]* and P.M. Wagh[2]

[1]Research Laboratory, Department of Botany,
Arts, Science and Commerce College, Naldurg,
Tq. Tuljapur, Osmanabad Dt. – 413 602, M.S., India
[2]Department of Biology,
S.S. and L.S. Patkar College of Arts and Science College, Goregaon (W),
Mumbai – 400 063, M.S., India

ABSTRACT

Phytotoxins are other important metabolites of fungi involved in the process of fruit pulp biodeterioration. Seeds of mataki (Vigna aconitifolia Jacq.) pre-soaked in cell free culture filtrate of post-harvest sapota fungi viz. Geotrichum candidum, Aspergillus niger and Rhizoctonia solani were evaluated. In present stud,y, impact of different media,incubation period, pH and temperature was studied. The results revealed that the toxicity of culture filtrates in case of all the three fungi was variable in 4 different media. Maximum toxic effect was observed in case of fungi from Richard's medium (66 per cent). It is clear from the result that with the increase in incubation period, toxicity also increased. Toxin production in all the tested fungi was observed at pH 3 and increased gradually

* *Corresponding Author.* E-mail: unbhale2007@rediffmail.com

with the increase in pH upto 6. Toxin production in all tested fungi was seen at 20 °C and increased upto 30°C. Optimum temperature for toxin production was between 30°C - 40°C in Aspergillus niger (70 per cent) while 30°C in Rhizoctonia solani (40 per cent) and Geotrichum candidum (60 per cent).

Keywords: *Achras sapota, Post-harvest pathogenic fungi, Vigna aconitifolia, Seed germination, Seedling parameters, Phytotoxicity.*

Introduction

Sapota (*Achras sapota* L.) belongs to the family sapotaceae and is an economically important edible fruit crop cultivated in tropical and subtropical regions of the world. A post-harvest fruit and food loss constitutes a vast complex of physical and biological changes due to microorganisms like fungi and bacteria. Mycotoxins are well known for their health–hazardous effects in human beings and animals (Probst *et al.*, 2007; Reddy and Raghavender, 2007). Fungal metabolites are substances discharged by fungi in their metabolic processes. The metabolites are products of some amino acids, cyclic peptides, aromatic, phenols, terpenoids and plant growth regulators (Griffin 1981; Madhosing, 1995). Fungi of the genera *Aspergillus, Fusarium, Penicillium* and *Rhizoctonia* are commonly known to produce toxic substances (Singh *et al.*, 1991). During the metabolic process, fungi secrete their metabolites in medium. These metabolites are of different types and are known to cause diseases in plants, animals and humans who use to eat this infected food (Schumann, 1991; Singh *et al.*, 1991).

Materials and Methods

Storage fungi: Storage fungi (*Geotrichum candidum, Aspergillus niger* and *Rhizoctonia solani*) were isolated from post-harvested sapota (*Achras sapota*) and maintained on PDA media.

a) Production of Phytotoxins

Evaluation of phytotoxin from the tested fungi was conducted in the present work. The test fungi isolated from sapota fruits were grown on Glucose nitrate medium. 25 ml of the medium was poured in 100 ml Erlenmeyer conical flasks and autoclaved at 15 lbs pressure for 20 minutes. The flasks on cooling were inoculated separately with 1 ml standard spore suspension of the test fungi prepared from 7 days old culture grown on PDA slants. The flasks were inoculated for 6 days at 25 ± 1°C. On the 7th day the flasks were harvested by filter the contents throughWhatman's filter No.l. The filtrates were collected in the pre-sterilized bottles and termed as crude toxin preparations, which is cell-free filtrate. These preparations were tested for their toxicity.

Assay of Phytotoxins (Umecharuba and Neachukwa, 1997)

The toxicity of culture filtrates was determined by using following method. Seed germination method- Surface sterilized 100 seeds of the test crops were soaked in crude toxin preparation for 24 hours. They were then placed on moist blotter in Petri plates. Seeds soaked similarly in freshly prepared uninoculated liquid medium served as control. Per cent germination or per cent inhibition of seed germination, root and shoot length of seedlings was measured after 7 days of incubation at room temperature.

Effects of different Sources on Toxin Production

Tested fungi on different media, they were cultured on four standard media (Glucose nitrate media, CMC media, Richard's and Czapek's media) and their effect was tested on per cent seed germination and root and shoot lengh of seeds of mataki (*Vigna aconitifolia* Jacq.). The fungi were grown for different period ranging from 5-25 days. The culture filtrates were collected and were tested for toxic effects on seeds. The tested fungi were grown at different pH values ranging from 3 to 9 and culture filtrates were tested for toxic effect on seeds. Three tested fungi were studied by growing them at different temperatures ranging from 10 °C to 50 °C. Culture filtrates obtained were studied for toxic effect on seeds. Untreated seeds served as control.

Results and Discussion

Effects of different media, incubation period, pH and temperature were studied on toxin production.

1) Effects of different Media on Toxin Production

The results are summarized in Table 15.1. The results revealed that the toxicity of culture filtrates in case of all the three fungi was variable in 4 different media. Maximum toxic effect was observed in case of fungi from Richard's medium (66 per cent) followed by Czapek's (64 per cent) and in case of Glucose nitrate medium (48 per cent) and in case of CMC media it was found to be least (40 per cent). Among the 3 fungi, culture filtrates of *Aspergillus niger* (66 per cent) showed maximum toxicity on seed germination in Richard's medium. Similar types of toxic effects are seen on root and shoot growth of germinating seeds.

2) Effects of Incubation Period on Toxin Production

From the data given in Table 15.2, it becomes clear that with increase in incubation period, toxicity also increased. It was noticed that after 5 days, *Geotrichum candidum* could produced toxin. The culture filtrates of three fungi harvested from 10th day onwards showed increase in toxicity with increase in the incubation period upto 25 days. On 25th day, *Aspergillus niger* (85 per cent), *Geotrichum candidum* (70 per cent) and *Rhizoctonia solani* (50 per cent) showed inhibition of seed germination.

3) Effects of pH on Toxin Production

The results given in Table 15.3 revealed that toxin production in all the tested fungi was observed at pH 3 and increased gradually with the increase in pH upto 6.Toxin production at pH 9 was not found in *Geotrichum candidum, Aspergillus niger* and *Rhizoctonia solani*. It was interesting to note that, optimum pH for toxin production was 6.0 in *Geotricum candidum* (60 per cent) and *Rhizoctonia solani* (50 per cent) whereas 4.5 in case of *Aspergillus niger* (60 per cent) for inhibition of seed germination.

4) Effect of Temperature on Toxin Production

It is evident from Table 15.4 that at 10°C, none of the fungus could produce toxin. Toxin production in all tested fungi was seen at 20 °C and increased upto 30 °C.The

Table 15.1: Effects of different Media on Toxin Production

Toxin Production in Sapota Fungi on different Media Fungi	Culture Filtrates from Media											
	Glucose Nitrate			Richards			Czapek's			CMC		
	Per cent Inhibition of Seed Germination	Root Length (cm)	Shoot Length (cm)	Per cent Inhibition of Seed Germination	Root Length (cm)	Shoot Length (cm)	Per cent Inhibition of Seed Germination	Root Length (cm)	Shoot Length (cm)	Per cent Inhibition of Seed Germination	Root Length (cm)	Shoot Length (cm)
Aspergillus niger	48	1.9	0.8	66	2.0	0.7	64	1.0	0.7	40	1.4	0.6
Geotrichum candidum	40	2.0	1.0	58	2.1	0.6	55	1.2	0.8	38	1.8	0.8
Rhizoctonia solani	45	2.2	1.2	60	2.8	0.9	58	0.9	0.4	34	1.2	0.9

maximum toxin production was at 30°C in all the fungi, *Aspergillus niger* (70 per cent), *Geotrichum candidum* (60 per cent) and *Rhizoctonia solani* (40 per cent). At 50 °C, there was low toxin production in *Geotrichum candidum* (20 per cent) and *Aspergillus niger* (25 per cent) while no toxin production in *Rhizoctonia solani*. Optimum temperature for toxin production was between 30°C - 40 °C in *Aspergillus niger* (70 per cent) while 30°C in *Rhizoctonia solani* (40 per cent) and *Geotrichum candidum* (60 per cent).

Table 15.2: Effect of Incubation Period on Toxin Production

Fungi	*Days*				
	5	*10*	*15*	*20*	*25*
	Per cent Inhibition of Seed Germination				
Aspergillus niger	20	70	75	80	85
Geotrichum candidum	15	55	60	65	70
Rhizoctonia solani	00	40	45	50	50
SEm±	6.0	8.66	8.66	8.66	10.13
CD (P=0.05)	19.10	27.53	27.53	27.53	32.23

Table 15.3: Effect of pH on Toxin Production

Fungi	*pH*				
	3	*4.5*	*6*	*7.5*	*9*
	Per cent Inhibition of Seed Germination				
Aspergillus niger	25	60	57	45	00
Geotrichum candidum	17	45	60	40	00
Rhizoctonia solani	10	17	50	25	00
SEm±	4.27	10.92	4.94	6.02	0.0
CD (P=0.05)	13.60	34.74	15.70	19.15	0.0

Table 15.4: Effect of Temperature on Toxin Production

Fungi	*Temperature (°C)*				
	3	*4.5*	*6*	*7.5*	*9*
	Per cent Inhibition of Seed Germination				
Aspergillus niger	00	60	70	70	25
Geotrichumcandidum	00	45	60	30	20
Rhizoctonia solani	00	30	40	25	00
SEm±	00	8.06	8.41	12.20	6.62
CD (P=0.05)	00	25.63	26.75	38.78	21.28

The role of toxic metabolites of pathogenic fungi in plant disease development has been reported by several workers. Harvested grains are colonized by various species of *Aspergillus*, under such conditions leading to deterioration and mycotoxin production (Reddy *et al.*, 2009). *Aspergillius* are the most common fungal species that can produce mycotoxins in food and feedstuffs. Mycotoxins are well known for their health–hazardous effects in human beings and animals (Probst *et al.*, 2007; Reddy and Raghavender, 2007). Christensen (1974) noted different types of seedling abnormalities and damages in seed health have been in different crop due to fungal toxin of seed-borne fungi. In present investigation, Richard's medium supported maximum toxinproduction in all the three tested fungi. The medium used for cultivation of fungi and its effect on toxin production has been studied by various workers (Vidyasekaran *et al.*, 1970; Mishra and Kanaujia, 1973). Charya and Reddy (1982) observed that, Richard's medium was found to be stimulatory than Czapek's medium for the maximum production of toxin by fungi. Enhancement of toxicity by Richard's medium was also reported by Mishra and Kanaujia (1973). It was reported that the filtrate of *A.niger, Penicillium chrysogenum* and *Macrophomina phaseolina* inhibited percentage seed germination, with increase in filtrate age, soaking time in all fungal filtrates It means that metabolites are discharged by the tested fungi in the media in which they were grown (Kakde *et al.*, 2010). *R. solani* toxin is heat stable because the culture filtrate still showed high toxicity after exposure to a temperature of 121°C for 30 min (Chen *et al.*, 1992). Fungi were grown for different incubation period from 5-25 days, increased incubation period, increased toxicity, but *Aspergillus niger* showed maximum inhibition of seed germination than *Geotrichum candidum and Rhizoctonia solani* (Wagh and Mangaonkar,2012) Sinha and Prasad (1981) recorded inhibition of seed germination in mung due to *Alternaria alternata, urvularia lunata* and *Macrophomina phaseolina.* Haikal (2008) reportedthat toxic metabolites secreted by *Aspergillus niger, Fusarium* percentage of seed germination of soyabean. The culture filtrates of pathogen *F.oxysporum f.sp. udum* were inhibited the seed germination and seedling growth of pigeonpea after seven days of incubation period (Jalonder and Gachonde, 2011). Recently, Chatage and Bhale (2013) reported *Bipolaris tetramera* was found high toxic on sorghum seed (10 per cent) and low toxic for sunflower seed (90 per cent). *Geotrichum candidus* was found high toxic on maize seed (10 per cent) while low toxic on fenugreek seed (80 per cent).

References

Charya, M.A.S. and Reddy, S.M. (1982). Toxic effects of *Fusarium oxysporum* on seed germination and growth of mung (*Vigna radiata*). *Indian Bot. Reptr.*, **1**(2): 169-170.

Chatage, V.S. and Bhale, U.N. (2013).Phytotoxic activity of fungal pathogens inciting ivy gourd Fruit. *International Journal of Biology, Pharmacy and Allied Sciences*, **2**(11): 2123-2129.

Chen, J., Yao, J.M., Liu, W.Z. and Song, Z.H. (1992).Preliminary study on toxin of *Rhizoctonia solani* in rice sheath blight, *J. Shenyang Agric. Univ.*, **23**(1): 19-22.

Christensen, C.M. (1974). Storage of cereal grains and their products. St. Paul, Minn.: Am.Assoc.*Cereal chem.*, **51**: 549.

Graffin DH.(1981). Fungal Physiology. John Wiley and Sons, New York, p 383.

Haikal, N.Z. (2008). Effect of filtrate of pathogenic fungi of soyabean on seed germination and seedling parameters. *J.App. Sci. Res.*, **4**(1): 48-52.

Jalonder, V and Gachonde, B.D.(2011). Effect of cutture filtrates of fusarium oxysporum on seed germination and seedl | ng growth of pigeonpea (*Cajanus cajan*) varities. *Bioscience Discovery*, **2** (2): 185-188.

Kakde, R.B., Gadgile, D.P., Pangrikar, P.P., Hatti, A.D., Gaikwad, R.S. and Chavan, A.M. (2010). Soyabean seed germination status under the influence of fungal metabolites. *Bionano frontier*, **3** (1): 116- 117.

Madhosing, C. (1995). Relative wilt-inducing capacity of the culture filtrates of isolates of *Fusarium oxysporum f.sp. radicis-lycopersrcr*, the tomato crown and root-rot pathogen. *J. Phytopathol*, **4**: 193-198.

Mishra, R.R. and Kanaujia, R.S. (1973). Studies on certain aspects of seedborne fungi-II.Seed –borne fungi of certain oil seeds. *Indian Phytopath.*, **26**: 284-294.

Pradnya, M. Wagh and Kiran V. Mangaonkar(2012). Toxic potentialities of fungal metabolites on germination status of Mataki seeds(*Vigna aconitifolia* Jacq.). *Elixir Appl. Biology*, **53** : 11898-11899.

Probst, C., Njapau, H., and Cotty, P. J. (2007). Outbreak of an acute aflatoxicosis in Kenya in 2004: Identification of the causal agent. *Applied and Environmental Microbiology*, **73**(8): 2762–2764.

Reddy, B. N., and Raghavender, C. R. (2007). Outbreaks of aflatoxicoses in India. *African Journal of Food, Agriculture, Nutrition and Development*, **7**(5): 1-15.

Reddy, K.R.N., C.S. Reddy, Muralidharan, K. (2009). Potential of botanicals and biocontrol agents on growth and aflatoxin production by *Aspergillus flavus* infecting rice grains. *Food control*, **20**: 173-178.

Schumann, G.L. (1991). Plant Diseases: their biology and social impact.*The American Phytopath. Soci., St.Paul, Minnesota*, p.377.

Singh, K, Frisvad, J.C., Thrane, U. and Mathur, S.B. (1991). An iltustroted Manual on tdentification of Some Seed-Borne Aspergilli, Fusario, Pencillia ond their Mycotoxins. AioTryk as Odense, Denmark, p. 133.

Sinha, R.K. and Prasad, T (1981). Effect of Fungal metabolites on seed germination, microbial association and seedling growth of mung. *Indian Phytopath.*, 34(4): 515-517.

Umecharuba, G.I. and Neachukwa, E.O. (1997). The effect of filtrates of seedborne fungi of African yam-bean on seed germination and seedling development. *Global J.Pure and Appl.Sci.*, **3**: 165-176.

Vidhyasekaran, P., C.L. Subramanian and Govindaswamy, C.V. (1970). Productionof toxin by seed-borne fungi and its role in paddy spoillage. *Indian Phytopath.*, **23**: 518-525.

2015, Major Constrains and Verdict of Crop Productivity *Pages 234–237*
Editor: Dr. U.N. Bhale
Published by: ASTRAL INTERNATIONAL (P) LIMITED, NEW DELHI

Chapter 16

Climatic Factors cause Plant Diseases in Crops

Veena Sharma* and A.K. Singh

Regional Agricultural Research Station, SKUAST-J, Rajouri – 185 131, J&K

Introduction

The earth's climate has always changed in response to changes in the cryosphere, hydrosphere, biosphere and other atmospheric and interacting factors. It is widelyaccepted that human activities are now increasingly influencingchanges in global climate (Pachauri and Reisinger, 2007).Greenhouse gas concentrations in the atmosphere are being altered by Human activities, thus causing global climate change. These activities, intensified after the Industrial Revolution at the end of the eighteenth century result from the use of natural resources such as fossil fuel burning, deforestation and other land use changes.

Several changes in the climate have been registered. Changes will probably continue to happen even if greenhouse gas concentrations stabilize, due to the system's thermal inertia and to the long period necessary for returning to a lower equilibrium (IPCC, 2007).The importance of the environment on the development of plant diseases has been known for over two thousand years. The impact of elevated carbon dioxide, elevated ozone and higher atmospheric temperatures on plant diseases could challenge crops in these changing conditions.Climate change poses a threat to the control of pest and disease invasions. These "pests and diseases" include insects, plant diseases. Climate change is affecting plants in natural and agricultural

* *Corresponding Author.* E-mail: veena_sharma29@rediffmail.com, ashoks286@gmail.com

ecosystems throughout the world (Stern, 2007). Severity of human, animal and plant disease epidemics is greatly affected by climatic factors, especially temperature and rainfall (Wint *et al.*, 2002; Thomson *et al.*, 2006). Human-driven changes in the earth's atmospheric composition are likely to alter plant diseases of the future.

Diseases are responsible for losses of at least 10 per cent of global food production, representing a threat to food security (Strange and Scott, 2005). Besides direct losses, the methods for disease control, especially the chemical methods, can result in environmental contamination and in residual chemicals in food, in addition to social and economic problems.

As climate variables continue to change in one place, new pests and diseases may become able to invade previously uninhabitable areas. Some crop diseases and pests are of one place, while others originate from other regions of the continent or world. In their native regions, potential pest species are kept under control by predators and other ecosystem processes. However, once introduced to a different region, these pests and diseases may no longer have natural predators or other environmental variables to control their population size. Climate change is aiding these invasions by widening the "invasion niche," or the set of environmental conditions under which pests can successfully invade.

The impacts of climate change can be positive, negative or neutral, since these changes can decrease, increase or have no impact on diseases, since there can be a decrease, an increase or no effect on the different pathosystems, in each region. The analysis of the potential impacts of climate change on plant diseases is essential for the adoption of adaptation measures, as well as for the development of resistant cultivars, new control methods or adapted techniques, in order to avoid more serious losses (Ghini, 2005).

Effect of Rising CO_2 Level

Researchers predict carbon dioxide will reach levels double those of the preindustrial era by the year 2050, complicating agriculture's need to produce enough food for a rapidly growing population. Elevated carbon dioxide levels are more likely to have a direct effect on plant diseases through changes to the plant hosts rather than the plant pathogens.

Both enhancement and reduction in disease severity under elevated CO_2 has been reported. Elevated CO_2 would increase canopy size and density of plants, resulting in a greater biomass production and microclimates may become more conducive for rusts, mildews, leaf spots and blights development. Decomposition of plant litter is important for nutrient cycling and in the saprophytic survival of many pathogens. Because of high C:N ratio of litter as a consequence of plant growth under elevated CO_2, decomposition will be slower. Increased plant biomass, slower decomposition of litter, and higher winter temperature could increase pathogen survival on over-wintering crop residues and increase the amount of initial inoculation available for subsequent infection.

Elevated carbon dioxide levels are more likely to have a direct effect on plant diseases through changes to the plant hosts rather than the plant pathogens. Plants growing in a high carbon dioxide environment tend to grow faster and larger, and they have denser canopies. The dense plant canopies favor the development of some diseases because the low light levels and reduced air circulation allow higher relative humidity levels to develop, and this promotes the growth and sporulation of many plant pathogens. At the same time, plants grown in high carbon dioxide environments also close their stomata, pores in the leaves that allow the plant to take in carbon dioxide and release oxygen, more often. Because plant pathogens often enter the plant through the stomata, the more frequent closing of the stomata may help prevent some pathogens from getting into the plant.

Malmström and Field (1997) observed that the CO_2 enrichment stimulated the growth of oats infected with the barley yellow dwarf virus (BYDV), attenuating the dwarfing symptom. Rice plants were grown in ambient and elevated CO_2 concentrations, and were artificially inoculated with *Magnaporthe oryzae*. Rice plants grown in an elevated CO_2 concentration were more susceptible to leaf blast than those in ambient CO_2 as indicated by the increased number of leaf blast lesions. Plants grown under elevated CO_2 concentration had lower leaf silicon content, which may have contributed to the increased susceptibility to leaf blast under elevated CO_2 concentrations. (Kobayashi *et al.*, 2006).

Manning and Tiedemann (1995) observed an upward trend in diseases from 1930 to 1993due to increase in CO_2 concentration. The potential effects of higher CO_2 concentration on plant diseases was analyzed based on the plant responses to the new environment. The increase in plant biomass production, *i.e.*, the increase in production of shoots, leaves, flowers and fruit, represents more tissue that can be infected by pathogens. Increased carbohydrate contents can stimulate the development of sugar-dependent pathogens, such as rusts and powdery mildews. Increases in canopy density and plant size can promote higher growth, sporulation and spread of leaf infecting fungi, which require high air humidity, but not rain, as rusts, powdery mildews and leaf necrotrophs. The increase in crop residues can represent better survival conditions for necrotrophic pathogens. The reduction in stomatal opening can inhibit stomata-invading pathogens, such as rusts, downy mildews and some necrotrophs. The shortened growth period and accelerated ripening and senescence can reduce the infection period for biotrophic pathogens, and increase the necrotrophic pathogen populations. The increase in root biomass increases the amount of tissue that could be infected by mycorrhyza or soil-borne pathogens, but can compensate the losses inflicted by the pathogens. Higher root exudation can stimulate both pathogens and antagonistic microbiota in the rhizosphere (plant growth promoters). Chakraborty and Pangga (2004) concluded that of the 26 diseases studied most of them increased in severity when in CO_2-enriched environments.

Effect of Rising Ozone

Many farmers are unaware that ozone is reducing their yields. Ozone enters the plant's leaves through its gas exchange pores (stomata), just as other atmospheric gases do in normal gas exchange. It dissolves in the water within the plant and reacts with other chemicals, causing a variety of problems. Plant physiologists are still trying to understand the specific pathways and locations of ozone's effects within plant cells. Some cell membranes become leaky, possibly because of ozone's ability to interact with lipid (fatty) components and/or membrane proteins. Photosynthesis slows, resulting in slower plant growth. Compounds resulting from oxidation by ozone interfere with the cell's energy production in the mitochondria. Such ozone-induced compounds also decrease the numbers of flowers and fruits a plant will produce, and they impair water use efficiency and other functions.Plants weakened by ozone may be more susceptible to pests, disease, and drought.

Elevated CO_2 alone or in combination with O_3 significantly increases brown spot severity in Soybean though the increase is small in magnitude (Eastburn, 2010).In elevated ozone, plant growth is inhibited and results in shorter plants with less dense canopies. This can slow the growth and reproduction of certain pathogens. However, ozone also damages plant tissues that can help pathogens infect the plant more easily. Elevated levels of carbon dioxide and ozone can make a plant more susceptible to some diseases, but less susceptible to others (Eastburn, 2010).

The effects of O_3 on leaf surface properties results in increased incidence of rust(Karnosky *et al.*, 2002). Percy *et al.* (2002) investigated in a FACE experiment the impacts of elevated CO_2 and O_3 on poplar leaf rust (*Melampsora medusae*) in natural disease outbreaks. In this study, rust infection increased fourfold under enhanced O_3, infection remained almost threefold higher as compared to the control leaves.

Effect of Elevated Ultraviolet B

There is considerable information on the effects of increased UV-B on crops and natural vegetation and on the growth and life cycle of pathogenic organisms such as fungi. Studies indicate that the UV-B component of solar radiation plays a natural regulation on plant diseases. Stimulatory effect of near-UV light on reproduction of many fungi, and spore production in Leptosphaerulina trifoli peaks at 287 nm are reported. Fungi differ in their sensitivity to UV-B. Some strains of *Septoria tritici* are more sensitive to UV-B than others and *S. nodorum*, as a species, is more sensitive than *S. tritici*. UV-B radiation can modify the relative composition of phylloplane organisms, such as pink and white yeast. Continued exposure to enhanced UV-B radiation lowers the level of antifungal compounds in foliar parts. UV-B has been shown to reduce tolerance of rice to blast (*Pyricularia grisea*) and although higher UV-B reduced plant biomass and leaf area; there was no increase in blast severity. There are some evidences that sunlight can influence pathogen by causing accumulation of phytoalexins or protective pigments in plant tissues. Therefore, UV-B may affect plant diseases directly via the pathogen or indirectly via the host (Usha and Parimal, 2008).

In four of ten studies, UV-B was found to counteract disease severity and in the other six studies, it promoted disease development (Manning and Tiedemann, 1995). The direction of the UV-B radiation effect on disease severity can also vary with the variety of the host. In a rust-resistant variety of wheat, additional UV-B radiation had little effect, but it promoted the rust infection in a rust-sensitive wheat variety (Manning and Tiedemann, 1995). It is not clear in many of these experiments whether the changes in disease severity were due simply to changes caused by UV-B radiation in the host plant, or whether direct UV-B radiation effects on the fungal or viral pathogens was involved. Cucumber plants first exposed to UV-B radiation were more susceptible to subsequent infection by two fungal pathogens if the host plants were exposed to UV-B radiation prior to infection; but UV-B irradiation after infection had no effect on disease severity (Orth *et al.*, 1990). Such an experiment suggests the effect of UV-B radiation was mediated through changes in the host plant tissues. There is also evidence from solar UV-B exclusion studies showing increased incidence of fungal disease when UV-B is removed (Gunasekera *et al.*, 1997).

Effect of Rising Temperature

The average global surface temperature has increased by 0.2°C per decade in the past 30 years (Hansen *et al.*, 2006). Eleven of the twelve warmest years ever registered by instruments since 1850 occurred between 1995 and 2006, with the exception of 1996. Alterations in the water cycle have also been observed.

Increases in temperature can modify host physiology and resistance. Both temperature and the length of exposure are important in determining the effect of climate change on disease severity. Temperature change might lead to appearance of different races of the pathogens hitherto not active but might cause sudden epidemic. Change in temperature will directly influence infection, reproduction, dispersal, and survival between seasons and other critical stages in the life cycle of a pathogen.

High temperature can cause sunscald in fruits and leaves, it can be broken down into 2 types, sub lethal (Figure 16.1) and lethal (Figure 16.2). Sub lethal sunscald can be described as a yellow, hard area usually on the shoulder of the fruit. This occurs when tissue temperature rises above about 86° F. The high tissue temperature will not allow the red pigment to develop nor the flesh to soften, but allows the yellow pigments to develop. With lethal sunscald, the tissue turns white and dies. Many times the dead tissue will turn black from fungi that are feeding on the dead tissue. Lethal sunscald occurs when tissue temperatures rise above 104° F. Damage usually occurs when fruits are suddenly exposed to sunlight. This most frequently occurs

Sunscald in Tomato Fruits.

Figure 16.1: Sub-lethal Sunscald.

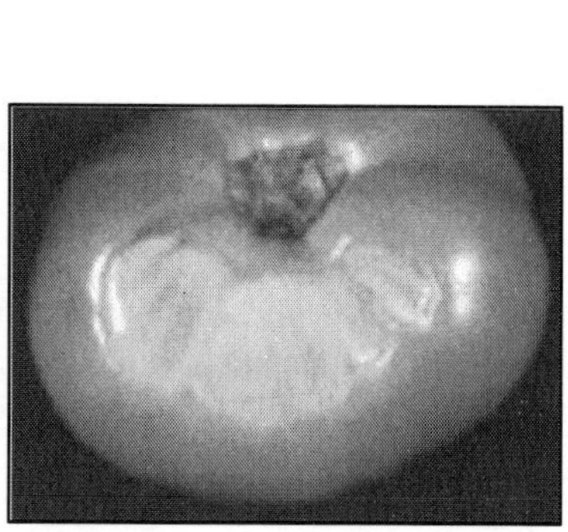

Figure 16.2: Lethal Sunscald.

Figure 16.3: Sunscald on Leaves.

after a harvest or a storm when leaves are moved around and fruit exposed. Over pruning can also increase sunscald problems especially with fruit in the upper part of the plant. Sometimes plant leaves may be effected Figure 16.3 good foliage cover can reduce the problem of fruits. Growers at times may use a sun screen material such as Snow or Surround to help reduce sunscald.

A rise in temperature above 20°C can inactivate temperature sensitive resistance to stem rust in oat cultivars. Increase in temperature with sufficient soil moisture may increase evapo-transpiration resulting in humid microclimate in crop canopy and may lead to incidence of diseases favoured under warm and humid conditions. Some of the soil-borne diseases may increase at the rise of soil temperature. If climate change causes a gradual shift of cropping regions, pathogens will follow their host.

Temperature is one of the most important factors affecting the occurrence of bacterial diseases such as Ralstonia solanacearum, Acidovorax avenae and Burkholderia glumea. Thus, bacteria could proliferate in areas where temperature-dependent diseases have not been previously observed (Kudela, 2009). As the temperature increases, the duration of winter and the rate of growth and reproduction of pathogens may be modified (Ladanyi and Horvath, 2010). In apple, due to lesser rainfall in the rainy season and more severe summers, incidence of different canker diseases due to fungal pathogens is increasing (Sharma, 2012).

A change in temperature may favor the development of different inactive pathogens, which could induce an epidemic. Due to changes in temperature and precipitation regimes, climate change may alter the growth stage, development rate, pathogenicity of infectious agents, and the physiology and resistance of the host plant (Chakraborty and Datta, 2003). In cool sub-tropical zones such as Japan and northern China, elevation of ambient temperature resulted in greater risk of blast epidemics. Situations in the humid tropics and warm humid subtropics were opposite

Figure 16.4: Frost Injury in Plant.

to those in cool areas. A lower temperature resulted in greater risk of blast epidemics. Frost heaving results from the alternate freezing and thawing of soils. Heaving exposes the roots to cold and desiccation and is especially a problem on newly planted or shallow-rooted plants. Fall planted ground covers, perennials and small container shrubs are highly vulnerable. Ground cover plugs and small container plants should be fully planted in soil. This may sound obvious, but with a tendency to plant high, these plants are often found planted in the mulch rather than completely surrounded by soil. In addition, a uniform mulch layer aids in preventing rapid soil temperature fluctuations. In larger container plants, planting high in the mulch layer may expose the upper level of the root system enough to influence plant quality the following spring. Proper planting technique is the best defense against frost heaving.

Effect of Rising Precipitation

A local increase in summer precipitation due to climate change has been found to be responsible for the increase and spread of needle blight in British Columbia

(Woods and Coates, 2005). In California and Oregon, sudden oak death mortality rates abruptly increase and then subside. The pattern is driven by heavy rains and extended wet weather during warm periods which create optimal infection conditions. Infected trees suffer a reduced capacity to manage water, but survive until high temperatures and extended dry periods overwhelm their vascular capability, resulting in death. Two cycles of this pattern have been noted in California during 1998 to 2001 and 2005 to 2008 (Frankel, 2007).

Increases in frequency and intensity of summer storms with high winds, rain, and hail will increase wounding of plants and result in increased transmission of viruses by mechanical means.Therefore, with predicted changes in climate, viral diseases of plants are expected to increase in importance.

Effect of Hurricanes

An increase in severe weather events such as hurricanes may also catalyze the spread of crop diseases such as soybean rust across continents or oceans (Rosenzweig *et al.*, 2001; Schneider and Hollier, 2005).

Agriculture is strongly influenced by weather and climate. While farmers are often flexible in dealing with weather and year-to-year variability, there is nevertheless a high degree of adaptation to the local climate in the form of established infrastructure, local farming practice and individual experience. Climate change can therefore be expected to impact on agriculture, potentially threatening established aspects of farming systems but also providing opportunities for improvements. Climate change directly impacts crops, as well as their interactions with microbial pests (Rosenzweig *et al.*, 2000).

Although there are a large number of studies which focus on the impact of a particular aspect of climate change in a specific location, there are relatively few studies which provide a global assessment. A comprehensive, internally consistent assessment of all potential direct and indirect effects of climate change on agricultural productivity has not yet been carried out.

Information derived from climate change studies will help us prepare for the changes ahead by knowing which diseases are most likely to become more problematic. Now is the time for plant pathologists, plant breeders, agronomists and horticulturalists to adapt disease management strategies to the changing environment.

Conclusion

Disease management strategies depend on climate conditions. Climate change will cause alterations in the disease geographical and temporal distributions and consequently the control methods will have to be adapted to this new reality. Changing disease scenario due to climate change has highlighted the need for better agricultural practices and use of ecofriendly methods in disease management for sustainable crop production. The adaptability of some agricultural systems can help to minimize the negative impacts of climate change with the adoption of new cultivars and other practices. In the changing climate and shift in seasons, choice of crop management practices based on the prevailing situation is important. Developing countries, though, will have more difficulties in adapting to climate change, due to the lower technological development and to the scarce resources available for the adoption of measures.

References

Chakraborty, S., and Datta, S. (2003). How will plant pathogens adapt to host plant resistance at elevated CO_2 under a changing climate? *New Phytol.*, **159**: 733-742.

Chakraborty, S. and Pangga, I.B.(2004). Plant disease and climate change.In: Gillings, M., Holmes, A. (Ed.) Plant microbiology. London: BIOS Scientific. p.163-180.

Eastburn, D.M., Degennaro, M.M., DeLucia, E.H., Dermody, O. and McElrone, A.J. (2010). Elevated atmospheric carbon dioxide and ozone alter soybean diseases at Soy FACE. *Global Change Biology*, **16**(1): 320-330.

Frankel, S.J. 2007. Climate change's influence on sudden oak death, PACLIM 2007. May 13-15. 2007, Monterey, CA.

Gunasekera, T.S., Paul, N.D. and Ayres, P.G. (1997). The effects of ultraviolet-B (UV-B: 290-320 nm) radiation on blister blight disease of tea (*Camellia sinensis*). *Plant Pathology*, **46**: 179-185.

Ghini, R. (2005). Mudançasclimáticasglobais e doenças de plantas.Jaguariúna: Embrapa MeioAmbiente.p104.

Hansen, J., Sato, M., Ruedy, R., Lo, K., Lea, D.W., Medina-Elizade, M. (2006).Global temperature change. Proceedings of the National Academy of Sciences.**103:** 14288-14293.

IPCC. Climate change (2007): the physical science basis. Geneva: IPCC, 2007, p. 996 (Assessment Report, 4).

Karnosky, D.F., Percy, K.E., Xiang, B.X., Callan, B., Noormets, A., Mankovska, B., Hopkin, A., Sober, J., Jones, W., Dickson, R.E., Isebrands, J.G. (2002). Interacting elevated CO_2 and tropospheric O_3 predisposes aspen (PopulustremuloidesMichx.) to infection by rust (*Melampsora medusae,f. sptremuloidae*). *Global Change Biology.*, **8**: 329-338.

Kobayashi, T., Ishiguro, K., Nakajima T, Kim, H. Y., Okada, M. and Kobayashi, K. (2006). Effects of Elevated Atmospheric CO_2 Concentration on the Infection of Rice Blast and Sheath Blight. *Ecology and Epidemiology*, **96** (4): 425-431.

Kudela, V. (2009). Potential impact of climate change on geographic distribution of plant pathogenic bacteria in Central Europe. *Plant Protect. Sci.*, **45**: S27-S32.

Ladanyi, M., and Horvath, L. (2010). A review of the potential climate change impact on insect populations – general and agricultural aspects. *Appl. Ecol. Environ. Res.*, **8**: 143-152.

Malmström, C.M., Field, C.B. (1997).Virus-induced differences in the response of oat plants to elevated carbon dioxide. *Plant, Cell and Environment.*, **20**: 178-188.

Manning, W.J., Tiedemann, A.V. (1995). Climate change: potential effects of increased atmospheric carbon dioxide (CO_2), ozone (O_3), and ultraviolet-B (UV-B) radiation on plant diseases. *Environmental Pollution.*, **88**: 219-245.

Orth, A.B., Teramura, A.H. and Sisler, H.D. (1990). Effects of ultraviolet-B radiation on fungal disease development in Cucumissativus. *American Journal of Botany* **77**: 1188-1192.

Pachauri RK, Reisinger A, (2007). Climate Change 2007: Synthesis Report. Contribution of Working Groups I, II and III to the Fourth Assessment Report of the Intergovernmental Panel on Climate Change. Geneva, Switzerland: IPCC.

Percy, K.E., Awmarck, C.S., Lindroth, R.L., Kubiske, M.E., Kopper, B.J., Isebrands, J.G., Pregitzer, K.S., Hendrey, G.R., Dickson, R.E., Zak, D.R., Oksanen, E., Sober, J., Harrington, R., Karnosky, D.F. (2002). Altered performance of forest pests under atmospheres enriched by CO_2 and O_3. *Nature.*, **420**: 403-407.

Rosenzweig, C., Iglesias, A., Yang, Y. B., Epstein, P. R., and Chivian, E. (2000). Climate Change and U.S. Agriculture: The Impacts of Warming and Extreme Weather Events on Productivity, *Plant Diseases and Pests*. Boston, MA, USA: Center for Health and the Global Environment, Harvard Medical School.

Rosenzweig, C. and A. Iglesias, (2001). "Climate Change and Extreme Weather Events, Implications for Food Production, Plant Diseases, and Pests." *Global Change and amp, Human Health*, **2**(2): 90-104.

Schneider, R. W., C. A. Hollier, (2005). "First Report of Soybean Rust Caused by Phakopsorapachyrhizi in the Continental United States." *Plant Disease*, **89**(7): 774-774.

Sharma, I. M. (2012).Changing disease scenario in apple orchards: Perspective, challenges and management strategies. In Proceedings of the National Symposium on Blending Conventional and Modern Plant Pathology for Sustainable Agriculture, Indian Institute of Horticultural Research, Bengaluru, 2012.

Stern, N. (2007) The economics of climate change: the Stern reviewCambridge, *UK: Cambridge University Press.*

Strange, R.N., Scott, P.R. (2005). Plant disease: A threat to global food security. *Annual Review of Phytopathology*.**43**: 83-116.

Thomson, M.C, Doblas-Reyes F.J, Mason S.J, Hagedorn R, Connor S.J, Phindela T, Morse A.P, Palmer T.N. (2006) Malaria early warnings based on seasonal climate forecasts from multi-model ensembles.*Nature.*, **439**: 576–579.

Usha, Mina and ParimalSinha.(2008).Effect of climate change on plant pathogens. *Environews.*, 14(460-66).

Woods, A., K. D. Coates, (2005). "Is an Unprecedented Dothistroma Needle Blight Epidemic Related to Climate Change?" *BioScience*, **55**(9): 761-769.

Wint, G.R.W, Robinson T.P, Bourn D.M, Durr P.A, Hay S.I, Randolph S.E, Rogers D.J 2002Mapping bovine tuberculosis in Great Britain using environmental data *Trends Microbiol.*, **10**: 441–444.

2015, Major Constrains and Verdict of Crop Productivity Pages 244–249
Editor: Dr. U.N. Bhale
Published by: ASTRAL INTERNATIONAL (P) LIMITED, NEW DELHI

Chapter 17

Impact of Cadmium Phytotoxicity on Photosynthetic Rate and Chlorophyll Content in *Triticum aestivum* L. DWR 225 Variety

V.S. Bheemareddy[1]* and H.C. Lakshman[2]

[1]Department of Botany,
J.S.S. Banashankari Arts, Commerce and S.K. Gubbi Science College,
Dharwad – 580 004, Karnataka, India
[2]P.G. Department of Botany,
Karnataka University Dharwad – 580 003, Karnataka, India

ABSTRACT

Higher plants take up nutrients present in the form of minerals. Many heavy metals like Cu, Zn, Ni and Co are essential for plants. Cadmium is a heavy metal constituting a nonliving component of the biosphere, and occurs naturally in soils and plants. It is a nonessential element ranked as one of the most hazardous heavy metal. Triticum aestivum *L. DWR 225 variety is widely cultivated as irrigated crop plant in North Karnataka. Present research work was aimed to evaluate the impact of Cd^{+2} toxicity levels on Photosynthetic rate and total chlorophyll content of DWR 225 variety at 50mg/kg and 100mg/kg soil. Cd^{+2} phytotoxicity harmfully affect both photosynthetic rate and chlorophyll content. Photosynthesis is harmfully affected by Cd^{+2} phytotoxicity at lowest concentration. It is well evident*

* *Corresponding Author.* E-mail: venpra1964@yahoo.co.in

from the results that, Cd^{+2} were found to be more lethal at 100mg/kg concentration to Triticum aestivum L. DWR 225 variety. Cd^{+2} phytotoxicity not only reduce total chlorophyll content but also the photosynthetic rate.

Keywords: *Cadmium, Phytotoxicity, Triticum aestivum* L., *Chlorophyll content and Photosynthetic rate.*

Introduction

Plants come across various kinds of environmental stresses. Currently heavy metal (HM) pollution is a serious environmental concern. Heavy metal pollution has increased due to mining activities (Ross, 1994), industries, automobile exhausts, and increased use of agrochemicals (Wagner, 1993). Cadmium, Zinc, Nickel,and Mercury are the heavy metals released in to enviornment. The presence of heavy metals (HM) in excess amounts is a global problem, Cadmium is a heavy metal (specific density exceeding 5 g/cm^3, and relative atomic mass above 40; Seregin and Ivanov, 2001) constituting a nonliving component of the biosphere, and occurs naturally in soils and plants (Page *et al.*, 2006). Cadmium (Cd^{+2}) is non essential metal present in the soil. It accumulates in soil due to various anthropogenic activities like industrial emissions, sewage pollution, and use of Cd^{+2} containing Phosphate fertilizers (Adriano 1986). Cadmium is also an environmental hazard, human exposures to environmental cadmium are primarily the result of fossil fuel combustion, cement production and related activities (Morrow,2010). It is a non essential element ranked as one of the most hazardous heavy metal (Kim *et al.*, 2003).

Plants are able to absorb and accumulate nutrients including those with unknown metabolic functions. Higher plants take up nutrients present in the form of minerals. Many heavy metals like Cu, Zn, Ni and Co are essential for plants. Cadmium present in soil is absorbed by the roots along with other minerals. It affects normal functioning in plants and known to induce many physiological changes such as leaf chlorosis, growth inhibition, disturbance of respiration, nitrogen metabolism, and reduction of photosynthesis (Wagner, 1993).

Bread wheat (*Triticum aestivum* L.) is a global food crop and it is one of the important crops in India. It is grown in different agro climatic regions. In Karnataka *Triticum aestivum* L. DWR 225 variety is widely cultivated. Phytotoxicity effect of cadmium at different concentrations is not yet studied in *Triticum aestivum* L. DWR 225 variety. The aim of present research work is to evaluate the impact of Cd^{+2} toxicity on Photosynthetic rate and total chlorophyll content in *Triticum aestivum* L. DWR 225 variety at 50mg/kg and 100mg/kg soil.

Materials and Methods

Plant Cultivation

Grains of DWR 225 variety were collected from wheat research station, University of Agricultural sciences Dharwad, India. Experiments were conducted in poly house using 30cm diameter earthen pots. About four kg soil was taken in each pot. Six grains were sown in each pot. Plants were regularly watered on alternate days till the harvest. Hoagland's solution was applied at every fifteen days till the time of harvest.

Treatments

Experimental plants were subjected to following treatments.

Treatment 1: Plants were cultivated in sterile soil taken in pots without Cadmium

Treatment 2: Plants were cultivated in sterile soil taken in pots containing Cadmium in the form of Cadmium Chloride ($CdCl_2$) at the concentration of 50 mg/kg soil.

Treatment 3: Plants were cultivated in sterile soil taken in pots containing Cadmium in the form of Cadmium Chloride ($CdCl_2$) at the concentration of 100 mg/kg soil.

Plants of the above treatments are maintained in triplicates and arranged in randomized design.

Photosynthetic Rate

Photosynthetic rate was studied by using IRGA (Infra Red Gas Analyzer, LICOR 6400, USA) in flag leaf at 90 DAS (Days After Sowing) at 1500 PAR (Photo synthetically Activated Radiation). Flag leaf is inserted in to the leaf cuvette of IRGA. Measurements were taken between 9 AM to 11 AM inside the laboratory at 27°C - 29°C room temperature. Infra red gas analyzer was installed with reference (incoming) and sample (Outgoing) CO_2 and water (H_2O) concentrations. Before measuring a fast check up is performed according to check list to confirm good condition of the instrument. An attached LED (Light Emitting Diode) light Source was used to ensure stable light source during measurements.

Total Chlorophyll Content

Total leaf chlorophyll content was determined with SPAD-502 Portable Chlorophyll meter (Minolta Camera Co.Ltd. Japan). First leaf (from the top) was used for the measurement of total chlorophyll content. The leaf was inserted into sample slot of the measuring head. Press and hold it until a beep sounds and the measurement value appeared in the displays.

This value will automatically store in the memory. The values measured by the SPAD-5O2 correspond to the amount of Chlorophyll present in the leaf. The values are calculated based on the amount of light transmitted by the leaf in two wave length regions in which absorbance of chlorophyll is different.

Results and Discussion

Results obtained from the experiments revealed that Cd^{+2} phytotoxicity harmfully affect both photosynthetic rate and chlorophyll content. Total chlorophyll content was reduced with increase of Cd^{+2} concentration. Reduction of 20 per cent chlorophyll content was observed in plants treated with 50mg Cd^{+2}/kg soil treatment. Drastic decrease of chlorophyll content was observed in plants treated with 100 mg Cd^{+2}/kg soil treatment. Almost 50 per cent reduction in the total chlorophyll content was

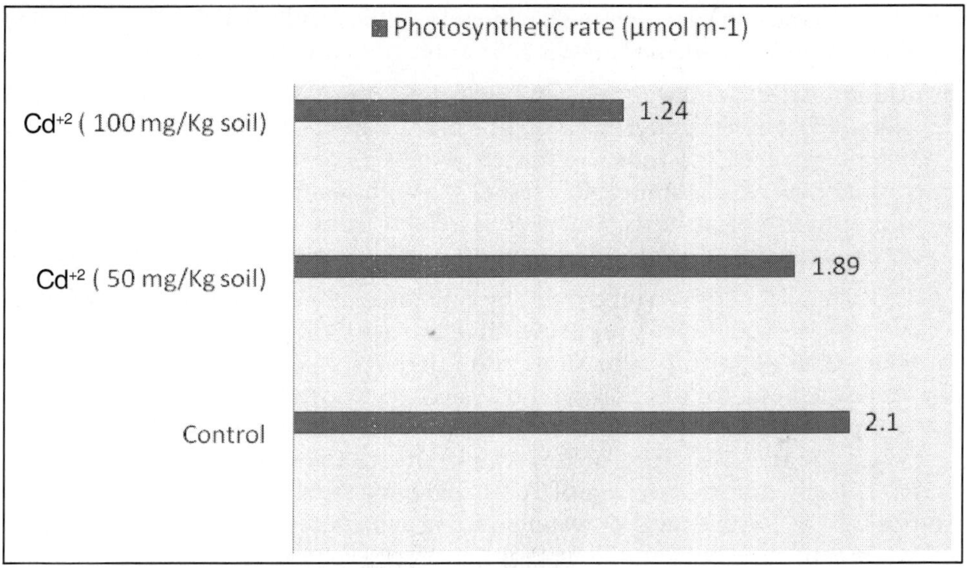

Figure 17.1: Effect of Cd^{+2} Phytotoxicity on Photosynthetic Rate.

observed in plants treated with 100 mg Cd^{+2}/Kg soil treatment compared to those grown under control conditions.

Photosynthesis is also affected by Cd^{+2} phytotoxicity at different concentration. There was reduction in photosynthetic rate with increase of Cd^{+2} concentrations. At 50mg Cd^{+2}/Kg soil, about 10 per cent decreases in the photosynthetic rate was observed. There was 40 per cent decrease in the photosynthetic rate at 100 mg

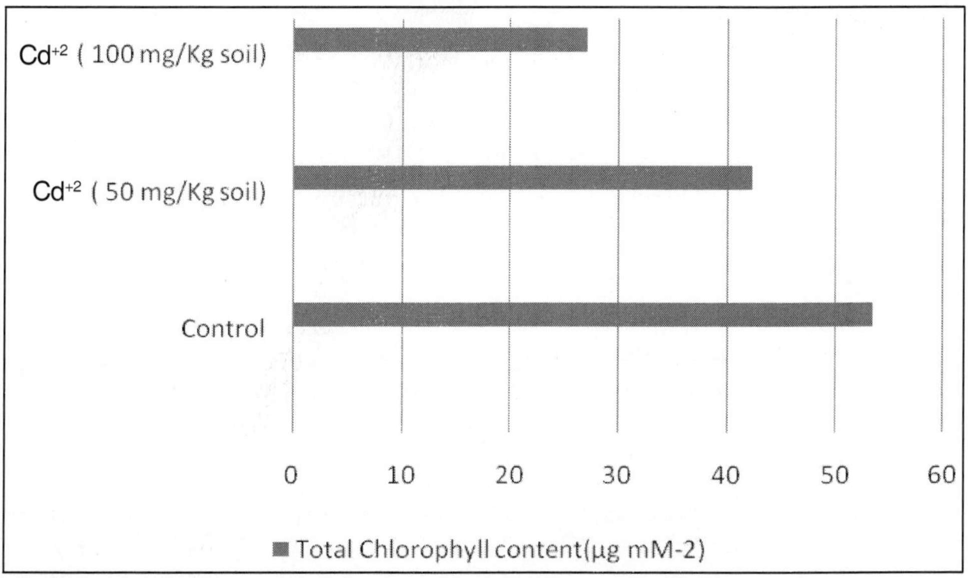

Figure 17.2: Effect of Cd^{+2} Phytotoxicity on Total Chlorophyll Content.

Cd^{+2}/kg soil concentration. It is well evident from the results that Cd^{+2} was found to be lethal to *Triticum aestivum* L. DWR 225 variety at 100mg/kg concentration.

Chlorophyll content was reduced due to toxicity of Cd^{+2}. Mysliwa- kurdziel and Strazatka, (2002) investigated the inhibition of chlorophyll biosynthesis. Stiborova *et al.* (1986) explained the reduction in chlorophyll content with increase in the concentration of Cd^{+2}. Muthuchelian *et al.*(2001) reported the inhibitory effect of Cd^{+2} on chlorophyll biosynthesis. Oncel *et al.* (2000) reported the effect of Cd^{+2} on biosynthesis in wheat seedlings. These findings strongly support the present findings.

Cd^{+2} altered the photosynthetic rate by inhibiting different reaction steps of Calvin cycle. Lucero *et al.* (1976) reported that at low concentrations Cd^{+2} acts as photophosphorylation inhibitor. Chug and Sawhney,(1999) explained that Cd^{+2} was very effective inhibitor of photosynthetic electron transport. Strzatka and Ketner, (1997) reported Cd^{+2} affect membrane bound protein complex of PS I, which is necessary for the oxidation of ferrodoxin under light reactions. Gregar and Ogren,(1991) investigated the role of Cd^{+2} in reducing the rate of photosynthesis. Cd^{+2} inhibits the PS II at the level of water splitting system resulting in the decrease of photosynthetic rate (Walker, 1987).

Malik *et al.* (1992) reported that Cd^{+2} inhibits the activity of RuBp carboxylase, further role Cd^{+2} in the inactivation of sulfohydryl proteins participating in photosynthesis. Present findings supports the research findings of earlier workers.

Conclusion

Cd^{+2} phytotoxicity not only reduce total chlorophyll content but also the photosynthetic rate. At 100 mg/kg soil concentration Cd^{+2} was found to be more lethal to *Triticum aestivum* L. DWR 225 variety.

Acknowledgements

We are thankful to Prof. B. N. Patil, Director, Wheat research station, University of Agricultural Sciences, Dharwad, for providing wheat germ plasm and Prof. S. B. Patil Dean, Agriculture research station, Hebballi for extending laboratory facilities. Our sincere thanks to Dr. S. N. Hegde Principal, J.S.S. College, Dharwad and Dr. Ajith Prasad Finance Officer Janata Shikshan Samiti, Dharwad for their encouragement to conduct this research.

References

Adriano, 1986. Trace Elements in the Terrestrial Environment. Springer, NewYork.

Chug, LK., Sawhney, S.K.(1999). Photosynthetic activities of *Pisum sativum* seedlings grown in presence of cadmium. *Plant Physiol Biochem*., **37**: 297-303.

Gregar, M., Ogren, E. (1991).Direct and indirect effects of Cd^{+2}on photosynthesis in sugar beet (*Beta vulgaris*). *Physiol Plant*,. **83**: 129-135.

Kim, C.G., Bell, J.N.B., and Power, S.A. (2003). Effects of soil cadmium on *Pinus sylvestris* L. seedlings. *Plant and Soil*, **257**: 443–449.

Lucero, H.A., Andreo, C.S., Vallejos, R.H. (1976). Sulfohydryl groups in photosynthetic energy conservation.II Inhibition of photophosphorylation in spinach chloroplastsby $CdCl_2$. *Plant Sci Lett.,* **6**: 309-313.

Malik, D., Sheoran, I.S., Singh, R. (1992). Carbon metabolism in leaves of cadmium treated wheat seedlings. *Plant Physiol Biochem.,* **30**: 223- 229.

Morrow, H. (2010). "Cadmium and Cadmium Alloys". *Kirk-Othmer Encyclopedia of Chemical Technology*. John Wiley and Sons. pp. 1–36.

Muthuchelian, K., Bertamini, M., Nedunchezhian, N. (2001).Triacontanol can protect *Erytrhina variegata* from Cd^{+2} toxicity. *Journal of Plant Physiology*, **158**: 1487-1490.

Mysliwa- Kurdziel B., Strazatka, K. (2002). Influence of metals on biosynthesis of photosynthetic pigments. In PRASAD MNV STRAZATKA K. (eds) physiology and biochemistry of metal toxicity and tolerance in plants.Kluwer, Dortrecht, pp. 201-227.

Oncel, I., Kele, S., Ustun, A.S. (2000).Interactive effects of temperature and heavy metal stress on the growth and some biochemical compounds in wheat seedlings. *Enviorn Pollut.,* **107**: 315-320.

Page, V., Le Bayon, R.C., and Feller, U. (2006). Partitioning of zinc, cadmium, manganese and cobalt in wheat (*Triticum aestivum*) and lupin (*Lupinus albus*) and further release into the soil. *Environmental and Experimental Botany, 58*: 269–278.

Ross, S.(1994). Toxic metals in soil plant systems. Wiley, Chichester, p. 469.

Seregin, I.V., and Ivanov, V.B. (2001). Physiological aspects of cadmium and lead toxic effects on higher plants. *Russian Journal of Plant Physiology,* **48:** 523–544.

Stiborova, M., Doubravova, M., Brezinova, A., Fredrich, A. (1986). Effect of heavy metal ions on growth and biochemical characteristics of photosynthesis of Barley. *Photosynthetica*., **20:** 418-425.

Strzatka, K., Ketner, P.(1997). Carbon dioxide. In: *Plant Ecophysiology*. Wiley New York, pp. 393-456.

Wagner, G.J. (1993) Accumulation of cadmium in crop plants and its consequences to human health. *Adv Agron,*. **51:** 173- 213.

Walker, D.A. (1987). The use of the oxygen electrode and fluorescence probes in simple measurements of photosynthesis, 145. U. K: Oxygraphics Limited, University of Sheffield Print Unit.

2015, Major Constrains and Verdict of Crop Productivity *Pages 250–264*
Editor: Dr. U.N. Bhale
Published by: ASTRAL INTERNATIONAL (P) LIMITED, NEW DELHI

Chapter 18

Fusarium Head Blight of Wheat and its Management

Vibha* and Susma Nema

Senior Scientist,
Department of Plant Pathology,
Jawaharlal Nehru Krishi Vishwa Vidhayalaya, Jabalpur, M.P., India

Introduction

India has recorded the self sufficiency in food grain production and; contribution of wheat production to this self sufficiency is one of the prime factors. The sustainable production of wheat is under heat of changing climatic conditions that is raising a big question mark to meet out the demand of ever increasing population of the world. The weather information throughout the world clearly indicating the global warming and; its consequence on emergence of new or minor diseases in more destructive forms are most likely to occur. *Fusarium* head blight caused by fungal pathogen is much dependent on weather conditions and are likely to become a major disease of all wheat growing areas of the world. Though, occurrence of this disease is sporadic but may become major threat in future in India, under changing climatic conditions.

Fusarium head blight (FHB) of wheat is one of the most important fungal diseases of all wheat growing areas of the world, including India. FHB infection is favored by warm and humid condition during flowering and early stages of kernel development (Gilbert and Takauz, 2000).A total of 17 species of *Fusarium* have been described to be potentially associated with FHB symptoms (Leonard and Bushnell, 2003).The disease is caused by the two major pathogenic genera of fungi: *Microdochium* and *Fusarium*

* *Corresponding Author.* E-mail: vibhapandey93@gmail.com

(Bottalico, 1998; Daamen and Langerak, 1991). However, out of five main species of *Fusarium* (*Fusarium graminearum, Fusarium triticum, Fusarium poae, Fusarium culmorum and Fusarium avenaceum*) and two sub- species of *Microdochinum nivale* (*M. nivale nivale* and *M. nivale majus*), the *F. graminearum* and *F. culmorum* are the most pathogenic (Champeil *et al.*, 2004). In India, *F. avenaceum*, F. *graminearum*, *F. moniliforme*, *F. subglutinans*, *Fusarium compactum* and *M. nivale* are found to be associated with head scab of wheat (Saharan *et al.*, 2004). Although, temperature and humidity are the main determining factor for its distribution and dominance, but are not the sole factor responsible for its population composition. Limited soil tillage or no tillage, crop succession, irrigation, nutrition, sowing date, weed management etc. are other factors responsible for disease establishment and development (Champeil *et al.*, 2009; Koch *et al.*, 2006; Schaafsma and Hooker, 2007).

FHB causes total or partial ear premature senescence with a consequent reduction in both crop yields and grain quality (Pirgozlien *et al.*, 2003). Moreover, the extent of loss magnifies many fold due to secretion of mycotoxin by the pathogen. The deposition of mycotoxins such as deoxynivalenal (DON) and its acetylated derivatives { (3- and 15- acetyldeoxynivalenol (3-and 15- A DON)}, results in the contamination of harvested grains (Xu and Berries, 2005). DON, a member of the trichothecene mycotoxins is primarily produced by the genus *Fusarium* (Sutton, 1982). The occurrence of DON in cereals grain is of great concern for human health, since trichothecenes are associated with dermal toxicity, toxic aleukia, food refusal, vomiting and depressed immune functions (Snijders and Perkowaski, 1990). Besides, mycotoxin causes two types of problem: firstly, it generates technological problem such as negative effect on malting of beer or on fermentation or even on production of high quality bread (Bakan, 1998). Secondly, it destroys starch grains, protein stores and cell wall (Parry *et al.*, 1995). The absorption of trichothecens has resulted in the poisoning of animals and humans in U S, Canada, the former Soviet Union and Japan (Vidal, 1990). Effects on human have been observed after the absorption of 300- 800 µg/kg deoxynivalenol in India (Battalico, 1998). Hence, its integrated management through disturbing primary source of inoculums, changing cropping system, crop rotation, nutrient management and judicious use of fungicides will certainly help in reducing to pathogen's population and further DON contamination in wheat grains.

Host Pathogen Interaction

Fusarium is known to cause vascular wilt; the fungus invades the roots and grows directly to vascular elements. Contrary to it, *F. graminearum* infects wheat spike tissues primarily through anthers that protrude from the glumes, and then spreads to the epidermis of glumes and to the ovaries (Xu and Hideki, 1989).The FHB- wheat pathosystem involves extreme variation in *Fusarium* pathogenicity and the existing differences in resistance among cultivars (Mesterhazy, 1987). The high levels of choline and betaine produced in extruded anthers stimulate the growth of *F. graminearum* and promote the infection of wheat spike by the pathogen at anthesis stage (Strange and Smith, 1978). These substances favor the extension of conidial hyphal but not the germination of spores of *F. avenaceum, F. culmorum* and *F. graminearum*. In addition, ear emergence (Rapilly *et al.*, 1973; www. synenta.ca, 2002) or the mid-grain-filling stage (Anderson, 1948) are consider as the most susceptible stage to attack in wheat, because susceptibility decreases strongly after the start of the dough stage (Lacey *et al.*, 1999). The extent of resistance to primary infection and development of fungal hyphae in different wheat cultivar is probably due to the coincidence of favourable climatic condition with critical receptive stage of the crop. Air temperature, humidity at receptive or before this stage, development of the inoculums, and the stage of maturity of the plant at the time of hyphal penetration are critical factors responsible for establishment of disease (Teich, 1989; Andersen, 1948; Pugh *et al.*, 1933). However, before penetration of host tissue, *Fusarium* species produce cell degrading enzymes and/or mycotoxins for weakening host defense mechanism and also aiding fungal penetration and colonization of host tissues (Kang and Buchenauer, 2000). *Fusarium graminearum* and *F. calmorum* are known to secrete enzymes that decomposes xylans, pectins, and cellulose of the host cell walls (Kang and Buchehauer 2000; Wanyoike *et al.*, 2002).Moveover, *Fusarium* species move towards the rachis or towards the young grains which they invade via the parenchyma of the pericarp close to the ovary (Kang and Buchenaur, 2000; Wanjiru *et al.*, 2002; Schroeder and Christensen, 1963). *F. graminearum* hyphae penetrate the thinner cell walls of the parenchyma tissue more easily than the thinner cell walls of other more specialized tissue (Pugh *et al.*, 1933). Upon penetration of the host, *F. graminearum* may form sub- cuticular hyphae between the cuticle and epidermal cell wall on the wheat spike (Pritsch *et al.*, 2000; Jansen *et al.*, 2005). The pathogen first penetrates host tissues 36-48 h. after inoculation (Kang and Buchehauer, 2000), whereas the penetration of the fungus into the ear is favored by relatively low temperature and high humidity (Andersen, 1948; Teich, 1989). Besides, birds also facilitate infection by creating lesions on the grains that favors the penetration of fungi. Although, infection may occur at temperatures of 20-30°C, but is negligible at temperatures below 15°C. A period of 4-8 days of low humidity reduces the incidence of infection, but does not entirely eliminate it (Andersen, 1948). The duration of required incubation period decreases with increasing humidity (Les Fusarioses, 1993). The germination of macroconidia, ascospore and chlamydospores is maximum between 0 and – 20 bar and inhibited between -60 and -80 bar (Sung and Cook, 1981). Ascospore germination is inhibited beyond a threshold of – 30 bars (-30 M Ps) of water potential after 8 hours of drought (Paulitz, 1996). In condition of high humidity for longer time, more spores are formed that may then result in infection to next crop.

There is no evidence for pathogenic specialization within *F. graminarum and F. culmorum* (Parry *et al.*, 1995) but genotypic diversity (Saharan *et al.*, 2004) and aggressiveness among isolates of *F. graminearum* has been reported by several workers (Bai and Shaner 1994, 1996; Parry *et al.*, 1995). To establish the variability among the isolates of *Fusarium* spp. and group them into their nearest lineage; band sharing coefficient was calculated through cluster analysis. Eight separate genetic lineages of *F. graminearum* were identified (O' Donnell *et al.*, 2004). In India, twenty isolates of different *Fusarium* species were identified with differential inter and intra species-specific aggressiveness. Out of them, *F. graminearum* isolates of Dalang Maidan and Wellington were more aggressive than Ludhiana and Gurdaspur isolates of *F. graminearum* as well as other *Fusarium* species/isolates (Saharan *et al.*, 2004).

Disease Symptoms

FHB is caused by complex of *Fusarium* species that can be isolated from the all parts of wheat including roots, stems, ears and leaves (Mudge *et al.*, 2006). The species *F. avenaceaum, F. culmorum, F. graminearum* and *F. tricinctum* colonizes the wheat leaves differ in their aggressiveness, that result in variability in necrotic lesion formation, which further limit their usage in disease diagnosis (Wagacha *et al.*, 2012). The fungi have brief biotrophic relationship with leaves depending upon the aggressiveness of isolates that causes discoloration around necrotic lesion, either to deep brown or deep red, by all *Fusarium* species except *F. poe* (Wagacha *et al.*, 2012). There is contradictory view on the role of stomata in *Fusarium* colonization. Wagacha and coworkers (2012) reported the occasional use of stomata for host penetration. In contrast, Bushnell *et al.* (2003) described a significant contribution of stomatal entry in the colonization of glumes, palea and lemma that enclose the floret.

Inter and intra-cellular growth of *Fusarium* hyphae in tissue surrounding necrotic leaf cells confirme its association on glumes, kernel and the rachis (Zange *et al.*, 2005). Water soaked spots at the base or middle of glume and discoloration of glume spot in all direction from the point of infection are commonly seen that are covered with salmon-pink fungal growth (Saharan *et al.*, 2004).Additionally, premature death or bleaching of spikelet is also a common symptoms and can be seen on emerged immature heads, where one or more spikelet's or the entire head may be affected (Wiese, 1987). In severe infection, the peduncle may turn dark brown (Parry *et al.*, 1995). Although, symptoms produced by *M. nivale* is almost invisible but sometimes it becomes difficult to distinguish from the symptoms produced by *F. culmorum, F. graminearum* and *F. avenaceum* infection (Cassini, 1981). The similarity between head blight caused by *M. nivale* and *F. graminearum* was first reported in Pacific Northwest of the USA by Inglis and Cook (1981).Systemic infection of wheat stem tissue by *F. culmorum, F. graminearum* and *M nivale* upto the second stem node (Clement and Parry, 1998) and; gradual upward movement of *Fusarium* species have been linked to a sequence of symptomless infection which develop following rain splash dispersal of conidia from lower to higher leaves (Zinkernagel *et al.*, 1997). Furthermore, grains infected with *F. graminearum* become smaller, shriveled and white to pale in color (Abramson *et al.*, 1987).

Epidemiology

Weather parameters influence the disease incidence and its severity. As discussed earlier, that the infestation of pathogenic *Fusarium* complex vary due to function of climate (Parry and Jenkinson, 1995; Scott *et al.*, 1998). The development of *Microdochium* is favored by lower temperature (Parry and Jenkinson, 1995; Pissinger, 1982) and rainfall (Caron, 2000) at flowering stage, while that of *Fusarium* at high temperature (Pissinger, 1982) and storms (Caron, 2000). *Fusarium culmorum, F. avenaceum* and *M. nivale* are predominant species of cooler regions, whereas *F. graminearum* is predominant in warmer regions (Parry *et al.*, 1995, Caron, 1993; Cromey *et al.*, 2002). Although, all species of *Fusarium* and *Microdochium* can survive saprophytically(Parry *et al.*, 1995) but these pathogens can survive longer on residues that do not degrade easily such as stem nodes (Sutton, 1982). Additionally, seeds are the major source of inoculums, making it possible for disease to begin in autumn from grain contaminated with mycelium (Cassini, 1970; Caron, 1993). Hence, seed borne infections are more likely to take hold over soil borne infections. Moreover, four types of asexual inoculums may be produced are; macroconidia, microconida, chlamydospore and hyphal fragments, depending on the species of pathogen (Messialen and Cassini, 1968). The macroconidia of *F. graminearum* may contain one to seven septa, most frequently three to seven and, measure 20 x105 μm in length (mostly 35-62 μm), and 2-56 μm in width (Sutton, 1982). This type of inoculums may be available throughout the crop cycle (Caron, 1993). Although, *M. nivale, F. graminearum* and *F. avenaceum* are capable of producing ascospore in natural (Parry *et al.*, 1995; Caron, 1993, Rapilly *et al.*, 1973) but, their size and form of ascospore are almost similar. These ascospore generally have three septa and measure about 17.5 – 26 μm x 3.5-5μm (Sutton, 1982). Ascospores take long time to mature and therefore contamination cannot occur before ear emergence (Caron, 1993). Light is required for the production and maturation of perithecia (Trail *et al.*, 1998) and ascospore (Tschanz *et al.*, 1976), whereas the process of ascospore release does not directly require light (Paulitz, 1996). The ascospore release began at temperature varying between 11 and 30°C (hold for 17 to 18 hr.) and relative humidity of 60 and 95 per cent (Pautitz, 1996). Daily minimum temperature below 9°C and maximum temperature greater than 26°C registered around anthesis may inhibit the growth of the fungus(Moschini and Fortugno, 1996) Moreover, aerial contamination by ascospores and conidia is thought to be principal source of ear contamination (Atanasoff, 1920) that involves two possible modes of dispersal: splashing and wind (Sutton, 1982; Mc Muller *et al.*, 1997). Splashing transports spores, macroconidia in particular, that is too heavy to be transported by wind (Gilbert and Tekauz, 2000). The relative importance of these two modes of dispersal depends on the climate and the species making up the pathogenic complex in the year and region studied and, the capacity of these species to produce ascospores.

Weeds such as *Locium, Digitaria* and *Setaria* may serve as primary source of inoculum for *F. graminearum* isolates that survive on these weeds during winter (Pereyra and Dill-Macky, 2008). *F. graminearum, M. nivale* and *F. avenaceum* produce ascospore, which contribute to the local inoculum, may travel longer distance to initiate disease in fields away from the primary source of inoculum (Kohl *et al.*, 2007). Finally, some *Fusarium* species can also persists in soil as saprophytic mycelium or as thick walled resting spores (chlamydospores). Landschoot *et al.* (2011) recorded the predominance of *F. culmorum* population was in crop residues and soil, *M. nivale* in weeds, crop residues and in soil in the month of November; while the population of *F. graminearum* and *F. poe* on wheat ear in July. They also suggested that the *Fusarium* population in July is more complex than the population at the beginning of the season, in November.

Factors Affecting Deoxynivalenol Contamination in Wheat

FHB infections can reduce grain quality by producing a variety of mycotoxins of which the trichothecene deoxynivalenol (DON) appears to be the most commonly encountered (Parry *et al.*, 1995). This mycotoxin is a protein synthesis disruptor and can cause serious health problems for both human of animals (D' Mello *et al.*, 1998).

Variation in DON content depends on weather conditions (Edwards, 2009), virulence of fungal isolates, tillage and other agronomic practices (Lori *et al.*, 2009), presence of resistance in different wheat cultivars (Wilde and Miedaner, 2006,) and use of fungicides (Ramirez *et al.*, 2004). Deoxynivalenol belongs to the trichothecene family which contains two types of toxins: type A{include toxins T—, HT, diacetoxyscirpenol(DAS) and scirpenol} and type B(include deoxynivalenol, vomitoxin and acetyle derivates of DON) (Vidal, 1990). Type A toxin acts on initiation of protein elongation, whereas type B acts on both elongation and termination steps of protein synthesis (McLaughlin *et al.*, 2000). In plants, deoxynivalenol retard the germination and growth of wheat, and inhibit the growth of grain and coleoptiles tissues (Snijders and Perkowski, 1990). However in adult plants, DON seems to circulate in the phloem, with the concentration of these molecules in the plant following a descending gradient from the rachis, through lemmas and grains to the peduncle (Sinha and Savard, 1997). Additionally, from the fourth day after inoculation, the flower parts, rachis and peduncle contain larger amount of deoxynivalenol below the point of infection than above it (Savard *et al.*, 2000).

Role of preceding crop is very important as they provide shelter to the pathogenic fungi to cause the infection and DON contamination in succeeding crop. There was six times higher concentration of DON was recorded in the wheat following maize rotation than those of wheat following another cereals (wheat, barley) or soybean (Teich and Hamilton, 1985). Similarly, it was recorded that cropping system involving ploughing (productive, integrated and organic) with direct drilling system over 2 years reported that levels of deoxynivalenol, zearalenone and nivalenol contamination were highest for the drilling system (Champeil *et al.*, 2004). The deoxynivalenol (DON) content of harvested grain was significantly correlated with FHB incidence and severity. The DON level in wheat following soybeans, averaged across tillage treatments, was 25 per cent lower than in wheat following wheat and 50 per cent of the level in wheat following corn. Furthermore, varietal maturity also plays significant role in DON accumulation, the early maturing varieties accumulates more deoxynivalenol than do late flowering varities(Jenney *et al.*, 2000). Although, very limited informations are available on effect of mineral nutrition on mycotoxin production, but top dressing of nitrogen increases grain-N content, simultaneously increases deoxynivalenol content and consequently risk to wheat crop(Thimmerman *et al.*, 2009; van der Burgt *et al.*, 2011). Additionally, potassium, phosphorus and pH did not appear to affect deoxynivalenol concentration significantly (Teich and Hamilton, 1985).

There is differential effect of DON production under the fungicide treatment for FHB disease management. The stimulation or reduction in DON production in the presence of fungicide is influenced by complex interaction between a_w, temperature, fungicide concentration and time of incubation (Ramirez *et al.*, 2004). In contrast to it, in condition of natural contamination, the application of fungicides does not significantly decrease the concentration of deoxynivalenol and nivalenol. There is also not necessarily a correlation between the application of fungicide and quantity of deoxynivalenol and/or nivalenol found in grain (Magan *et al.*, 2002).

Management

As a consequence of climate change and global warming, emergence of mycotoxin producing virulent species of *Fusarium* are becoming a new threat to quality and safety of wheat grains. Hence, integrated management of this disease through cultural, biological and chemical means will help in mitigating the risk to crop. The extent of infection can be reduced by taking proper tillage operation (including minimum soil erosion), but simultaneously exposes the residue of previous crops that harbor the pathogen to adverse climatic conditions. Similarly, maize, soybean, sorghum and wild oats should not be taken after wheat as they serve as alternate host of *F. graminearum*. Besides, cultivation of those crop that produce volatiles organic compounds such as chickpea can reduce the pathogenic population of *Fusarium* (Cruz *et al.*, 2012). Additionally, uses of non-conventional chemicals/plant hormones are effective in disease control as they play critical role in plant defence against pathogens. Salicyclic acid (SA), jasmonic acid and ethylene are recognized as key players (De Vos *et al.*, 2005: Koorneef and Pieterse, 2008) in plant defence system. As salicyclic acid is predominantly associated with resistance against biotrophic and hemi-trophic pathogens, and with triggering systemic acquired resistance (SAR) in many species, including *Arabidopsis thaliana* and wheat (Gorlach *et al.*, 1996; De Vos *et al.*, 2005). Reduction in efficiency in germination and growth at higher concentration of SA has been recorded in F. graminearum (Qi *et al.*, 2012). Several investigators have focused on finding affordable and environmentally compatible biocontrol agents for the management of FHB. Biocontrol agents could play an important role in organic cereal production. In conventional production, such agents may extend protection of spikes past the flowering stage after fungicides can no longer be applied. Certain strains of spore- producing bacteria (such as *Bacillus* species) and yeasts (such as *Cryptococcus flavescens*) show some promise for the control of FHB and the reduction of mycotoxin contamination (SchmaleIII and Bergstrom, 2003).Most bacterial and yeast strains for FBH are presumed to inhibit the disrease through the production of antibiotics or through competition for nutrients (Da Luz *et al.*, 2003). Use of endophytic fungi and bacteria, found on graminaceous host, has ability to confer ecological benefits to its host could be exploited for disease control. *Acremonium zeae*, a corn endophytic, was found to inhibit kernel rot and mycotoxin production by *Aspergillus flavus* and *Fusarium verticillioides* (Wicklow *et al.*, 2005), and, thus, it is conceivable that endophytic micro-organisms could be found with the ability to inhibit *F. graminearum* while growing in cereal tissues. Additionally, use of enzymes found in soil micro-oganisms could be exploited for detoxification and degradation of DON. *Agrobacterium- Rhizobium* group strain E3-39(Shima *et al.*, 1997) and mixed microbial culture(Volki *et al.*, 2004) are reported to possess such DON detoxification and degradation enzymes. Plants also can be source of such enzymes; aUDP-gluycosyltransferase that can detoxify DON was reported from *Arabidopsis thaliana* (Poppenberger *et al.*, 2003). Genetic manipulation of key genes or enzymes involved in DON biosynthesis is another option to save the wheat grains from DON contamination. The gene *TR16* encodes a transcription regulator of trichothecene biosynthesis, while *TR15* encodes sequiterpene cylase, a critical enzyme for cyclization of farnesylpyrophosphate into trichodiene. Inactivation of *TR15* or *TR16* eliminated trichothecene production (Proctor *et al.*, 1995a and b).

The effects of fungicides (benomyl, chlorothalonil, fenbuconazole, flusilazole, myclobutanil, potassium bicarbonate, propiconazole, tebuconazole, thiabendazole, triadimefon and mancozeb) on the severity of *Fusarium* head blight of wheat (caused primarily by *Fusarium graminearum* [*Gibberella zeae*]), deoxynivalenol [vomitoxin] level, yield and test weight were evaluated under conditions of high disease pressure in the field. These fungicides did not reduce head blight incidence or vomitoxin level and did not increase yield or test weight (Milus and Parsons, 1994). However, the trizole group (tebuconazole and propiconazol) and imidazole derivatives (prochloraz) have been reported to reduce the severity of disease (Galich, 1996; Ramirez *et al.*, 2004). These types of fungicides interfere with the metabolism of fungal pathogens, mainly by inhibition of ergosterol biosynthesis (Hewitt, 1998). Fungicide treatments for head blight depend not only on active ingredient but also on the method and date of application of fungicide. However, this may also be accounted for the multitude of species present in field and the relative specificity of fungicide treatments for only one or few species (Champeil *et al.*, 2004). In condition of natural contamination, the application of the fungicide does not significantly decreases the concentration of deoxynivalenol and nivalenol and there is not necessarily a correlation between the application of fungicides and the quantity of deoxynivalenol and/or nivalenol found in the grains (Parry *et al.*, 1995; Champeil *et al.*, 2004). Contradictions in effectiveness of fungicide have been observed in laboratory studies with pure cultures of phyto-pathogens and field trials with crop plants (Ramirez *et al.*, 2004). More generally, there is also growing concern that sub-lethal doses of some fungicides may lead to an increase in mycotoxin production by *Fusarium* species (Milus and Parson, 1994; D'Mello *et al.*, 1998). Such effects have been observed in several in-vitro studies with some compound inhibiting the growth of *Fusarium* species but causing increased trichothecene production (Hope *et al.*, 2002).The mechanisms by which the fungicides stimulate toxin production in *Fusarium* species are not known. It may be assumed that in the presence of sub-lethal concentration of certain fungicides the fungal strains respond to this stress by increased production of secondary metabolites including mycotoxins, as a possible mechanism (Ramirez *et al.*, 2004).

The risk of mycotoxin contamination can be reduced by developing resistant/transgenic crop as another stable method of control. It encompasses the three possible approaches (Dvick, 2001): (i) trasgene-mediated control of the ability of the ability of fusarium to infect and colonise the ear, through the overproduction of specific antifungal proteins and metabolites, or by increasing the plant's own defence system in kernel tissues; (ii) the prevention of mycotoxin biosynthesios, or the detoxification of mycotoxins in plants: and (iii) the development of more resistant plants based on transgenic crops engineered to produce a *Bacillus thuringiensis*(Bt) toxin. However, the most reliable type of resistance is physiological resistance to initial infection, referred to as Type 1, and to the spread of the pathogen to nearby florets via the rachis, or Type 2 (Bai and Shaner, 2004). While physiological resistance does not confer immunity to *F. graminearum*, it appears to be stable. It is difficult, however, to breed resistance to *F. graminearum* into cereals while maintaining desired agronomic, baking (wheat) and brewing (barley) qualities, along with resistance to other diseases. Wheat breeders have relied primarily on resistance derived from 'Sumai 3', a cultivar from

China (Van Ginkel *et al.*, 1996), but more recently have also used other sources of resistance, *e.g. Triticum dicocoides* (Mergoum *et al.*, 2005a). While newer wheat cultivars with improved resistance have indeed shown to reduce FHB severity and DON concentrations in comparison to than older, more susceptible cultivars, DON levels in more resistant cultivars can still be problematic under high inoculum conditions (Mergoum *et al.*, 2005b).

Conclusion

Sporadic occurrence of FHB in wet and warm conditions during flowering that ended up by contaminating wheat grains with mycotoxins, is a serious threat to ever bulging population of developing nations. A total eight species of pathogens have been identified to be associated with disease. Although, *Fusarium* spp. are well known to cause vascular diseases, but infestation at flowering and subsequent stages causes more economic losses to growers and the food processing industries. Apart from climate, agronomic practices, weed management, nutrient management etc. decides the severity of disease. As the fungus produces the deoxynivalenol and its acetyl derivates which may be implicated with human and animal health disorders. *Fusarium* spp. associated with FHB produces mycotoxins exhibit dermal toxicity, toxic aleukia, and vomiting, depressed immune system, necrosis of the throat and stomach, decrease in red blood cell count. Hence, several countries have formulated the regulations on DON toxins limit in food and feed. Since, wheat is an important food crop of worldwide; therefore its integrated disease management through cultural, biological and chemical is urgently needed to work on. Meanwhile, new sources of resistance should be sought to be introduced into commercial cereals through conventional breeding methods to augment resistance in existing cultivars. To restrain the mycotoxin to enter into the food chain, emphasis on development of methods for inhibiting DON synthesis or detoxifying DON in fungus-infected grain are critical to address.

References

Abramson, D., R.M. Clear and Nowicki T.W.(1987). *Fusarium* spp. and trichothecene mycotoxins in suspect samples of 1985 Manitoba wheat. *Can. J. Plant Sci.*, **67**: 611-619.

Andersen, A.L.(1948). The development of *Gibberella zeae* head blight of wheat. *Phytopathol.*, **38**: 595–611.

Atanasoff, D. (1920). *Fusarium* blight (scab) of wheat and other cereals. *J. Agric. Res.*, **20**: 1–32.

Bakan, B. (1998). Approche physiologique de la biosynthèse des trichothécènes par *Fusarium*, Thèse INA-PG, p. 150.

Battalico, A. (1998). *Fusarium* diseases of cereals: species complex and related mycotoxin profiles. *Eur. J. Plant Pathol.*, **80**: 85–103.

Bai, G.(1996). Variation in *Fusarium graminearum* and cultivar resistance to wheat scab. *Plant Dis.*, **80**: 343-348.

Bai, G. and Shaner G.(1994). Scab of wheat: prospects for control, *Plant Dis.*, **78**: 760–766.

Bushnell, W.R., B.E. Hazen and Pritsch, C. (2003). Histology and physiology of Fusarium head blight. In: Leonard J, Bushnell WR(eds.), Fusarium Head Blight of Wheat and Barley. APS Press, St. Paul, MN, pp. 44-83.

Caron, D. (2000). Fusarioses des épis. Sait-on prévoir leur développement? Perspect. *Agricoles*, **25**: 356–62.

Caron, D., Les Fusarioses.(1993). In: Maladies des blés et orges, ITCF, pp. 30–39.

Cassin, R. I.(1970). Facteurs favorables ou défavorables au développement des fusarioses et septorioses du blé, in: Proceedings of the Meeting of Sections Cereals and Physiology, Eucarpia, Dijon, pp. 271–279.

Cassini, R.(1981). *Fusarium* disease of cereals in Western Europe; *In Fusarium: Disease, Biology and taxonomy* (Eds.Nelson *et al.*). Pennysylvania State Univ.Press, pp. 56-63.

Champeil, A, Doré T and Fourbet, J.F.(2004).*Fusarium* head blight: epidemiological origin of the effects of cultural practices on head blight attacks and the production of mycotoxins by *Fusarium* in wheat grains. *Plant Sci.*, **166**: 1389-1415.

Champeil, A, Fourbet J F and Doré, T.(2009). Influence of cropping system on *Fusarium* head blight and mycotoxin levels in winter wheat. *Crop Protec.*, **23**: 531-537.

Clement, J.A. and Parry, D. (1998). Stem-base disease and fungal colonization of winter wheat grown in compost inoculated with *Fusarium culmorum, F. graminearun*, and *Microdochium nivale. Eur. J. Plant Pathol.*, **104**: 323-330.

Cromey, M.G., Shorter S.C., Lauren D.R. and Sinclair, K.I. (2002). Cultivar and crop management influences on *Fusarium* head blight and mycotoxins in spring wheat (*Triticum aestivum*) in New Zealand. *N. Z. J. Crop Horticultural Sci.*, **30**: 235–247.

Cruz, A.F., C. Hamel, C. Yang, T. Matsubara, Y. Gan, A.K. Singh, K. Kuwada and Ishii, T.(2012). Phytochemicals to suppress Fusarium head blight in wheat-chickpea rotation. *Phytochem.*, **78:** 72-80.

Da Luz, W.C., C.A. Stockwell and Bergstrom, C.A. (2003). Biological control of Fusarium graminearum. In: Leonard, K.J., Bushnell, W.R. (Eds.), Fusarium Head Blight of Wheat and Barley. Am. Phytopathol. Society Press, St. Paul, MN, pp. 381–394.

Daamen, R.A., C.J. Langerak and Stol, W.(1991). Surveys of cereal diseases and pests in The Netherlands. III. *Monographella nivalis* and *Fusarium* spp. in winter wheat fields and seed lots. *Neth. J. Plant Pathol.*, **97:** 105–114.

Dick, J. 2001. Prospects for reducing fumonisin contamination of maize through genetic modification. *Environ. Health Perspect*, **109**: 337-342.

Edwards, S.G. (2009). *Fusarium* mycotoxin content of UK organic and conventional wheat. *Food Add. Contam*. **26:** 496-506.

De Vos, M., V.R. Van Oosten, R.M.P. Van Poeke, J.A. M.J. Van Pelt, Pozo, M.J. Muller, A.J. Buchala, J.P. Metraux, L.C. Van Loo, M. Dicke and Pieterse, C.M.J.(2005).Signal structure and transcriptome changes of *Arabidopsis* during pathogen and insect attack. *Molecular Plant Microbes Interactions*, **18:** 923-937.

D'Mello, J.P.F., A.M.C. Macdonald, D. Postel, W.T.P. Dijksma, A. Dujardin and Plancinta, C.M.(1998). Pesticide use and mycotoxin productionin *Fusarium* and *Aspergillus* phytopathogens. *Eur. J. Plant Pathol.*, **104**: 741-751.

Galich, M.T. (1996). *Fusarium* Head Blight in Argentina. In; Duvin, H J, Gilchrist L, Reeves J, Mc Nab A(Eds.), *Fusarium* Head Scab: Global Status and Future Prospects. Proceeding of Workshop Held at CIMMYT, Mexico.

Gilbert, J. and Tekauz A. (2000). Recent developments in research on *Fusarium* head blight of wheat in Canada. *Can. J. Plant Pathol.*, **22**: 1–8.

Gorlach, J., S.A. Volrath, G. Knauf-Beiter, G. Henry, U. Beckhove, K. Kogel, M. Oostendrop, T. Staub, E. Ward, H. Kessmann and Ryals, J. (1996). Benzothiadiazole, a novel class of inducers of systemic acquired resistance, activates gene expression and disease resaistance in wheat. *Plant Cell*, **8**: 629-643.

Hewitt, A.F.(1998). Fungicides in Plant Protection. CAB International, Wallingford.

Hope, R. J., Colleate A., Baxster E.S. and Magan, N.(2002). Interactions between environmental stress and fungicides effect on growth and mycotoxin production by *Fusarium culmorum* isolates from wheat grain. *Eur. J. Plant Pathol.*, **108**: 685-690.

Inglis, D.A. and Cook, R J.(1981). *Calonectria nivalis* cause scab in the Pacific Northwest. *Plant Dis.*, **65**: 923-924.

Jansen, C., D. W. von Wettstein, Scharfer, K.H. Kogel, A. Felk and Maier F.J. (2005). Infection pattern in barley and wheat spikes inoculated with wild-type and trichodiene synthase gene disrupted *Fusarium graminearum. Proceeding of National Academy of Sciences of United States of America*, **102**: 16892-16897.

Jenny, E., A. Hecker, P. Kessler, C. Kulling and Forrer, H. R. (2000). Cereal fusorioses: varietal resistance and toxin content. *Agraforschung*, **7:** 270-273.

Kang, Z. and Buchenauer, H.(2000). Ultrastructural and immunocytochemical investigation of pathogen development and host responses in resistant and susceptible wheat spikes infected by *Fusarium culmorum*. *Physiol. Mol. Plant Pathol.*, **57**: 255–268.

Koch, H. J., C. Pringas and Maerlaender, B.(2006). Evaluation of environmental and management effects on Fusarium head blight infection and deoxynivalenol concentration in the grain of winter wheat. *Eur. J. Agron.*, **24**: 357-366.

Kohl, J., B. de Haas, Kastelein P., S.L.G.E. Burgers, Waalwijk, C. (2007). Population dynamics of *Fusarium* spp. and *Microdochium nivale* in crops and crop residues of winter wheat. *Phytopathol.* **97**: 971-978.

Kornneef, A. and Pieterse, C.M.J.(2008). Cross talk in defence signaling. *Plant Physiol.*, **146**: 836-844.

Lacey, J., G.L. Bateman and Mirocha, C. J. (1999). Effects of infection time and moisture on development of ear blight and deoxynivalenol production by *Fusarium* spp. in wheat, *Ann. Appl. Biol.*, **134**: 277–283.

Landschoot, S., K. Audenaert, W. Waegman, B. Pycke, B. Bekaert, B. D. Baets and Haesaert, G. (2011). Connection between primary fusarium inoculums on gramineous weeds, crop residue and soil samples and the final population on wheat ears in Flanders, Belgium. *Crop Protec.*, **30:** 1297-1305.

Leonard, K.J. and Bushnell, W.R. (2003). Fusarium Head Blight of Wheat and Barley. APS Press, St. Paul, Minnesota, p. 512.

Lori, G.A., M.N. Sisterna, S.J. Sarandon, I. Rizzo and Chidichimo H.(2009). *Fusarium* head blight in wheat: Impact of tillage and other agronomic practices under natural infection. *Crop Protec.*, **28**: 495-502.

Magan, N., R. Hope, A. Colleate and Baxter, E.S. (2002). Relationship between growth and mycotoxin production by *Fusarium* species, biocides and environment. *Eur. J. Plant Pathol.*, **108**: 685–690.

McLaughlin, C.S., M.H. Vaughan, L.M. Campbell, C.M. Wei, M.E. Stafford and Hansen, B.S. (2000). Inhibition of protein synthesis by trichothecenes. In: J.V. Rodricks, C.W. Hesseltine, M.A. Mehlman (Eds.), Mycotoxins in Human and Animal Health, Park Forest South Pathotox, 1977, pp. 262–272.

McMuller, M., R. Jones and Gallenberg, D. (1997). Scab of wheat and barley: a reemerging disease of devastating impact. *Plant Dis.*, **81**: 1340–1348.

Mergoum, M., R.C. Frohburg, J.D. Miller and Stack, RW. (2005a). Registration of ' Steele-ND wheat. *Crop Sci.*, **45**: 1163-1164.

Mergoum, M., R.C. Frohburg and Stack, R.W. (2005b).'Glenn' hard red spring wheat cultivar: a new step in combatingF head blight disease. In: canty, SM, Boring t, wardwellJ, Siler L, Ward RW (Eds.), Proceeding of the National *Fusarium* Head Blight Forum, 2005 Dec. 11-13. Michigan State University, East Lansing, Milwaukee, WI, pp. 69-71.

Milus, E. A and Parsons, C.E.(1994). Evaluation of foliar fungicides for controlling *Fusarium* head blight of wheat. *Plant Dis.*, **78**: 697-699.

Moschini, R and Fortugno, C.(1996). Predicting wheat head blight incidence using models based on meteorological factors in Perrgamino, Argentina. *Eur. J. Plant Pathol.*, **102**: 211-218.

Mudge, A.M., R. Dill-Macky, Y. Dong, D.M. Gardiner, R.G. White and Manners, J.M.(2006). A role for the mycotoxin deoxynivalenol in stem colonization during crown rot disease of wheat caused by *Fusarium graminearum* and *Fusarium pseudograminearum*. *Physiol. and Mol.Plant Pathol.*, **69:** 73-85.

O, Donnell, K., T.J. Ward, D.M. Geiser, H.C. Kistler and Aoki, T.(2004). Genealogical concordance between the mating type locus and seven other nuclear gene supports formal recognition of nine phylogenetically distinct species within the *Fusarium graminearum* Clade. *Fungal Genet. Biol.*, **41**: 600-623.

Parry, D.W., P. Jenkinson and McLeod, L.(1995). *Fusarium* ear blight (scab) in small grain cereals–a review, *Plant Pathology*. **44** : 207–238.

Paulitz, T. (1996). Diurnal release of ascospores by *Gibberella zeae* in inoculated wheat plots, *Plant Dis.*, **80**: 674–678.

Pereyra, S.A. and Dill-Macky, R. (2008). Colonization of the residues of diverse plant species by *Gibberella zeae* in wheat residue. *Plant Dis.*, **92:** 800-807.

Pirgozliev, S.A., S.G. Edwards, M.C. Hare and Jenkinson, P.(2003). Strategies for the control of the Fusarium head blight in cereals. *Eur. J. Plant Pathol.*, **109**: 731-742.

Pissinger, P. (1982). *Fusarium* infection of wheat seeds in the Vas province, Novenyvedelem 18, 83–85 (Abstract, in Hungarian).*Plant Dis.*, **78**: 697-699.

Poppenberger, B., Berthiller F., Lucyshyn D., Sieberer T., Schumacher R., Krska R., Kuchler K., Glossl J., Luschnig C. and Adam, G.(2003). Detoxification of the *Fusarium* Mycotoxin deoxynivalenol by a UPD-gl;ucosyltransferase from *Arabadopsis thaliana*. *J. of Biol. Chem.*, **278**: 47905-47914.

Pritsch, C., G.J. Muhlbauer, W.R. Bushnell, D.A. Somers and Vance, C.P. (2000). Fungal development and induction of defence response genes during early infection of wheat spikes by *Fusarium graminearum*. *Mol. Plant Micro. Interact.*, **8**: 593-601.

Proctor, RH, Hohn TM, McCormick SP. (1995a). Reduced virulence of *Gibberella zeae* caused by disruption of a trichothecene toxin biosyntheticgene reversion and mutant complementation. *Mol. Plant Micro. Interact.*, **8**: 593–601.

Proctor, R.H., T. M. Hohn, S.P. McCormick and Desjardins, A.E.(1995b). TRI6 encodes an unusual zinc finger protein involved in regulation of trichothecene biosynthesis in Fusarium sporotrichioides. *Appl. Environ. Microbiol.*, **61:** 1923–1930.

Pugh, G.W., H. Johann and Dickson, J.G. (1933). Factors affecting infection of wheat heads by *Gibberella saubinetii, J. Agric. Res.*, **46**: 771–797.

Qi P.F., A. Johnston, M. Balcerak, H. Rocheleau, L. J. Harris, X. Y. Long, Y. M. Wei and Zeng Y L.(2012).Effect of salicyclic acid on *Fusarium graminearum*, the major casual agent of *Fusarium* head blight in wheat. *Fungal Biol.*, **116**: 413-426.

Ramirez, M. L., S. Chulze and Magan, N.(2004).Impact of environmental factors and fungicides on growth and deoxynivalenol production by *Fusarium graminearum* isolates from Argentinian wheat. *Crop Protect.*, **23**: 117-125.

Rapilly, F., J.M. Lemaire, R. Cassini and Les Fusarioses. (1973). In: Les maladies des céréales–Principales maladies cryptogamiques, INRAITCF, pp. 119.

Saharan, M.S., J. Kumar, Sharma A.K. and Nagarajan, S.(2004). Fusarium Head Blight (FHB) or Head Scab of Wheat-a review. *Proc. Indian Natl. Sci. Acad.*, **B70**: 255-268.

Savard, M.E., R.C. Sinha, W.L. Seaman and Fedak G. (2000).Sequential distribution of the mycotoxin deoxynivalenol in wheat spikes after inoculation with *Fusarium graminearum*. *Can. J. Plant Pathol.*, **22**: 280–285.

Schaafsma, A.W. and Hooker, D C.(2007). Climatic model to predict occurrence of *Fusarium* toxin in wheat and maize. *Int. J. Food Microbiol.*, **119**: 116-125.

Schmale, III, D.G. and Bergstrom, G.C. (2003). Fusarium head blight in wheat. *The Plant Health Instructor*, 0612-01.

Schroeder, HW and Christensen, J.J.(1963). Factors affecting resistance of wheat to scab caused by *Gibberella zeae*. *Phytopathol.*, **53**: 831–838.

Scott, L, S. Leath, W. Hagler and Murphy, J. (1998).Variation in *Fusarium graminearum* associated with wheat scab in North Carolina, In: National *Fusarium* Head Blight Forum. Chapter1: Epidemiology and Disease Management, Michigan State University, USA, p.15

Shima, J., S. Takase, Y. Takahashi, Y. Iwai, H. Fujimoto, M. Yamazaki and Ochi, K.(1997). Novel detoxification of the trichothecene mycotoxin deoxynivalenol by a soil bacterium isolated by enriched culture. *Appl. Enviro. Microbiol.,* **63**: 3825-3830.

Sinha, R.C. and Savard, M.E.(1997). Concentration of deoxynivalenol in single kernels and various tissues of wheat heads. *Can. J. Plant Pathol.*, **19**: 8–12.

Snijders, CHA, Perkowski, J.(1990).Effects of head blight caused by *Fusarium culmorum* on toxin content and weight of wheat kernels, *Phytopathol.*, **80:** 566–570.

Strange, R.N and Smith, H. (1978). A fungal growth stimulant in anthers which predisposes wheat to attack by *Fusarium graminearum. Physiol. Plant Pathol.*, **1**: 141-150.

Sung, J.M. and Cook, R.J. (1981). Effects of water potential on reproduction and spore germination by *Fusarium roseum 'F. graminearum', 'F. culmorum'*, and *'F. avenaceum'*, *Phytopathol.*, **71**: 499–504.

Sutton J.C.(1982). Epidemiology of wheat head blight and maize ear rot caused by *Fusarium graminearum*, *Can. J. Plant Pathol.*, **4:** 195–209.

Teich, A.H.(1989). Epidemiology of wheat (*Triticum aestivum* L.) scab caused by *Fusarium* spp., In: J. Chelkowski (Ed.), *Fusarium*: Mycotoxins, Taxonomy and Pathogenicity, vol.2, Elsevier, Amsterdam (NLD), Chapter 16, pp. 269–282.

Teich, A.H. and Hamilton, J.R. (1985). Effect of cultural practices, soil phosphorus, potassium, and pH on the incidence of *Fusarium* head blight and deoxynivalenol levels in wheat, *Appl. Environ. Microbiol.,* **79**: 1429–1431.

Timmermans, B.G.T., A.M. Osman and van der Bugt, G.J.H.M. (2009). Difference between spring wheat cultivars in tolerance *Fusarium* seedling blight under organic field conditions. *Eur. J. Plant Pathol.*, **125:** 377-386.

Trail, F., Gadoury D. and Loranger, R. (1998). Environmental parameters of ascospore discharge in *Gibberella zeae*, in: National *Fusarium* Head Blight Forum. Chapte 1: Epidemiology and Disease Management, Michigan State University, USA, p. 11–13.

Tschanz, A.T., R.K. Horst and Nelson, P.E.(1976). The effect of environment on sexual reproduction of *Gibberella zeae*, *Mycologia,* **68**: 327–340.

Van Ginkel, M., W. van der Schaar, Y. Zhuping and Rajarm, S.(1996). Inheritance of resistance to scab in two cultivars from Brazil and China. *Plant Dis.,* **80:** 863-867.

van der Burgt, G.J.H.M., B.G.H. Timmermans, J.M.S. Scholberg, Osman, A.M. (2011). *Fusarium* head blight and deoxynivalenol contamination in wheat as affected by nitrogen fertilizer. *NJAS- Wageningen J. Life Sci.*, **58**: 123-129.

Vidal, D.R.(1990). Propriétés immunosuppressives des mycotoxines du groupe des trichothécènes, *Bull. Inst. Pasteur*, **88**: 159–192.

Volki, A., B. Vogler, M. Schollenberger and Karlovsky, P.(2004). Microbial detoxification of mycotoxin deoxynivalenol. *J. of Basic Microbiol.*, **44**: 147–156.

Wagacha, J.M., E.C. Oerke, H.W. Dehne and Steiner, U.(2012) .Colonization of wheat seedling leaves by *Fusarium* species as observed in growth chambers: a role as inoculum for head blight infection? *Fungal Ecol.*, **5:** 581-590.

Wanjiru, W.M., K. Zhensheng and Buchenauer, H.(2002). Importance of cell wall degrading enzymes produced by *Fusarium graminearum* during infection of wheat heads, *Eur. J. Plant Pathol.*, **108:** 803–810.

Wanyoike, M.W., F. Walker and Buchenauer, H.(2002). Relationship between virulence, fungal biomass and mycotoxin production by *Fusarium graminearum* in winter wheat head blight. *Zeitschrift fur Pflanzenkrankheiten and Pflanzenschutz*, **109**: 589-600.

Wicklow, D.T, S. Roth, S.T. Deyrup and Gloer, J.B.(2005). A protective endophyte of maize: Acremonium zeae antibiotics inhibitory to Aspergillus flavus and Fusarium verticillioides. *Mycol. Res.*, **109:** 610–618.

Wiese, M.V.(1987).Compendium of wheat Disease*: American Phytopathol. Soc.*, St. Paul, MN.

Wilde, F. and Miedaner, .(2006). Selection of *Fusarium* head blight resistance in early generations reduces the deoxynivalenol(DON) content in grain of winter and spring wheat. *Plant Breed.*, **125:** 96-98.

Xu, X.M. and Berrie, A M. (2005). Epidemiology of mycotoxigenic fungi associated with *Fusarium* ear blight and apple blue mould: a review. *Food Add. Contaminat.*, **22**: 290-301.

Xu, Y.G. and Hideki, N. (1989). The infection process of wheat scab pathogen. *J. of Nanjing Agriculture University* **12**: 33-38.

Zange, B.J., Z. Lang and Buchenauer, H.(2005). Effect of Folicur® on infection processs of *Fusarium culmorum* in wheat spikes. *J. Plant Dis and Prot.*, **112**: 52-64.

Zinkernagel, V., B. Adolf and Habermeyer, J.(1997). The spread of *Fusarium* spp. from the above ground level to the ears of wheat. *Cereal Res. Commun.*, **25**: 677-679.

2015, Major Constrains and Verdict of Crop Productivity *Pages 265–291*
Editor: **Dr. U.N. Bhale**
Published by: **ASTRAL INTERNATIONAL (P) LIMITED, NEW DELHI**

Chapter 19

Insect Pests of Rice and their Integrated Pest Management

Venkat Reddy*, R. Sunitha Devi, S. Malathi and B. Dileep Kumar

Regional Agricultural Research Station, ANGRAU, Warangal – 506 007, A.P., India

Introduction

Over the next three decades, production of food grains in India has to increase at least 2 million tonnes a year to meet the food demand of the growing population. In the past, agricultural production increased through area expansion and increasing use of high yielding varieties, synthetic fertilizers, chemical pesticides and irrigation water. Prior to the introduction of modern varieties during 1960's, farmers used to grow varieties with different genetic background, resulting in the existence of several varieties in the field during the crop season. This together with low or no fertilizer use probably was the major reason for the maintenance of pest population at below economic injury level. However, during the mid 1960's, modern high yielding varieties (HYVs) were developed and introduced with a narrow genetic base and fertilizer responsiveness. These changes in cultivation resulted in an altered microclimate, which led to the accentuation of insect pest problems. Now, a chance of increasing agricultural production through expansion of area and application of technologies appears to be severely limited. Land frontiers are closing down, and there is little scope to bring additional land under cultivation. Green revolution technologies have now been widely adopted and the process of diminishing returns to additional input usage has set in.

---

* *Corresponding Author.* E-mail: venkatreddyakkati1968@yahoo.com

Agricultural production continues to be influenced by a number of biotic and abiotic factors. For instance, insect pests, diseases and weeds cause considerable damage to potential agricultural production. Studies indicate that pests cause 25 per cent loss in rice, 30 per cent in pulses and 50 per cent in cotton. It is estimated that herbivorous insects eat about 26 per cent of the potential food production. Emerging problems of insecticide resistance, secondary pest out breaks and resurgence further add to the cost of plant protection. Annual crop losses due to insect pests and diseases in India are estimated to be 18 per cent of the agricultural output. New pests have appeared due to the changes in the cropping patterns and intensive agricultural practices. Though these losses cannot be eliminated altogether, they can be reduced. Until recently, synthetic pesticides were increasingly relied upon to limit the production losses.

Till the beginning of 20th century, farmers relied exclusively on cultural practices such as crop rotation, healthy crop variety, manipulation in sowing dates, etc., to manage the pests. Use of pesticides, although began in 1870's with the development of arsenical and copper based insecticides, discovery of pesticides properties of DDT during World War II revolutionized the pest control. In India pesticide use has been increasing at an annual rate of 2.5 per cent since early 1970's. About 96,000 tonnes of technical grade pesticides are currently produced in the country, of which two-thirds are used in agriculture. Pesticides often kill natural enemies along with pests. With the natural enemies eliminated, it is difficult to prevent recovered pest populations from exploding to higher and more damaging levels, and often developing resistance to chemical pesticides. At present, resistance to insecticides is common and more than 500 insect species have acquired resistance to insecticides. Repeated applications of chemical pesticides only repeat this cycle. However, as yields started increasing, pesticide use started becoming wide spread. Their adverse effects on the environment and human health also soon became apparent (Dhandapani and Murugan, 2004).

The solution to the pesticide hazards lies in the implementation of Integrated Pest Management (IPM) which combines the use of different pest control strategies. Integrated pest management is a broad spectrum ecological approach for pest management which employs all available skills, technologies and practices such as cultural, genetic, mechanical and biological methods including application of chemical pesticides as a last resort in a harmonious and compatible manner with a view to suppress pest population below economic injury level, based on regular crop pest surveillance and monitoring (Dhaliwal and Ramesh, 2001). The IPM is a dynamic approach and process which varies from area to area, time to time, crop to crop and pest to pest and aims at minimizing crop losses with due consideration to human and animal health besides safety to environment. "Live and let live" is the philosophy behind IPM. IPM approach has been globally accepted for achieving sustainability in Agriculture. Agro Eco System Analysis (AESA) is an approach that can be used in IPM to analyze field situation with regard to pests, natural enemies, soil condition, plant health, the influence of climatic factors and their inter relationship for growing healthy crop.

Tools of IPM

1. **Monitoring:** Crop monitoring, that keeps track of the pest and their potential damage is the foundation of IPM. Pheromone traps have got advantage over other tools such as light and sticky traps.
2. **Pest resistant varieties (Host Plant resistance) :** Breeding for pest resistance is a continuous process. It is the most effective, economical, practical and easiest means of controlling the pests. Further, it is compatible with other methods of pest control.
3. **Cultural control:** It includes crop production practices that make crop environment less susceptible to pests. Crop rotation, fallowing, manipulation of planting/sowing and harvesting dates, manipulation of row and plant spacing, destruction of old crop debris, summer deep ploughing, balanced use of fertilizers as per the recommendations are the important cultural practices.
4. **Physical or Mechanical control:** These are based on the knowledge of pest behaviour. Removal and destruction of pest infested plant parts, clipping of leaf tips of rice seedlings, collection of egg and larvae of pests, installation of bird perches are some of the physical or mechanical control practices.
5. **Biological control:** These include augmentation and conservation of natural enemies of pests such as parasitoids, predators, parasitic nematodes, fungi, bacteria and viruses. In IPM programme, native natural enemy population are conserved and non-native agents are released with utmost care (Gupta *et al.*, 1987).
6. **Chemical control:** Pesticides are used to keep pest populations, below economically damaging levels, when the pests cannot be controlled by other means. Pesticides include synthetic pesticides, plant derived pesticides and bio-pesticides. These are very easy to use, fast-acting and relatively inexpensive. Ideally, pesticides should be used as a last resort in IPM because of their potential negative effect on the environment. Fortunately, new generation pesticides, with novel mode of action and low environmental effects are developed that are short lived and act on few specific target organisms.

Although IPM has been accepted as the most attractive option for crop protection, implementation at the farmer's level has been limited. For an effective implementation strategy, it is necessary to identify the obstacles for its dissemination. Some of which are :

1. Low awareness and innovativeness of extension personnel and target groups (farmers).
2. Inadequate interaction between research and extension agencies.
3. Problem of timely and adequate supply of quality IPM inputs.
4. Complexity of IPM Vs simplicity of pesticides.
5. Dominant influence of pesticide industry.
6. Non-availability of local specific IPM modules for many crops.

Rice

Rice is the staple food of atleast half of the world population and is grown in approximately 148 million ha of land globally. Nearly 90 per cent of this area falls in Asian region. Rice is a major field crop of India covering the largest area of 44 million hectares with annual production of about 95 million tonnes. In order to meet the growing demand of ever increasing population, we need to produce 1 million tonnes of more rice every year. But the rice production is limited by both biotic and abiotic stresses of which insect pests alone cause 25 per cent losses amounting to Rs. 2,40,138 millions. Hence, we need to reduce the yield losses due to major pests. Different insect pests cause damage to rice crop resulting in annual loss of about 10 per cent output worth Rs. 5,000 corer (Beevi *et al.*, 2000).

Prior to the introduction of modern varieties during 1960's the rice crop survived for centuries with traditional varieties with robust plant type but low yield. However, during the mid 1960's modern high yielding varieties (HYVs) were developed and introduced. Single or a couple of such varieties with a narrow genetic base started occupying vast areas. These were fertilizer responsive and therefore, farmers started applying higher doses of fertilizers in general and nitrogen in particular. These changes in rice cultivation resulted in an altered microclimate, which led to buildup of insect problems.

Over 200 species of insects attack rice from nursery to maturity. Of these, a dozen are consistently reported as major pests of economic importance. Rice is attacked by insect-pests, diseases, weeds, rodents and nematodes. Some of the pests are of national significance, while others are pests of regional significance. More than 200 species of insect pests attack rice crop in India, but about 20 insect pests cause severe damage in rice.

Insect pests like gallmidge (*Orseolia oryzae*), white backed plant hopper (*Sogatella furcifera*), yellow stem borer (*Scirpophaga incertulas*), leaf folder (*Cnaphalocrosis medinalis*) and brown plant hopper (*Nilaparvatha lugens*) are pests of national significance.

Leaf hoppers, plant hoppers and leaf folder, which were of minor importance, have assumed the status of major pests. Gallmidge has become a serious problem in many areas and has also extended its activity to dry season (*rabi*), particularly in coastal areas. Sporadic pests like hispa, climbing cut worm and gundhi bug have been causing severe damage, intermittently. India has a history of pest out breaks resulting in extensive losses in rice production systems.

A number of factors have contributed to pest out breaks of which major ones are : cultivation of modern high yielding varieties in large areas, cultivation of susceptible varieties, cultivating rice throughout the year, use of high levels of nitrogenous fertilizers and increased use of insecticides.

Yield losses ranging from 21 to 51 per cent have been estimated due to moderate to severe incidence of insect pests in the country. The development and implementation of suitable Integrated Pest Management (IPM) strategies is highly essential to overcome biotic constraints mainly, insect pests for realizing yield potential of rice (Anonymous *et al.*, 1992). In view of the past, that farmers have been mostly relying on chemical control for managing the pests, now it has become necessary to implement holistic system of tackling pests, which is environment-friendly, economically viable and socially acceptable (Pasalu, 2002).

Insect Pests of Rice

Plant Hoppers

Among the plant hoppers, brown plant hoppers (BPH), *Nilaparvatha lugens* (Stal) and white backed plant hopper (WBPH), *Sogatella furcifera* (Horvath) are of importance in India. These two belong to the order Homoptera, family Delphacidae.

Brown Plant Hopper (BPH)

Marks of Identification and Biology

Adult BPH has brownish body from light brown to dark brown. Adult females are longer than males and bear dark brown ovipositor at the base of the abdomen. Two distinct wing forms are seen in adult stage. Large winged or macropterous forms have wings extended beyond the tip of the abdomen while short winged or brachypterous forms have smaller wings that do not extend beyond the tip of the body. Macropterous forms are adopted for long distance flights and these forms are the first to appear in the newly planted fields and begin colonization. Brachypterous females lay about 300-500 eggs while macropterous forms lay less number of eggs. The eggs are inserted in groups in a straight line in the tissue of the lower part of the plant, not only near mid rib of leaf sheaths but also in leaf blades. The eggs are covered with dome shaped egg plug secreted by the female. When the adult population is high, eggs are found in the upper parts of rice leaves. The eggs are banana shaped and 0.99 mm long. Newly laid eggs are whitish but turns darken when about to hatch. Before hatching, two distinct eye spots appear. Eggs hatch in 7-11 days. The nymphs are creamy white with a pale brown tinge. The nymphs undergo four moults with intervening five nymphal instars. The nymphal period varies widely depending on the food condition, density and temperature and is generally completed within 12-20 days. The adults survive for 10 to 25 days.

White Backed Plant Hopper (WBPH)

Marks of Identification and Biology

The adult hopper is 3.5-4.0 mm long. Forewings are almost uniformly sub-hyaline with dark veins. There is a prominent white band between the junctions of the wing. The body is creamy white. The hoppers especially adults prefer to stay at the upper portions of rice plant. The eggs are laid in longitudinal rows within the leaf midribs. On an average; a female hopper lays 150-300 eggs. Eggs are similar to those of BPH but have a longer, more pointed egg plug. Eggs laid in batch are not joined at the tip. The eggs hatch in about 8-12 days. Nymphs reach adulthood in 12-17 days passing through five instars.

Symptoms of Damage

The hopper damage is more often seen in well irrigated, densely planted fields with high doses of nitrogen and frequent insecticide application. The nymphs and adults colonize at the base of rice plant and suck the plant sap. As a result of feeding, plants turn yellow and dry up rapidly. This condition is called 'hopper burn'. This hopper is a vector of the rice ragged stunt virus (RRSV) and rice grassy stunt virus (RGSV). Upto 60 per cent yield loss is common in susceptible rice cultivars.

There are however, subtle differences in population dynamics and damage symptoms by these two hoppers. WBPH, generally prefers young stage of the crop and is seen in greater numbers during early tillering stage. BPH, on the other hand, builds up later during reproductive phase of the crop. WBPH preferentially feeds at the base of upper leaves, while BPH prefers base of the plant. Oviposition marks on leaf sheaths are less distinct in case of BPH while plant tissue is slit and an open window is seen when eggs are laid by WBPH. Initial feeding by WBPH leads to leaf turning to orange before turning yellow and drying, while exclusive feeding by BPH turns leaf yellow directly (Lim *et al.*, 1980).

Excessive use of urea as nitrogen fertilizer and insecticides can lead to outbreak by increasing fecundity and by reducing populations of natural enemies. Insecticides like abamectin and cypermethrin cause outbreak of hoppers. There is evidence that some insecticides increase the protein content of BPH male accessory glands and thereby increase fecundity. Some increase amino acids and sucrose in the phloem and thereby increase BPH survival.

The temperature is a critical factor in the life activities of hopper. The egg hatchability and survival rate are highest around 25°C. The population growth is maximum at 28-30°C. Excessive use of N fertilizer and insecticides can lead to outbreaks by increasing the fecundity by reducing populations of natural enemies.

Gall Midge

Marks of Identification and Biology

The rice gallmidge *Orseolia oryzae* (Wood-Mason) belongs to the family Cecidomyiidae of the order Diptera. Though several related species of *Oreseolia* are known to survive on grasses in and around rice fields, the rice gallmidge has limited host range besides rice. The species can be identified by the presence of unequal bifid cephalic horns at the anterior tip of the pupa Ambika *et al.* (1998).

The life cycle of the insect consists of egg, maggot, pupa and adult stage. Eggs are laid singly on leaf and other parts of the plant and take 4 days to hatch. Maggots wriggle and crawl down the plant and reach apical meristem. Larval feeding results in a gall chamber around it. Maggots undergo three instars before pupating. Pupal duration is about 6-8 days. During this time, the gall elongates and protrudes above the surrounding water as silver shoot. The mature pupa wriggles up the gall tube, drills a hole, pushes itself out and eclodes to emerge as adult fly. Male adult is smaller and brownish while female is large and pinkish. Adult males have antennae with 23 segments while female have only 13 segments. Adults live only for 24-48 hours and are crepuscular in being active during dawn and dusk. Rice gallmidge outbreaks tend to follow periods of high rainfall early in the season.

Symptoms of Damage

The maggots wriggles out from egg and crawl down from the plant and reach apical meristem of just protruding tiller. The larval feeding on apical portion results in transformation of tiller into a gall or onion leaf like structure called as "silver shoot".

Such tillers will not bear panicles. Delayed planting, growing of susceptible varieties, congenial weather like high rainfall followed by prolonged drought are the important factors for buildup of gallmidge.

Stem Borers

Several species in families Pyralidae and Noctuidae of order Lepidoptera have been reported as important stem borer pests of rice in India. Of these, yellow stem borer (YSB) *Scirpophaga incertulas, S. innotata* (Walker) is the predominant species. White stem borer (WSB) *Scirpophaga* spp., Dark headed stem borer (DHB), *Chilo polychrysus* (Meyrick), Striped stem borer (SSB), *Chilo suppressalis, Chilo auricilius* (Walker), Pink stem borer (PSB), *Sesamia inferens* (Walker) are the other species reported (Table 19.1).

Table 19.1: Duration of different Stages of Life Cycle of Stem Borer

Sl.No.	*Species*	*Life Cycle (days)*				
		Egg	*Larva*	*Pupa*	*Adult*	*Total Duration*
1.	YSB	5-10	30-46	6-10	4-5	45-71
2.	DHB	4-7	16-43	4-8	2-5	26-63
3.	PSB	6-10	28-56	8-11	4-6	46-83
4.	SSB	5-6	20-48	5-9	3-5	41-70

Marks of Identification and Biology of YSB

The female moth is bigger than the male and its forewings are bright yellowish brown with a distinct black spot in the centre. The abdomen is wide and covered with a distinct anal tuft of hairs. The male moth is pale yellow with inconspicuous spots on forewings. The eggs of yellow stem borer and white stem borer are laid in masses on the leaf tips and covered by a tuft of scales. The tufts of scales of YSB egg mass are buff coloured whereas those of WSB are pale buff to white colour. Eggs of PSB are naked and laid in two rows on the leaf sheath. DHB eggs are flat, scale like and laid in clusters on leaf surface in longitudinal overlapping rows each cluster with 80 eggs. Usually one egg mass contains 50 to 80 eggs. When humidity is below 70 per cent, maturity of eggs is high.

Symptoms of Damage

The newly hatched larva of YSB scrapes on the leaf surface and later disperses and bores into the plant. YSB larvae are solitary and mostly enter the plant just above the water. Larvae of PSB, SSB and DHB are gregarious. The damage symptoms caused by all the stem borers are similar except that they differ in their intensities. As larva bores into the plant and feeds on the growing point, it results in the death of growing tip. This is known as "dead heart". When this dead portion is pulled out it comes out easily and smells foul. When the larva infests the plant at booting stage, the damage manifests as a 'white ear' causing total grains chaffiness. It is a serious pest of rice during *rabi* season. Larvae bore into the stem and feed on internal tissue. Final instar larvae make an exit hole and pupate within the stem at the base of the plant in a silken cacoon. Their feeding causes crop loss of upto high as 95 per cent.

Leaf Folder

Leaf folder complex in rice in Asia comprises of ten species. In India, four species of leaf folders *i.e., Cnaphalocrosis medinalis* (Guenee), *Marasmia patnalis* (Bradley), *Marasmia exigua* (Butler) and *Brachmia arotraea* (Meyrick) are reported on rice. Among these, *C.medinalis* is the most predominant and wide spread species. Different species of leaf folder adults could be distinguished based on wing markings. These belong to order Lepidoptera and family pyralidae.

Marks of Identification and Biology

The adult female moth is golden yellow in colour with dark brown margins on both the wings. There are three lines traversing entire forewing of which ante median and post median lines are prominent. Median line is short, comma like and curved outside. Male moths have prominent patch of dark brown andraconial scales along the mid costa of forewings. The adult appears triangular in shape when at rest. The larva is green in colour and can be easily distinguished by the presence of a pair of black spots/patches behind the head. The females deposit eggs singly or in semi clusters of 3-8 on the upper surface of the leaf. Fecundity ranges from 135-175 eggs. Eggs hatch in 3-4 days. Immediately after hatching, neonate larvae move to the unopened leaf at the centre of the plant and feed gregariously by scraping the green matter. From second in star onwards, the larvae fold the leaves by stitching with silky threads keeping open both the ends of the fold. The pupal period ranges from 6-9 days. The total developmental period ranges from 22-25 days.

Symptoms of Damage

Larval damage is found in all the stages of the crop growth. Larva folds the leaf stitches the two edges together and feed by scraping the green matter from within the fold. This feeding results in longitudinal white stripes and under severe infestation while field appears scorched. The damage to the flag leaf is of more serious concern as there will be a direct impact on grain yield.

Rice Whorl Maggot

Marks of Identification and Biology

The rice whorl maggot, *Hydrellia Philippines* belongs to order Diptera and family Ephydridae. The adults are dull grey and prefer to lay eggs on young plants planted with wider spacing and standing water that reflects light. The female lays 20 eggs. Eggs are laid singly on leaves. Egg period is 2-6 days. The freshly hatched larvae are transparent to light green colored. The larval period lasts for 10-12 days. The pupal period lasts for 5-10 days. The total developmental period is normally 25-28 days.

Symptoms of Damage

Whorl maggot, a dipteran pest, often feeds on an opened leaves during the vegetative stage of the crop. The damage symptoms are visible as distorted leaf margins and shot holes on the lamina of young unfolded leaves with yellow hallow around. This is a pest in the field immediately after planting in tillering stage.

Rice Hispa

Marks of Identification and Biology

The rice hispa, *Dicladispa armigera* (oliver) belongs to order Coleoptera and family Chrysomelidae. The adult is a small, shiny, blue black beetle with a spiny body. The eggs are laid singly near the tip of the leaf blade, generally embedded in leaf tissue. A single female lays 55 eggs. The eggs hatch in 4-5 days. The hatched grub slits open the two epidermal layers of the leaf and mines the lamina and feeds. Three larval instars are observed and entire development takes place inside the leaf.

Symptoms of Damage

Both larvae and adult are cause damage. Adult scrape the chlorophyll of the leaf giving the characteristic appearance of white, parallel streaks. Adults are often present on the damaged leaves. The grubs feed on the green tissue from within the mine and produce irregular white hollow blotches. The larvae can be seen by holding the damaged leaf against sunlight and can be detected by passing the fingers along the tunnel.

Case Worm

Marks of Identification and Biology

The rice case worm, *Paraponyx stagnalis* belongs to order Lepidoptera and family Pyralidae. The adult moths are snowy white with white wings marked with a few light brown to black specks. The adults are nocturnal. A single female lays an average of about 50 eggs. Five larval instars are completed in an average of 20 days. The full grown larva is pale green with brown head and the larva breathes through its gills.

Symptoms of Damage

The case worm occurs only in rice fields with standing water. During the day, larvae float on the water, at night they climb plants to cut off leaves to make new tabular cases. The damage is caused by larvae feeding and cutting of the leaf tips for making cases. Damage is characterized by ladder like appearance of the removed leaf tissue.

Rice Thrips

Marks of Identification and Biology

The rice thrips, *Stenchaetothrips biformis* (Bagnall) belongs to family Thripidae and order Thysanoptera. Adult thrips are dark brown, slender insects. Their antennae are 7 segmented. Adults can be either winged or wingless. Thrips prefers young rice seedlings. Female thrips lay on an average of 20 eggs singly and the total life span is about 20 days.

Symptoms of Damage

Both larvae and adults are pests from tillering to booting stage. Due to sucking of sap from the tips of leaves, the tips roll and gives a pointed appearance and the field gives wilted appearance. When opened, different stages of the insect are seen. Newly planted crop is more affected. Thrips outbreaks are associated with drought.

Panicle Mite

Steneotarsonemus spinki (Smiley) also known as panicle rice mite, spinki mite and rice tarsonemid mite is a serious rice pest in Asian tropical regions, Central America and the Caribbean. This belongs to class Arachnida, order Trombidiformis and family Tarsonemidae.

Marks of Identification and Biology

The panicle mite is not visible to the naked eye. A minimum 20 x hand lense is required to observe it on the inside of the leaf sheath. The mites are clear to straw colored and are approximately 250 mm in length. The males are elongated with rear legs containing a pair of elongated spines. The legs are carried above the body. Males are highly active and can be seen moving on the surface of the leaf. Females are ovoid-shaped. Rice mites are parthenogenetic (virgin females can produce male off springs). The female will then mate with the male off spring and produce eggs. A mated female produce an average of 55 eggs in her life time. The life cycle can vary from 3 days to 20 days. High temperature and low rainfall, continuous rice cultivation and sharing of implements between fields are conducive to building damaging population. Panicle rice mite, when found in association with bacterial panicle mite (*Burkholderia glumae*) and sheath rot (*Sarocladium oryzae*) pathogens cause substantial economic losses.

Symptoms of Damage

Feeding takes place behind the leaf sheath. The feeding lesion can be detected by cinnamon to chocolate-brown discoloration of the leaf sheath. When a new leaf begins development, a female will move to the leaf sheath to produce male offspring and then establish a new feeding lesion. They also feed on developing panicles from the boot stage to the milky stage of heading. They cause damage to plants by directly feeding on leaf tissue in the leaf sheath and developing grains at milky stage and indirectly by transmitting fungal pathogens. During feeding they inject toxic saliva. The damage to grains results in sterility and deformed grains, straight-head and parrot-beaking of grains. Crop losses ranging from 5 per cent to 90 per cent have been recorded.

Leaf Mite

Marks of Identification and Biology

Paddy leaf mite, *Olygonychus oryzae* is a small microscopic spider mite. Adults are sexually mature when they emerge and mate as soon as possible. Egg laying begins in 1-3 days and singly in rows along the leaf midribs and veins. Incubation lasts 4-9 days this is followed by the three immature stages-larva, protonymph and deutonymph. Entire life cycle can take 8-12 days.

Symptoms of Damage

This mite is minute and spins a very delicate web on the lower surface of leaf blade. It remains on the lower side of leaf blade and appears like a powdery substance. That's why this mite infestation symptom is popularly known as Boodida rogam. The adults are visible to naked eyes. Infested plant can easily be identified from a distance due to its peculiar symptom of chlorophyll less leaves. They appear light yellow or pale. In the beginning white patches appear and later on incidence increases all the patches joined and entire leaf blade appear light yellow or pale. Severely infested leaves fall down. Usually the symptoms appear near the bunds first on the weeds, Haryali grass (*Cynedon dactylan*) in particular is its alternative host. They survive on this weed during off-season and immediately after planting of rice, will transfer to their preferred host. However, yield loss is imminent if plants are infested at flag leaf to grain filling stage. Their incidence is more prevalent during September to October.

Gundhi Bug

Marks of Identification and Biology

Rice gundhi bug or Stink bug, *Leptocorisa acuta* belongs to order Hemiptera and family Coridae. Adult is grey green color and measures about 16-19mm in length and can be identified by their characteristic foul odor produced by the scent glands present on the abdomen. It is diurnal in habit but active during early morning and evening. Females are strong fliers. Female lay 250-300 eggs in two or three straight rows along side mid rib on upper surface of leaf blade.Eggs are disc shaped, dorsally flat and elliptical. Incubation period is one week. Bugs complete 1Or 2 generations on grasses and then migrate to rice fields. Newly hatched nymphs are green but turn brown as they grow. Nymphal period varies from 25-30 days with 5 instars (Misra, 1999).

Symptoms of Damage

Both the adults and young ones suck the milky grains leaving dull whitish spots on the grain. The grain remains unfilled. Infested paddy straw contains foul smell. Young succulent leaves and shoots are also attacked before grain formation stage. Infestation is characterized by the presence of some empty or ill-filled grains in the panicles. A hole is left on the grain at the point of puncture and around that hole a brownish spot develops causing discoloration of the panicles.

The population buildup is usually noticed at the end of rainy season and declines rapidly during dry months when temperatures are unfavorable. Maximum population is observed during Setember-Nevember months.

Climbing Cutworm

Marks of Identification and Biology

The rice climbing cutworm, *Mythimna seperata* belongs to order Lepidoptera and family Noctuidae. The adult moth is stout and red brown in color and have a hairy body covered with dark specks and patches. Whitish to pale yellow eggs are laid between the leaf sheaths and stem in clusters. The full grown larva is long, stout and greenish black with longitudinal stripes

Symptoms of Damage

It is a polyphagous pest. The cut worm is a problem generally in upland rice only because it needs dry soil for pupation and for completion of its life cycle. Low land rice usually suffers heavy damage from larvae migrating from adjacent grassy areas.The damage is prominent at ear head stage. The full grown caterpillar climbs the plant and cut off the ear head or rachis of the ear head. The larvae hide at the base of the plant during day time and feed voraciously at night. The infested crop presents a dried look and plants easily break at the nodes. Heavy late rains followed by prolonged drought are favorable for its multiplication.

Green Leaf Hopper

Marks of Identification and Biology

Rice green leaf hopper, *Nephotettix virescens* belongs to order Homoptera and family Cicadellidae. The adults and nymphs are light green to dark green in color with two distinct black spots on fore wings of the adult. Pre mating period ranges from 1-2 days and eggs are laid in small slits made in all soft parts of the plant in batches of around 30 eggs. The incubation ranges from 6 to 12 days. The nymphal period lasts for 14-37 days.

Symptoms of Damage

The nymphs and adults cause direct damage to rice crop by sucking sap from leaf sheaths and blades. The feeding marks predispose plants to fungal and bacterial infections. The effected leaves and plant turns to yellow color and growth is retarded. They indirectly acts as vectors by transmitting virus diseases such as Tungro and Yellow dwarf. The insects are usually more abundant during dry season than during wet season. Rice plants at tillering and panicle initiation stage are more vulnerable for the rapid buildup of pest population (Table 19.2).

Integrated Pest Management (IPM) in Rice

Host Plant Resistance (Resistant Varieties)

Growing of pest resistant varieties are the most effective, economical, practical and easiest way of controlling insects. Most of the modern high yielding varieties widely cultivated in the pest prone areas possess resistance to at least one insect pest or disease. In India, approximately more than 570 commercial rice varieties were released. Some of the important pest resistant donors and pest resistant varieties are listed below in the Table 19.3 (Chellaiah *et al.*, 1989). The only limiting factor in host plant resistance is the breakdown of resistance due to selection pressure on the pest owing to large scale and continuous growing of particular resistant variety. In the last two decades, insect biotypes have caused serious concern as an important biotic factor affecting the host plant resistance and its durability. Biotypes are discrete

Table 19.2: Economic Thresholds for Important Pests of Rice

Sl.No.	*Pest*	*Economic Thresholds*
1.	Stem borer	10 per cent dead hearts or I egg mass or1 adult moth/m^2
2.	Plant hoppers	10 insects/hill at vegetative stage and 20-25 insects/hill at reproductive stage.
3.	Gallmidge	5 per cent silver shoots
4.	Leaf folder	3 damaged leaves/hill
5.	Gundhi bug	1-2 nymphs or adults/hill
6.	Cut worm and Army worm	1 larvae/hill or 4 larvae/m2
7.	Green leaf hopper	2 insects/hill in Tungro virus endemic areas and 20-30 insects/ hill in other areas
8.	Hispa	1 Adult/leaf or 5 per cent damaged leaves
9.	Whorl maggot	25 per cent damaged leaves
10.	Green leaf hopper	10-20 insects/hill and 2 insects/hill in virus endemic areas

Table 19.3. Released Varieties and Resistance Donors against Major Insect Pests of Rice

Sl.No.	*Insect Pest*	*Released Varieties*	*Resistance Donors*
1.	Plant hopper	Chaitanya, Vajram, Pratibha, Vijetha (MTU-1001), Krishnaveni, Makom, Pavizham, Manasarovr, Co-42, Chandana, Nagarjuna, Sonasali, Rasmi, Jyothi, Bhandra, Neela, Aunanga, Daya, Aruna, Kanaka, Remya, Karthika, HKR120	Ptb 10,18,21,33, ARC 5984, 650, Karivennel, Leb Mue Nhang, Manoharsali, Oorapandy, Triveni
2.	Gallmidge	Kakatiya, Pothana, Kavya, Erramallelu, Orugallu, Surekha, Divya, Phalguna, JGL-1798, JGL-384, JGL-3844, Sneha, Rajendradhan, Karna, Ruchi, Samridhi, Usha, Asha, MDU3, Bhuban, Abhaya, Shakti, Surksha, Daya, Pratap, Udaya, IR-36, Shaktiman, Tara, Kshira, Sarasa, Neela, Lalat, Mahaveer, Vibhava, Dhanyalaxmi, Vikram, Kunti.	CR 143, Eswarakorra, Leuang 152, Ob 677, Ptb 10, 18, 21, Siam 29
3.	Stem borer	Ratna, Vikas, Sasyasree	TKM-6
4.	Green leaf hopper	Vikramarya, Lalat, Kharia, Nidhi	Ptb 2, W1263

population or individual distinguished from other population or individuals of its species by some morphological/physiological trait. In rice, biotypes of gallmidge and brown plant hopper have emerged as serious threat to success of host plant resistance. For example, in gallmidge, so far six biotypes were reported from different parts of India. The gallmidge biotypes so far identified can be very well distinguished by the reaction pattern displayed by the differentials (Table 19.4) (Pasalu and Katti, 2003). Although rice varieties resistant to brown plant hoppers have been successfully developed, the pest has evolved prolific biotypes that can destroy these varieties. So far, four biotypes of BPH are known. Biotype 1 is widely distributed in South and South East Asia, biotype 2 originated in Philippines, biotype 3 was produced in laboratories in Japan and Philippines and biotype 4 only in South Asia (Table 19.5).

Important Gall Midge Resistant Varieties Related from ANGRAU

Table 19.4: Reaction Pattern of differentials against Gall Midge

Group	*Differentials*	*Reaction against Biotype*					
		1	*2*	*3*	*4*	*5*	*6*
I	Eswarakorra/W 1263/ARC 6605	R	S	R	S	R	R
II	Siam 29/ARC 5984/Phalguna	R	R	S	S	R	S
III	Velluthacheera/Aganni/CR-MR 1523/Ptb 10/T-1477	R	R	R	R	S	S
IV	TN1	S	S	S	S	S	S
	Distribution in states of India	A.P, Orissa, M.P.	Orissa	A.P	A.P, Mahara-shtra	Kerala	Manipur

R: Resistant; S: Susceptible.

Table 19.5: Reaction Pattern of differentials against Brown Plant Hoppers

Test Entry	*Resistance Gene*	*Reaction to PBH Biotype*		
		Pantnagar	*Cuulong (Vietnam)*	*Biotype 2*
94-42-5-1	Unknown	HR	HR	HR
TN1	None	S	S	S
Mudgo	Bph 1	S	S	S
ASD 7	Bph2	S	S	R
Rathu Heenali	Bph3	MR	MS	R
Ptb 33	Bph 2, Bph 3	R	MR	R
Babawee	Bph 4	S	MR	R
ARC 10550	Bph 5	S	S	S
Swarnalatha	Bph 6	MS	S	S
T12	Bph 7	S	S	S
China-saba	Bph 8	S	S	S
Pokkali	Bph 9	S	–	–

*: HR: Highly Resistant; R: Resistant; MR: Moderately Resistant; MS: Moderately Susceptible and S: Susceptible.

Cultural Control

Cultural practices are normal agronomic practices that are followed to increase crop productivity and at the same time are useful in pest suppression. The cultural practices offer great scope in rainfed rice where scope for application of insecticides is less due to greater uncertainty. Some of the cultural control practices are:

1. Early and synchronous sowing and planning often escapes pest attacks like gallmidge and hoppers. This needs community action and often depends on weather factors.
2. Application of recommended dosage of balanced fertilizers especially Nitrogen fertilizers in split doses avoids buildup of insects like hoppers and leaf folder. Application of potash fertilizers confers tolerance to insect pests.
3. Water management practices like frequent drying and wetting of field, flooding field up to crop height followed by spreading of kerosene-saw dust mixture reduces automatically hopper population.
4. Harvesting crop up to ground level and ploughing or burning stubbles effectively to prevent the carryover of yellow stem borer and gall midge during *Rabi* season.
5. Formation of alleyways of 30 cm width after every 2 meters in hopper endemic areas effectively keeps hopper population under check.
6. Crop rotation is advisable in pest endemic areas where mono-cropping of rice is practiced.

Biological Control

Biological control may be defined as the action of predators, parasitoids, pathogens, antagonists or competitive population to suppress a pest population making it less abundant and less damaging than it would be otherwise. Rice insect pests have had long, close association with their natural enemies, allowing stable relationships to develop. For example, predator and parasitoid guilds recorded in five irrigated rice ecosystems belong to 10 orders and 57 families of predators and 3 orders comprising 40 families of parasitoids indicating great diversity and richness of the natural enemies' community in rice. The collective importance of natural enemies to pest regulation is clearly demonstrated by observing out breaks of insect pests after removal of natural enemies by use of broad spectrum insecticides. Thus conservation of natural enemies through modification of the environment or habitat in combination with judicious use of pesticides appears to be very important (Ridgeway and Vinson, 1976).

Insecticides are more harmful to natural enemies than to pests. In order to conserve indigenous natural enemies and enhance natural biological control, the following strategies for insecticide use are recommended.

1. Preventive or calendar –based use of insecticides should be stopped.
2. Early season spraying during the first 40 days after transplanting should be avoided.
3. Selective insecticide should be used only when pest population density reaches a damaging level.

4. Use appropriate formulation in the right dose. Apply at the optimum time of intervention following proper application methods.
5. Broad spectrum insecticide use should be avoided.

Some habitat management activities that have positive influence on the abundance of the natural enemies of rice insect pests are described below:

1. Vegetated rice bunds support many herbivore species and many species of natural enemies. Bunds provide shelter at critical times during harvesting, land preparation and between seasons. Manipulation of bunds vegetation, particularly the enhancement of broad leaf weeds, can provide vital source of natural sources.
2. Some non-rice crops such as cow pea, mung bean, maize, bell pepper, garlic, onion, soy bean etc serve as important reservoirs of natural enemies of rice insect pests during non-rice season.
3. From plant nutrient and natural enemy conservation point of view, it is better to incorporate crop residues into the soil rather than burning them.
4. Conservation of predatory birds, frogs and toads by preventing capture from rice environment can also enhance biological control.

Augmentation refers to inundative or inoculative release of mass reared natural enemies. It seems that, mass rearing and release is not economical in rice. Other augmentation methods include the use of bird perches for insect predatory birds, use of ducks, and rice-fish culture. The black drongo bird is a purely insectivorous bird and very common in rice environment of South Asia. More than one hundred species of natural enemies attack rice stem borers. Among all the pests, stem borers are most vulnerable particularly at the egg, neonate larval and adult stages. Use of biological agents to control insects is a key component of IPM. About 60 per cent of natural control of insects in rice is due to biological control agents, which have to be protected and conserved by avoiding unnecessary use synthetic broad spectrum insecticides. However, bio control agents provide adequate solution for only one or few pest species unlike pesticides which are broad spectrum in activity. The optimum natural enemy pest ratio of 2:1 should be considered while taking up chemical pest control decisions. Use of bio control agents through inundative or inoculative releases in rice ecosystem has provided limited success. Hence, maximizing the impact of *in situ* natural enemies as an essential part of IPM program needs emphasis. Among the natural enemies, inundative release of egg parasitoid, *Trichogramma japonicum* @ 1 lakh egg/ha for four releases at 10 days interval is successful in reducing stem borer and leaf folders. Several natural enemies have been indentified from different rice growing areas of the country (Table 19.6).

19 species of spiders were reported in rice ecosystem. The nymphs and adult spiders can consume 5 insect pests per day depending on their size. Green mirid bug consumes 7-10 eggs or 1-5 hoppers, water striders can predate upon 5-10 prey daily. The ground beetles and rove beetles can consume 3-5 larvae/day. The grubs of coccinellid beetle can consume 5-10 eggs, nymphs, larvae or adults/day. The predatory grasshoppers are known to predate 3-4 egg masses of yellow stem borer.

Table 19.6: Important Natural Enemies of Major Insect Pests of Rice *Telenomus* spp.

Insect Pest	*Natural Enemies*	*Stage of the Pest Attacked*
1. Stem borer	**a) Egg parasitoids**	Egg
	1. *Trichogramma* spp.	
	2. *Telenomus* spp.	
	3. *Tetrastichus* spp.	
	b) Larval parasitoids	Larval
	1. *Stenobracon nicevillei*	
	2. *Amauromorpha accepta*	
	c) Pupal parasitoids	Pupal
	1. *Xanthopimpla flavoloneata*	
	2. *Temelucha* spp.	
	d) Pathogens	Larval
	1. *Bacillus thuringiensis*	
	2. *Beauveria bassiana*	
	3. *Nomuraea rileyi*	
	4. *Cordyceps* spp	
	5. Unidentified NPV	
	e) Nemotodes	Larval
	1. *Rhabditis* spp.	
	2. *Steinernema thermophilum*	
	f) Predators	Adults
	1. Long horned grass hopper (*Conocephalus longipennis*)	
	2. Crickets (*Metrioche vittaticollis*)	
2. Gallmidge	1. *Platygastor oryzae*	Larva/pupa
	2. *Neanastatus grallarius*	
3. Plant hoppers	**a) Parasitoids**	
	1. *Anagrus* spp.	Egg
	2. *Oligostia* spp.	Egg
	b) Predators	
	1. *Gonatopus*	Nymph
	2. *Cyrtorhinus lividipennis*	Egg/nymph
	3. (Spiders) *Lycosa* spp., *Pardosa* spp., *Oxyopes* spp.	Adults/Nymphs
	c) Pathogens	Adults/Nymphs
	1. *Metarhizium* spp.	
	2. *Beaveria bassiana*	
	3. *Fusarium* spp.	
	4. *Hirsutella citriformis*	

Contd...

Table 19.6–*Contd...*

Insect Pest	*Natural Enemies*	*Stage of the Pest Attacked*
4. Leaf folder	**a) Parasitoids**	Egg
	1. *Trichogramma* spp., *Copidiosomopsis nacoleiae*	
	2. *Apanteles* spp., *Macrocentrus*	Larval
	3. *Xanthopimpla, Brachymeria, Goniozus*	Pupal
	b) Predators	
	1. *Ophionea* spp.	Larval
	2. *Nabis* spp.	
	3. *Paederus* spp.	
	c) Pathogens	
	1. *Beaureria bassiana*	Larval
	2. *Nomuraea rileyi*	
5. Hispa	**a) Parasitoids**	
	1. *Apanteles* spp.	Egg
	2. *Bracon* spp.	Larval
	b) Predators	Adults
	1. Orbweb spider (*Agriope* spp.)	
6. Whorl maggot	**Predators**	Larval/Adults
	1. Damsel flies	
	2. Cricket	
7. Rice gundhi bug	**a) Parasitoids**	
	1.Gryon nixoni	Egg
	b) Predators	
	1. Meadow grass hopper (*Conocephalus longipennis*)	Egg
	2. Assasin bug (*Nabis* spp.)	Nymphs/Adults
	3. Spiders (*Neoscona thesis*, *Agriope* spp.)	
8. Climbing cut worm	**a) Parasitoids**	
	1. Scelionid wasps	Egg
	2. Braconid wasps	Larva
	b) Predators	
	1. Grass hopper	Larva
	c) Nematode Pathogen	
	(*Neoaplecta* spp.)	

Behavioral Control

Sex Pheromones

Sex pheromones play a vital role in communication between male and female insects and this can be exploited in IPM (Tables 19.7 and 19.8). They control the insect through capture and annihilation by either mass trapping or disrupting mating communications. These have been attempted for monitoring of yellow stemborer and leaf folder populations as alternatives to light traps. Mass trapping by installing 20 traps per hectare each with 5 mg pheromone impregnated lures could reduce damage due to stem borer by about 70 per cent. A single application of slow release formulation of pheromones @ 40 g *a.i*/ha within 15 days after planting results in season long control of stem borer (Hall *et al.*, 1998, Hazarika and Puzari, 1997, Krishnaiah *et al.*, 2000 and Krishnaiah, 1995).

Table 19.7: Mixture of Sex Pheromones Used in YSB Management

Formulation No.	*Composition*	*Ratio*
1.	Z9H : Z11H	1:3
2.	Z9H : Z11H	1:10
3.	Z9H : Z11H : Z9O	1:10:0.75
4.	Z9H : Z11H : Z9O	1:10:1

Table 19.8: Mixture of Sex Pheromones Used in Leaf Folder Management

Pheromone	*Ratio*
Hexadecye acetate	10
(Z)-9-Hexadecenal	15
(Z)-11- Hexadecenal	60
(Z)-11-Hexadecen-1-ol	10
(Z)-9-Octadecenol	6

The female sex pheromones of yellow stem borer have been reported to be a blend of several chemicals such as (Z)-9-hexadecenal (Z9H), (Z)-11-haxadecenal (Z 11 H), (Z)-9-Octodecenal (Z9O) and (Z)-13-Octadecenae (Z13O) (Katti *et al.*, 2001 and Katti and Pasalu, 2001).

Light Traps

Use of light traps in one of the non-chemical eco-friendly method of control against the insect species known to be positively phototropic in adult stage. Light trap is a device used at night in the field to collect moths and other flying insects such as army warms, bugs, cut worms, leaf and plant hoppers, stem borers and gallmidge etc. Install trap five meters from the ground and put the light trap from early evening until early morning. Adults of most stem borers are nocturnal, positively phototropic and most attracted to ultraviolet and green fluorescent light. These are strong fliers capable of flying 5 to 10 miles.

Types of light sources used:

1. Incandescent tungsten filament lamp of 100-150 watts.
2. Compact fluorescent lamp (CFL).

3. High intensity discharge lamps of M.V.lamps and sodium lamps.
4. Black light or U.V lamps of 15-20 watt.

Among the three light traps, mercury light trap shows maximum ability followed by black light trap and UV light trap. The mercury light is more efficient for Lepidoptera, Hemiptera, Hymenoptera, Odonata and Diptera, Black light for Coleoptera, Orthoptera, Isoptera and Dictyoptera and UV light for Dermaptera, Orthoptera and Isoptera.

Mechanical Control

The suppression of rice insect population by means of manual devices is referred to as mechanical control. Removal and destruction of egg masses, caterpillars and adults of insects during surveillance reduce the pest number and clipping of rice seedlings tips before transplanting. Dragging a rope across the crop to dislodge the larvae of leaf folder, case worms, adults of hispa to kerosenized water is helpful. Pour kerosene on a flooded field and drag a rope across the foliage bending it into kerosene film on the water. The insects dislodge into the oil and get killed. The field is drained off after 6 hours to prevent phyto toxicity. Setting yellow sticky traps to catch the flying insects like leaf and plant hoppers. In upland situations, hanging fermented snails/frogs/rotten fish during milky stage of crop attracts the rice gundhi bug to the trap, thereby preventing damage to rice crop. Providing perching sites to avian predators also aids them to search their prey (Manimaran and Vasagam, 2000).

Chemical Control

Farmers opt for chemical control as one of the effective and quick method of reducing insect populations particularly in times of sudden pest out break. However, for better results care should be taken that the correct insecticide, suitable formulation, correct method of application and correct dose of the chemical is followed. Further, understanding the potential environmental hazards of chemicals used is essential not only to the user and consumer but also the whole farming system.

In pest endemic areas every care should be taken while selecting pesticides to overcome the problem of resurgence, particularly in case of BPH and leaf folder. In case of hoppers, synthetic pyrethroids *i.e.* cypermethrin, deltamethrin and some OP compounds like chlorphyriphos, quinalphos and methyl parathion have been observed to cause resurgence. Application of granular insecticides like phorate and carbofuran in early stage can lead to resurgence of leaf folder in later stages (Heinrichs and Mochida, 1984).

Botanical pesticides are plant derived chemicals that repel approaching insects, deter feeding and oviposition on the plants, disrupt behaviour and physiology of insects. The replellant and antifeedant affects of neem have been reported against hoppers and cutworms with concentration ranging from 0.001 to 0.4 per cent of neem seed kernel extract.

For management of rice insect pests several insecticides both granules, spray formulations and dusts have been found effective against different insect pests (Table 19.9).

Table 19.9: Spectrum of Toxicity of Spray Formulations against Insect Pests of Rice

Sl.No.	*Insecticide*	*Formulation Dosage (ml or g)/Lt*	*Dosage (a.i/ha)*	*Stem Borer*	*Leaf Folder*	*Hispa*	*BPH*	*WBPH*	*Cut Worm*	*GLH*
1.	Quinalphos	2.0	500	**	***	**	*	**	*	*
2.	Monocrotophos	1.6	400	***	***	**	***	***	**	***
3.	Chlorpyriphos	2.5	500	***	***	**	*	**	**	*
4.	Carbaryl	3.0	750	*	**	**	***	***	*	**
5.	Fenitrothion	2.0	500	**	**	*	*	*	*	*
6.	Phosphamidan	2.0	500	**	***	**	***	***	*	**
7.	Diohlorvos	1.0	500	*	**	*	**	*	***	*
8.	Endosulfan	2.0	600	***	**	*	*	*	***	*
9.	Ethofenprox	1.5	75	*	*	*	***	***	*	***
10.	Cartap hydrochloride	2.0	300	***	***	*	*	*	**	**
11.	Fipronil	2.0	50	***	***	**	***	***	**	***
12.	Triazophos	1.5	400	*	***	**	*	*	**	*
13.	Profenephos	2.0	400	**	***	**	*	*	**	*
14.	Imida cloprid	0.25	25	*	*	*	***	***	*	***
15.	Thiamethoxam	0.25	25	*	*	*	***	***	*	***

*: Moderately effective; **: Effective; ***: Highly effective.

Table 19.10: Spectrum of Toxicity of Granular Insecticides against Insect Pests of Rice.

Sl.No.	*Insecticide*	*Formulation Dosage (ml or g)/Lt*	*Dosage (a.i/ha)*	*Stem Borer*	*Gall Midge*	*Whorl Maggot*	*Leaf Folder*	*Hispa*	*BPH, WBPH*	*GLH*
1.	Carbofuron 3G	750	25.0	***	**	***	*	**	***	**
2.	Phorate 10 G	1250	12.5	***	***	*	*	*	**	**
3.	Cartap hydrochloride 4G	750	20.0	***	*	*	***	*	*	***
4.	Fipronil 0.3 G	75	20.0	***	***	**	**	**	***	**

*: Moderately effective; **: Effective; ***: Highly effective.

Seed Treatment

Seed treatment with Fipronil 5 per cent FS @ 25 g/kg is effective against stemborer in nurseries. Sprouted seeds soaked with imidacloprid 200 SL @ 0.05 per cent for three hours are effective against gallmidge up to one month.

Seedling Root Dip

Seedling root dip of uprooted seedlings ready for transplanting with 0.2 per cent Chlorphyriphos solution overnight provides protection against insect pests in initial stages of crop growth.

Soil Application in Nursery

An effective and economical way is to apply granules of Carbofuran 3G or Phorate 10 G or Cartap hydrochloride 4G @ 1.5 kg a.i/ha or Fipronil 0.3G @ 75 g a.i/ha mixed with sand 5-7 days before uprooting seedling gives protection to crop up to 3 weeks against stem borers and gall midge.

Soil Application in Main Field

In the main field, the crop can be protected either through application of granules like carbofuran, phorate, cartap hydrochloride @ 1 kg *a.i*/ha or Fipronil @ 75 g *a.*i/ha or need based sprayings with insecticides (Table 19.10).

The emphasis in recent years is to find out eco-friendly chemicals belonging to novel groups like neonicotinoids, nereis toxin, phenyl pyrazoles and other derivatives. These are comparatively safer to mammals and effective at 10-15 times lower doses than the traditional compounds (Tables 19.11 and 19.12).

Table 19.11: Pest-wise Recommended Effective Insecticides in Rice

Sl.No.	*Name of the Pest*	*Insecticides Recommended*	*Dosage (ml or g/lt)*
1.	Stem borer	1. Cartap hydrochloride	2.0 g
		2. Acephate	1.5 g
		3. Profenophos	2.0 ml
		4. Cartap hydrochloride 4G	20.0 kg/ha
		5. Chlorpyriphos	2.5 ml
		6. Phosphomidan	2 ml
2.	Plant hoppers	1. Acephate	1.5 g
		2. Monocrotophos	2.2 ml
		3. Ethofenprox	2.0 ml
		4. Fenobucarb	2.0 ml
		5. Imidacloprid	0.25 ml
		6. Thiamethoxam	0.2 g
		7. Buprofezin	1.6 ml

Contd...

Table 19.11–*Contd...*

Sl.No.	*Name of the Pest*	*Insecticides Recommended*	*Dosage (ml or g/lt)*
3.	Leaf folder	1. Cartrap hydrochloride	2.0 g
		2. Acephate	1.5 g
		3. Prefenophos	2.0 ml
		4. Chlorpyriphos	2.5 ml
4.	Gallmidge	1. Phorate 1G	12.5 kg/ha
		2. Carboforan 3G	25.0 kg/ha
5.	Hispa	1. Profenophos	2.0 ml
		2. Chlorpyriphos	2.5 ml
		3. Monocrotophos	1.6 ml
6.	Cutworms	1. Dichlorovos + Endosulfan	1 ml + 2.0 ml
		2. Dichlorovos + Chlorpyriphos	1 ml + 2.5 ml
7.	Panicle mite	1. Profenophos	2.0 ml
		2. Dicofol	5.0 ml
8.	Leaf mite	1. Wettable sulphur	3.0 g
		2. Dicofol	5.0 ml
		3. Spiromisifen	0.6 ml
9.	Gundhi bug	1. Malathion dust	20Kg/ha
		2. Endosulfan dust	25Kg/ha
		3. Quinalphos dust	25Kg/ha
10.	Green leaf hopper	1. Monocrotophos	1.6ml
		2. Ethofenprox	2.0ml
		3. Fipronil	2.0ml
		4. Imidacloprid	0.25ml
		5. Thiomethoxam	0.25ml
11.	Whorl maggot	1. Corbofuran 3G	25Kg/ha
		2. Fipronil 0.3G	20Kg/ha
12.	Case worm	1. Monocrotophos	1.6ml
		2. Chlorpyriphos+Nuvan	2.5ml +1ml
13.	Thrips	1. Acephate	1.5g
		2. Dimethoate	2.0ml
		3. Monocrotophos	1.6ml

Acknowledgements

We are greatly motivated by Dr. D. Vishnu Vardhan Reddy, Associate Director of Research, Regional Agricultural Research Station, Warangal to bring out this publication. We thank all the scientists of RARS, Warangal who have helped in bringing out this publication. Our special thanks to Dr. B. Bhaskar Reddy, Professor

Table 19.12: Active Insecticide Compounds Recommended for Use on Rice in India as per the Information Available on National Portal of India (15 February, 2011)

Sl.No.	*Name of the Active Compound*	*Sl.No.*	*Name of the Active Compound*
1.	Acephate	2.	Azadirachtin
3.	Buprofezin	4.	Carbaryl
5.	Carbofuran	6.	Carbosulfan
7.	Cartap hydrochloride	8.	Chlorantraniliprole
9.	Chlorpyriphos	10.	Clothianidin
11.	Cypermethrin	12.	Deltamethrin
13.	Dichlorovos	14.	Endosulfan
15.	Ethofenprox	16.	Fenitrothion
17.	Fenobucarb	18.	Fipronil
19.	Flubendiamide	20.	Formothion
21.	Imidacloprid	22.	Lambda cyhalothrin
23.	Lindane	24.	Malathion
25.	Monocrotophos	26.	Oxydemeton methyl
27.	Methyl parathion	28.	Phenthoate
29.	Phorate	30.	Phosalone
31.	Phosphamidan	32.	Quinalphos
33.	Thiacloprid	34.	Triazophos

(Agronomy, Retd.), Dr. Shyam Sundar Reddy of IIIT, Hyderabad for the photographs provided. We extend our thanks to Dr. K. Narendar Reddy, Sr. Scientist (Ento.), Dr. D. Jagdishwar Reddy, Professor (Ento.), Dr. B. Ram Prasad, Scientist (Ento.), ANGRAU for critically going through the manuscript and making suggestions for its improvement. We appreciate assistance from Mrs. T. Laxmi, RARS, Warangal, who typed this manuscript.

References

Ambika, Devi D., Nair K.P.V, George Thomas and Rama Devi. (1998). Status of rice gall midge, *Orseolia oryzae* (Wood-mason) in Kuttand (abst.) National Symposium on Rice Research for 21st Century, Challenges, Priorities and Strategies, organized by the Association of Rice Research Workers, Cuttack, India. Feb 5-7, 1998., pp.96-97.

Anonymous. (1992). Manual on Integrated Pest Management in Rice Directorate of Plant Protection Quarantine and Storage, Faridabad, India, pp.99.

Beevi, S.P., Lyla K.R and Prabhakaran, P.V. (2000). Quantification of pests and natural enemies of rice ecosystems. In: Proceedings of the Twelfth Kerala Science Congress, January 2000, Kumily., pp.617-621.

Chellaiah, S.S., Bentur, J.S. and Prakasa Rao P.S. (1989). Approaches in rice pest management: Achievements and Opportunities. *Oryza*, **26**: 12-26.

Dhaliwal, G.S. and Arora, Ramesh. (2001). Integrated Pest Management: Concepts and Approaches, Kalyani Publishers, New Delhi, p. 109.

Dhandapani N. and Murugan M. 2004. Role of predatory fauna in integrated management of rice epsts. *IPM Mitr.*, **3**: 99-122.

Gupta M., Rao P.P. and Pawar A.D. (1987). Role of parasites and predators in rice ecosystem. Proceedings of the National Symposium on Integrated Pest Control: Progress and Perspectives, (Eds. N. Mohan Das and George Koshy)., p.545.

Hall, D.R., Cork A., Krishnaiah K. and Hendarish S. (1998). Identification of pheromones for rice pests and their use in control by mating disruption. In: Research highlights under the IPM strategy area (eds. R.J. Hillocks and S.J. Eden Green).

Hazarika, L.K. and Puzari, K.C. (1997). Field efficacy of white muscardine fungus (*Beauveria bassiana*) on rice hispa (*Discladispa armigera*). *Indian Journal of Agriculture Science*, **67**(10): 463-465.

Heinrichs, E.A. and Mochida, O. (1984). From secondary to major pest status. The case of insecticide-induced rice brown planthopper *Nilaparvata lugens* resurgence. *Prot Ecol.*, **7**: 201-218.

Katti, G., Pasalu, I.C., Varma, N.R.G. and Krishnaiah, K. (2001). Integration of pheromone mass trapping and biological control as an alternate strategy for management of yellow stem borer and leaf folder in rice (paper presented). Abstract No.38., p 38 In: Advance abstracts – Proceedings of the eighty eight session of the Indian Science Congress, New Delhi, 2001, Indian Science Congress, January 3 to 7, 2001, Indian Agricultural Research Institute, New Delhi.

Katti, G. and Pasalu, I.C. (2001). Quantification of pest incidence *vis-à-vis* natural biological control in rice ecosystem for possible exploitation in rice IPM (paper presented). Abstract No. 60, p 45 in: Abstract supplements – Proceedings of the eighty eighth session of the Indian Science Congress, New Delhi.

Krishnaiah, K. (1995). Insect pheromones in integrated pest management of cereals. *Proc. Natl. Workshop, Pheromones in IPM*, pp. 33-39.

Krishnaiah, K., Rama Gopala Varma, N., Pasalu I.C. and Syed Zainulabedin. (2000). Development of mass trapping technique for management of yellow stem borer (Scirpophaga incertulas, Walkar) inrice. pp. 33-40, In: Innovative pest and disease management in horticultural and plantation crops. (Eds. S. Narasimhan, G. Suresh and Weisley Daniel). Allied Publishers, Chennai.

Krishnaiah, K., Reddy, A.P.K., Krishnaiah, N.V. and Pasalu, I.C. (1999). Current problems and future needs in plant protection in rice. *Indian J. Plant Prot.*, **27** (1 and 2): 47-64.

Lim, G.S., Ooi, P.A.C. and Koh, A.K. (1980). Brown planthopper outbreaks and associated yield losses in Malayasia. *Int. Rice Res. Newsl.*, **5**: 15-16.

Manimaran, D and Manicka vasagam S. (2000). Light tram catches of hopper and mirid in rice. *Insect Environm*ent, **5** (4): 156-157.

Misra, H.P. (1999). Control of gundhi bug in upland rice. *Indian J.* Ent., **61**(2): 169-172.

Pasalu, I.C. and Katti, G. (2003). Biotypes of insect pests of rice: an emerging and challenging problem. In: *Prceedings of the National Symposium on Frontier Areas of Entomological Research*, Nov.5-7, 2003, New Delhi, India, pp.64-83.

Pasalu, I.C. (2002). Advancese in IPM Technology in Rice. In: *Compendium of short course training on Advances in Implementable Pest Management Technology.* Directorate of Oil seed Research, Hyderabad, AP-500 030, pp. 136-142.

Ridgeway, R.L. and Vinson, S.B. (1976). Biological control by augmentation of natural enemies. Plenum Press, New York, pp.480.

2015, Major Constrains and Verdict of Crop Productivity *Pages 292–302*
Editor: **Dr. U.N. Bhale**
Published by: **ASTRAL INTERNATIONAL (P) LIMITED, NEW DELHI**

Chapter 20

Enhancement of Seedling Emergence and Biomass Production of Crop Plants by *Trichoderma* spp.

J.N. Rajkonda[1] and U.N. Bhale[2]*

[1]Department of Botany, Yeshwantrao Chavan Mahavidyalaya, Tuljapur, Dist. Osmanabad – 413 601, M.S., India
[2]Research Laboratory, Department of Botany, Arts, Science and Commerce College, Naldurg, Dist. Osmanabad – 413 602, M.S., India

ABSTRACT

Trichoderma species are used extensively as a biocontrol agent for the management of many soil and seed borne diseases. Five Trichoderma species viz. Trichoderma viride, T. harzianum, T. koningii, T. pseudokoningii and T. virens were treated with seeds of various crop plants. Seed germination, biomass production and vigour index of treated seeds and seedling were evaluated and found to be significant. Biomass production and vigour index of wheat (Triticum aestivum L.) seedlings were enhanced by T. virens while the rate of germination also increased by T. harzianum and T. pseudokoningii. Trichoderma viride showed better results on the paddy (Oryza sativa L.) seedlings with respect to germination, biomass production and vigour index. Germination rate of pigeonpea (Cajanus cajan L.) seeds were enhanced by T. harzianum and T. virens. Trichoderma harzianum enhanced the biomass production and seedling vigour index of pigeonpea (Cajanus cajan L.).

* *Corresponding Author.* E-mail: unbhale2007@rediffmail.com, jnrajkonda@gmail.com

Germination, biomass production and vigour index of Gram (Cicer aurietinum L.) due to T. koningii. Maximum germination and seedling vigour index of groundnut (Arachis hypogea L.) was recorded better by T. pseudokoningii and T. virens. Its biomass production was accelerated by T. koningii. Trichoderma virens was found to be superior in case of soybean (Glycine max L.) with respect to germination rate, biomass production and seedling vigour index. Trichoderma harzianum showed promoting effect on the germination rate and biomass production of spinach while maximum seedling vigour index was due to the effect of T. koningii. Germination rate and biomass production of shepu (Anethum graveolens L.) was better under the treatment of T. koningii and T. virens, while seedling vigour index was enhanced due to T. viride. Trichoderma species were responsible for healthy and disease free crops and products.

Keywords: *Trichoderma species, Crop plants, Biomass production.*

Introduction

Trichoderma species are biological control organisms against a wide range of soil borne pathogens and have plant growth promotion capacity. While using *Trichoderma* species as a biocontrol agent for the management of plant diseases, it should be necessary to determine its effects on the crop plants with respect to seed germination, biomass production and vigour index. The interaction between *Trichoderma* species and plants are intriguing. The *Trichoderma* species can be beneficial to the plants causing growth stimulation (Kleifeld and Chet, 1992; Ousley *et al.*, 1994), or through biocontrol of plant diseases (Askew and Laing, 1994; Devan and Sivasithamparam, 1988; Lo *et al.*, 1996; Yang *et al.*, 1995).

The effect of *Trichoderma* species on the plant growth and their development was studied by many workers and researches regarding this aspect were revealed that *Trichoderma* species increases the growth of the plants. Windham *et al.* (1986) reported that addition of *Trichoderma* species to autoclaved soil increased the rate of emergence of tomato and tobacco seedlings over that of control. *Trichoderma harzianum* when applied to sterilized soil induced an emergence of seedlings, plant height, leaf area and dry weight (Kleifeld and Chet, 1992). Seed germination was maximum when *T. polysporum* was treated; fresh weight of eggplant was more with *T. viride* (Watanabe, 1993). Effect of biocontrol agent when primed was effective with respect to increased germination, biomass production of crop plants and healthy crop production. The effect of *T. harzianum* and *T. viride* on cumin seeds by biopriming method (Sharma *et al.*, 2009) indicated that germination was increased; shoot: root ratio and seedling height was also increased. It has been shown that *Trichoderma* species stimulated the growth of tomato plants (Datnoff *et al.*, 1995; Ozbay *et al.*, 2004).

The effects of *Trichoderma* isolates on plant growth and development have important economical implications such as shortening the plant growth period and time in nursery, as well as improving plant vigour to overcome biotic and/or abiotic stresses, resulting in increase plant productivity and yields. Therefore, the use of *Trichoderma* isolates in plant growth improvement is crucially important in sustainable agriculture system, because chemical fertilizer is not economical in the long run due to their cost and environmental pollution. Accordingly, reduction or elimination of synthetic fertilizer applications in agriculture is highly desirable. There are many abiotic and biotic factors such as plant species, the strain of *Trichoderma* used, the form of applied inoculum and its concentration and the soil environment as well as the rhizosphere microflora that may have an influence on *Trichoderma* activity. Therefore, it is very important to collect information about these factors. Recently, several attempts have been undertaken to survey *Trichoderma* spp. promotion of seedling establishment, enhancement of plant growth and elicit plant defense reaction in some crops such as cotton (Shanmugaiah *et al.*, 2009), vegetables (Celar and Valic, 2005; Rabeerdran *et al.*, 2000; Lynch *et al.*, 1991), bean (Hoyos-Carvajal *et al.*, 2009) and corn (Windham *et al.*, 1989).

The physical interaction between *Trichoderma* and the plant appears that this interaction evolves in to a symbiotic rather than a parasitic relationship between the fungus and the plant. (Yedidia *et al.*, 1999). It Elicitors from *Trichoderma* activate the expression of genes involved in the plant defence response system and promote the growth of the plant, root system and nutrient availability. This effect inturn augments the zone for colonization and the nutrients available for the biocontrol fungus, subsequently increasing the overall antagonism to plant pathogens (Yedidia *et al.*, 2003; Honson and Howell, 2004;). Many species of *Trichoderma* are not only able to control the pathogens that cause plant disease, but are also able to promote plant growth and development.

Therefore in the present investiation an effort was taken to determination of effect of *Trichoderma* species on the germination, biomass production and vigour index of crop plants.

Materials and Methods

Isolation of *Trichoderma* Species

Five *Trichoderma* species *viz. T. viride, T. harzianum, T. koningii, T. pseudokoningii* and *T. virens* were isolated from the rhizosphere soil of Marathwada region of Maharashtra state. Isolated *Trichoderma* species were maintained on PDA slants for the study.

Treatment with Seeds of Crop Plants

The effect of culture filtrate of *Trichoderma* species on germination, biomass production and vigour index of crop plants were determined by pot culture method. Healthy seeds of crop plants such as wheat (*Triticum aestivum* L.), rice (*Oryza sativa* L.), pigeonpea (*Cajanus cajan* L.), gram (*Cicer aurietinum* L.), groundnut (*Arachis hypogea* L.), soybean (*Glycine max* L.), spinach (*Spinacia oleraea* L.) and shepu (*Anethum graveolens* L.) were selected. Twenty healthy seeds were taken and sterilized by treatment with Mercuric chloride($HgCl_2$) (0.1 per cent) solution. The seeds were washed with sterile distilled water and treated with culture filtrate of isolated *Trichoderma* species. The treated seeds were sown in pot containing sterilized 2 kg black soil. Seeds without treatment were sown in separate pot and considered as a control. Water was added periodically at regular time intervals for 30 days. The pots were allowed to maintain at natural conditions.

After 30 days, number of germinated seeds were counted and recorded in terms of percentage root length, shoot length, number of leaves fresh, weights, dry weight were measured. The vigour index [Vigour index = [Mean root length + Mean Shoot length] X Germination (per cent)] of crop seedlings was calculated as per Abdul Baki and Anderson (1973).

Results and Discussion

Effect of isolated *Trichoderma* species on the growth of crop plants were tested by seed treatment with culture filtrate and growing such seeds in pot. The treatment of *Trichoderma* species with the seeds significantly increased the germination rate, biomass production and vigour index of the seedlings.

Results from the Table 20.1 indicated that germination of wheat was maximum under the treatment of *T. harzianum* (80 per cent) and *T. pseudokoningii* (80 per cent). Number of leaves recorded more in case of all treated seeds and was maximum when *T. virens* (4.6) used. Root length of root recorded maximum in the treatment of *T. virens* (8.9 cm) and shoot length was found to be maximum by *T. harzianum* (9.1 cm). Biomass production of wheat seedlings with respect to fresh weight and dry weight was recorded maximum in the treatment of *T. virens*. However, the vigour index of wheat seedlings was recorded maximum when *T. harzianum* (1368) was treated. Germination of rice seeds was increased when they treated with *T. viride* (90 per cent) and *T. pseudokoningii* (90 per cent) followed by *T. harzianum* (80 per cent). The effect of *Trichoderma* species on seedlings with respect to number of leaves were little significant. Root length of paddy seedlings were enhanced by *T. viride* (5.5 cm) and *T. harzianum* (5.5 cm) while shoot length was maximum due to the effect of *T. pseudokoningii* (3.0

Table 20.1: Effect of *Trichoderma* Species on Germination, Biomass Production and Vigour Index of Cereal Crop Plants

Crop Plant	*Treatment*	*Germ. (%)*	*No. of Leaves*	*Root Length (cm)*	*Shoot Length (cm)*	*Fresh Weight (gm)*	*Dry Weight (gm)*	*Vigour Index*
Wheat	Control	75	3.9	7.6	6.5	0.62	0.090	980
	T. viride	50	4.0	8.3	7	0.78	0.100	765
	T. harzianum	80	4.1	8.0	9.1	0.8	0.100	1368
	T. koningii	75	4.0	7.8	6.7	0.7	0.080	1087
	T. pseudokoningii	80	4.1	8.4	7	0.72	0.060	1232
	T. virens	70	4.6	8.9	6.8	1.24	0.180	1089
Rice	Control	65	1.1	2.4	1.7	0.03	0.010	164
	T. viride	90	1.3	5.5	2.4	0.06	0.018	711
	T. harzianum	80	1.4	5.5	2.7	0.05	0.017	656
	T. koningii	50	1.2	3.2	2.1	0.04	0.013	265
	T. pseudokoningii	90	1.5	4.1	3	0.05	0.017	639
	T. virens	65	1.3	3.8	2.5	0.05	0.024	346
Mean±		72.5	2.7	6.1	5.19	0.42	0.059	775

cm) and *T. harzianum* (2.7cm). Biomass production was maximum in the treatment of *T. viride* which also increased vigour index of paddy seedlings (711).

Seeds of pulse crops were treated with *Trichoderma* species and the results were determined. The effects were found to be superior with respect to germination, biomass production and vigour index of seedlings of pulse crop plants. Table 20.2 showed that pigeonpea seeds when treated with *T. harzianum* (100 per cent) and *T. virens* (100 per cent) recorded highest per cent germination followed by *T. viride* (90 per cent). Numbers of leaves were recorded maximum in the treatment of *T. harzianum* (11). Maximum root length of seedlings was observed in the treatment of *T. harzianum* (11.2 cm) and shoot length was better in *T pseudokoningii* (25.6cm). *Trichoderma harzianum* also showed maximum biomass production of pigeonpea. Effect of *T. harzianum* on the seedling vigour index was significant (3180) followed by *T. pseudokoningii* (2720) and *T. virens* (2520). Germination of gram seeds were maximum under the treatment of all *Trichoderma* species (100 per cent). Numbers of leaves were maximum by *T. koningii* (14). *Trichoderma koningii* was the species that found to be effective with respect to root length (14.70 cm), shoot length (12.50 cm), fresh weight (25.86 gm), dry weight (7.06gm) and vigour index (2720).

Table 20.2: Effect of *Trichoderma* Species on Germination, Biomass Production and Vigour Index of Pulse Crop Plants

Crop Plant	Treatment	Germ. (%)	No. of Leaves	Root Length (cm)	Shoot Length (cm)	Fresh Weight (gm)	Dry Weight (gm)	Vigour Index
Pigeonpea	Control	80	8	8.20	12.4	11.32	1.15	1648
	T. viride	90	8	9.40	16.5	11.80	1.18	2331
	T. harzianum	100	11	11.20	20.6	14.04	1.40	3180
	T. koningii	75	8	8.30	20.6	11.48	1.15	2167
	T. pseudokoningii	80	7	8.40	25.6	12.01	1.19	2720
	T. virens	100	5	8.20	17.30	11.42	1.16	2520
Gram	Control	90	8	13.15	10.3	17.70	4.58	2110
	T. viride	100	10	14.20	12.3	21.98	5.85	2660
	T. harzianum	100	11	14.47	11.5	25.55	6.96	2597
	T. koningii	100	14	14.70	12.5	25.86	7.06	2720
	T. pseudokoningii	100	11	14.10	10.0	18.93	6.12	2410
	T. virens	100	12	14.27	11.0	21.50	5.44	2527
Mean±		93	9	11.77	15.02	16.96	3.60	2465

Among the oil seed crops, seeds of groundnut and soybean were treated with the *Trichoderma* species to determine their effects on germination, biomass production and vigour index. The results were obtained more superior and recorded. Table 20.3 showed that *T. pseudokoningii* (95 per cent) was the species effective for germination of groundnut seeds followed by *T. virens* (85 per cent). *Trichoderma koningii* caused the increased number of leaves (8.33) followed by *T. pseudokoningii* (8.22). Maximum induction of root length was occurred by treatment of *T. harzianum* (11.10 cm) and *T. virens* (11.00 cm) while shoot length by *T. koningii* (7.00 cm). Maximum biomass production was observed under the treatment of *T. koningii*. Vigour index of seedlings were maximum when *T. pseudokoningii* (1520) and *T. virens* (1437) were treated. Germination of seeds of soybean were maximum under the treatment of *T. virens* (90 per cent), *T. viride* (85 per cent), *T. harzianum* (85 per cent) and *T. pseudokoningii* (85 per cent). Maximum numbers of leaves were observed when treated with *T. viride* (6.12) and *T. harzianum* (6.00). Root length and shoot length of seedlings were increased but root length was maximum due to *T. virens* (14.09 cm) while shoot length by *T. pseudokoningii* (16.65 cm). Biomass production of the seedlings was highest in the treatment of *T. virens*. However, seedling vigour index was highest when treated with of *T. virens* (2610) followed by *T. pseudokoningii* (2468) and *T. viride* (2055).

Table 20.3: Effect of *Trichoderma* Species on Germination, Biomass Production and Vigour Index of Oilseed Crop Plants

Crop Plant	*Treatment*	*Germ. (%)*	*No. of Leaves*	*Root Length (cm)*	*Shoot Length (cm)*	*Fresh Weight (gm)*	*Dry Weight (gm)*	*Vigour Index*
Groundnut	Control	45	6.67	6.52	4.81	8.37	0.32	510
	T. viride	65	6.77	7.57	5.7	8.62	0.35	862
	T. harzianum	50	7.40	11.10	6.16	8.71	0.37	863
	T. koningii	55	8.33	10.81	7	14.73	0.70	980
	T. pseudokoningii	95	8.22	9.74	6.26	9.30	0.38	1520
	T. virens	85	7.53	11.00	5.91	9.02	0.34	1437
Soybean	Control	55	5.81	09.46	11.67	6.08	0.24	1162
	T. viride	85	6.12	10.84	13.34	7.61	0.37	2055
	T. harzianum	85	6.00	10.37	13.07	6.16	0.28	1992
	T. koningii	75	5.91	10.03	13.65	7.71	0.31	1776
	T. pseudokoningii	85	5.83	12.39	16.65	7.87	0.33	2468
	T. virens	90	5.81	14.09	14.91	7.96	0.38	2610
Mean±		73	6.70	10.32	9.93	8.51	36.00	1520

Seeds of spinach treated with the culture filtrate of *Trichoderma* species and the results were recorded in the Table 20.4. The germination of the seeds were increased and maximum due to *T. harzianum* (90 per cent) followed by *T. virens* (80 per cent), *T. koningii* (85 per cent) and *T. pseudokoningii* (80 per cent). Number of leaves recorded maximum by the treatment of *T. viride* and *T. harzianum*. The root length was more under the treatment of *T. koningii* (7.01 cm) and *T. virens* (6.27 cm). The shoot length was increased due to *T. harzianum* (0.99 cm), *T. koningii* (0.98 cm) and *T. pseudokoningii* (0.94 cm). Biomass production with respect to fresh weight and dry weight was found to be maximum due to effect of *T. harzianum* and *T. viride* followed by *T. koningii* and *T. pseudokoningii*. *Trichoderma koningii* (639) and *T. virens* (609) were effective in increased seedling vigour index. The germination of shepu was observed maximum due to the effect of *T. koningii* (80 per cent) and *T. virens* (80 per cent). Numbers of leaves were increased slightly and maximum under the treatment of *T. koningii* (4.62) and *T. virens* (4.61). Maximum root length (11.01 cm) and shoot length (2.11 cm) was recorded by the effect of *T. viride*. Fresh weight of seedlings was observed in the treatment of *T. koningii* (0.055 gm) and *T. harzianum* (0.052 gm). Dry weight was found to be more under the treatment of *T. koningii* (0.011 gm) and *T. harzianum* (0.011 gm). Significant increase in the seedling vigour index was recorded due to the treatment of *T. viride* (984).

Table 20.4: Effect of *Trichoderma* Species on Germination, Biomass Production and Vigour Index of Vegetable Plants

Crop Plant	*Treatment*	*Germ. (%)*	*No. of Leaves*	*Root Length (cm)*	*Shoot Length (cm)*	*Fresh Weight (gm)*	*Dry Weight (gm)*	*Vigour Index*
Spinach	Control	60	5.33	4.33	0.90	0.260	0.028	314
	T. viride	70	12.30	5.30	0.90	3.290	0.370	434
	T. harzianum	90	10.55	5.34	0.99	3.330	0.391	570
	T. koningii	80	8.80	7.01	0.98	1.630	0.178	639
	T. pseudokoningii	80	7.62	5.11	0.94	1.130	0.115	484
	T. virens	85	7.00	6.27	0.90	0.780	0.098	609
Shepu	Control	70	4.07	3.46	1.30	0.029	0.007	333
	T. viride	75	4.10	11.01	2.11	0.034	0.008	984
	T. harzianum	75	4.40	6.07	1.97	0.052	0.011	603
	T. koningii	80	4.62	4.31	1.90	0.055	0.011	497
	T. pseudokoningii	75	4.26	4.38	1.73	0.042	0.009	458
	T. virens	80	4.61	4.39	1.92	0.043	0.009	505
Mean±		77	6.47	5.58	1.38	0.890	0.102	536

Trichoderma species are the useful biocontrol agents against number of plant pathogens. When the bioagents are applied, their effects on the growth and development of plants should be determined. Therefore in the present investigation, effect of *Trichoderma* species on the various crop plants had been carried out. The obtained results showed that *Trichoderma* species exhibited promoting effects on the crop plants with respect to germination, seedling length, number of leaves, biomass production and vigour index. There were no negative effects of the *Trichoderma* species on the growth of crop plants.

During the development process, initial growth of seedling is necessary for their establishment in the surroundings. Many seedlings were failed to proper growth and due to lack of proper growth caused low productivity. The increased seedling length, number of leaves, biomass production and seedling vigour index also indicated that the survivals of crop plants were increased. This is because the *Trichoderma* species probably caused increased metabolism in the crop plants. The healthy seedlings were also the sign of increased immunity of the crop plant seedlings. The seeds of pulse crops, oil crops and vegetables germinated actively under the treatment of *Trichoderma* species.

Precolonization provides the biocontrol agent with a competitive advantage over attacking pathogens and often provides superior seed protection (Harman and Taylor, 1988; Harman *et al.*, 1989; Harman, 1991). *Trichoderma* species can compete with other microorganisms for key exudates from seeds that stimulate the germination of propagules of plant pathogenic fungi in the soil (Howell, 2003) Application of *T. harzianum* at the time of sowing resulted significant increased groundnut pod yield and reduced the incidence of stem and pod rot (Jadeja and Ralcholiya, 2009). Seed treatment and soil treatment with *Trichoderma* species resulted healthy and disease free crop plants. This indicated that these plants have increased disease resistance which was induced by *Trichoderma* species. However, the induction of disease resistance in plants by *Trichoderma* species has been poorly studied because earlier works focused on factors like mycoparasitism and antibiosis. The first clear demonstration of induced resistance by *Trichoderma* was published in 1997 by Bigirimana who showed that treating soil with *T. harzianum* (T-39) made leaves of bean plants resistant to diseases caused by *B. cinerea* and *Colletotrichum lindemuthlanum*. Similar studies have been carried out with wide range of plants, including both monocotyledons and dicotyledons. In cotton, *T. virens* induced disease resistance against *R. solani* by inducing fungitoxic turpenoid phytoalexins (Howell *et al.*, 2000).

Earlier research works confirmed that induced germination of seeds by *Trichoderma* species. Wright (1954) found that when *Trichoderma* added to soil, germination of tomato and eggplant occurred two to three days earlier and showed better plant growth over control. The increased growth response induced by *Trichoderma* species has been reported for many crop plants such as beans, cucumber, pepper, maize and wheat (Lo and Lin, 2002). Seed pelleting of sunflower and mung bean with *T. harzianum, T. polysporum, T. pseudokoningii* and *T. virens* enhanced the plant growth as compared to control (Yaqub and Shahzad, 2008). There are several reports where *Trichoderma* species effectively increased the plant growth, weight and root mass (Cole and Zvenyika, 1986; Paulitz *et al.*, 1985; Kumar *et al.*, 2007) that supports the results of present research work.

Recently chilli seeds coated with spore suspension of strains of *T. harzianum, T. pseudokoningii* and *T. virens* resulted in to increased germination, reduced the delay of germination and seedling parameters in both laboratory and field conditions (Islam *et al.*, 2011). Seed dressing and soil application with *T. virens, T. harzianum* and *T. viride* were found to be superior in controlling wet root rot of mungbean caused by *Rhizoctonia solani* and improvement of the yield (Dubey *et al.*, 2011). Increased growth of shoot and root caused by *Trichoderma* implied that there was beneficial effect of inoculation on plant growth and development since root collar and stem diameters were a measure of survivability of seedlings (Okoth *et al.*, 2011).

Earlier works suggested that the *Trichoderma* species are naturally occurring soil fungi that colonize roots and stimulate plant growth. Such fungi have been applied to a wide range of plant species for the purpose of growth enhancement, with a positive effects on plant weight, crop yield and disease control. Their agricultural use could be expanded if the mechanisms of growth enhancement were known.

Conclusion

It is concluded that the *Trichoderma* species likely act as plant growth promoting microorganisms because the given experiments showed enhanced root growth along with shoot growth and biomass production of crop plants. The data obtained in the result suggested an important role of *Trichoderma* species in plant growth regulation. The results showed great promise for the use of *Trichoderma* species as inoculants for plant improvement under controlled and field conditions. The introduction of *Trichoderma* species in the soil with or without pathogen did not affect the population of existing beneficial microorganisms. Use of *Trichoderma* species as seed treatment for better crop productivity could be suggested.

Acknowledgement

Authors are thankful to principal Dr. S. D. Peshwe, A. S. C. College, Naldurg and principal Dr. J. S. Mohite, Y. C. M. Tuljapur for providing necessary laboratory facilities. And also thankfully acknowledged to UGC (WRO) Pune and UGC, New Delhi for providing financial assistance.

References

Abdul Baki and Anderson (1973). Vigour determination in soybean by multiple criteria. *Crop Science,* **13**: 630-633.

Askew, D. J. and Liang, M. D. (1994). The *in vitro* screening of 119 *Trichoderma* isolates for antagonism to *Rhizoctonia solani* and an evaluation of different environmental sites of *Trichoderma* as sources of aggressive strains. *Plant and Soil,* **159**: 277-281.

Bigirimana, J. (1997). Induction of systemic resistance on bean (*Phaseolus vulgaris*) by *Trichoderma harzianum*. *Med. Fac. Landbouww Uni. Gent.,* **62**: 1001-1007.

Celar, F. and Valic, N. (2005). Effects of *Trichoderma spp* and *Glicladium roseum* culture filtrates on seed germination of vegetables and maize. *J. Plant Dis. Prot.,* **112**(4): 343-350.

Cole, J. S. and Zvenyika, P. (1986). Integrating *T. harzianum* and triadimenol for the control of tobacco sore shin in Zimbabwe. *Bull. Inf. Coresta,* p.68.

Datnoff, L. E., Sand, N. and Pernezny, K. (1995). Biological control of *Fusarium* crown and root rot of tomato in Florida using *Trichoderma harzianum* and *Glomus intraradices. Biol. Cont.,* **5**: 427-431.

Devan, M. M. and Sivasithamparam, K. (1988). Identity and frequency of occurrence of *Trichoderma* spp in roots of wheat and rye-grass in Western Australia and their effects on root rot caused by *Gaeumannomyces graminis* var *tritici. Plant and Soil,* **109**: 93-101.

Dubey, S. C., Ranganaicker, B. and Singh, B. (2011). Integration of soil application and seed treatment formulations of *Trichoderma* species for management of wet root rot of mungbean caused by *Rhizoctonia solani. Pest Management Science,* **67**(9): 1163-1168.

Harman, G. E. (1991). Seed treatments for biological control of plant disease. *Crop Prot.*, **10**: 166-171.

Harman, G. E. and Taylor, A. G.(1988). Improved seedling performance by integration of biological control agents at favourable pH levels with solid matrix priming. *Phytopathology*, **78**: 520-525.

Harman, G. E., Taylor, A. G. and Stasz, T. E. (1989). Combining effective strains of *Trichoderma harzianum* and solid matrix priming to improve biological seed treatments. *Pl. Dis.*, **73:** 631-637.

Honson, L. E. and Howell, C. R. (2004). Elicitors of plant defencse responses from biocontrol strains of *Trichoderma virens*. *Phytopathology*, **94**(2): 171-176.

Howell, C. R. (2003). Mechanisms employed by *Trichoderma* species in the biological control of plant disease; the history and evolution of current concepts. *Plant Dis.*, **87**: 4-10.

Howell, C. R., Hanson, L. E., Stipanovic, R. D. and Pukhaber, L. S. (2000). Induction of turpenoid synthesis in cotton roots and control of *Rhizoctonia solani* by seed treatment with *Trichoderma virens*. *Phytopathology*, **1990**: 248-252.

Hoyos-Carvajal, L., Ordua, S. and Bissett, J. (2009). Growth stimulation in bean (*Phaseolus vulgaris* L.) by *Trichoderma*. *Biol. Control*, **51**: 409-416.

Islam, M. S., Rahman, M. A., Bulbul, S. H. and Alam, M.F. (2011). Effect of *Trichoderma* on seed germination and seedling parameters in chilli. *Int. J. Expt. Agric.*, **2**(1): 21-26.

Jadeja, K. B. and Rakholiya, K. B. (2009). Efficacy of soil amendments with *T. harzianum* on stem and pod rot incident and yield of groundnut. (Abs.) *J. Mycol. Pl. Pathol.*, **39**(3): 569.

Kleifeld, O. and Chet, I. (1992). *Trichoderma harzianum* – interaction with plants and effect on growth response. *Plant and Soil*, **144**: 267-272.

Kumar, S., Arya, M. C. and Singh, R. (2007). Efficacy of *P. fluorescence* and *T. harzianum* as bioenhancers in tomato at high altitude in Central Himalayas. *Indian J. Crop Sciences*, **2**(1): 79-82.

Lo, C. T. and Lin, C. Y. (2002). Screening strains of *Trichoderma* spp for plant growth enhancement in Taiwan. *Plant Pathology Bull.*, **11**: 215-220.

Lo, C. T., Nelson, E. B. and Harman, G. E. (1996). Biological control of turfgrass diseases with a rhizosphere competent strains of *Trichoderma harzianum*. *Plant Dis.*, **80**: 736-741.

Lynch, J. M., Wilson, K. L., Ousley, M. A. and Wipps, J. M. (1991). Response of lettuce to *Trichoderma* treatment. *Lett. Appl. Microbiol.*, **12**: 59-61.

Okoth, S. A., Otaodh, J. A. and Ochanda, J. O. (2011). Improved seedling emergence and growth of maize and beans by *Trichoderma harzianum*. *Tropical and Subtropical Agroecosystems*, **13**: 65-71.

Ousley, M. A., Lynch, J. M., and Whipps, J. M. (1994). Potential of *Trichoderma* spp. as consistent plant growth stimulators. *Biol. Fertil. Soils,* **17**: 85-90.

Ozbay, N., Newman, S. E. and Brown, M. (2004). The effect of *Trichoderma harzianum* strains in the growth of tomato seedlings. Proc. XXVI IHC Manage, *Acta Hort.,* **635**: 131-135.

Paulitz, T., Windham, M. and Baker, R. (1985). Enhanced plant growth induced by *Trichoderma* amendments. *Phytopathology*, **75**: 1302.

Rabeerdran, N., Moot, D. J., Jones, E. E. and Stewart, A. (2000). Inconsistent growth promotion of cabbage and lettuce from *Trichoderma* isolates. *New Zealand Plant Prot.,* **53**: 143-146.

Shanmugaiah, V., Balasubramanian, N., Gomathinayagam, S., Monoharan, P.T. and Rajendran, A. (2009). Effect of single application of *Trichoderma viride* and *Pseudomonas fluorences* on growth promotion in cottonplants. *Afr. J. Agric. Res.,* **4**(11): 1220-1225.

Sharma, Y. K., Anwer, M. M., Lodha, S. K., Sriram, S. and Ramanujam, B. (2009). Effect of biopriming with biological control agents on wilt and growth of cumin. *J. Mycol. Pl. Pathol.,* **39**(3): 567.

Watanabe, N. (1993). Promoting effect of *Trichoderma* spp on seed germination and plant growth in vegetables. *Memories of the Institute of Science and Technology, Meiji Uni.,* **32**: 9-17.

Windham, G. L., Windham, M. T. and Williams, W. P. (1989). Effects of *Trichiderma* spp. on maize growth and *Meloidogyne arenaria* reproduction. *Plant Dis.,* **73**: 493-495.

Windham, M. T., Elad, Y. and Baker, R. (1986). A mechanism for increased plant growth induced by *Trichoderma* spp. *Phytopathology,* **76**: 518-521.

Wright, J. M. (1954). The production of antibiotics in soil 1 production of gliotoxin by *Trichoderma viride*. *Ann. Appl. Boil.,* **41**: 280-289.

Yang, D., Bernier, L. and Dessureault, M. (1995). *Phaeotheca dimorphospora* increases *Trichoderma harzianum* density in soil and suppresses red pine damping off caused by *Cylindrocladium scoparium*. *Can. J. Bot.,* **73**: 693-700.

Yaqub, F. and Shahzad, S. (2008). Effect of seed pelleting with *Trichoderma* spp and *Gliocladium virens* on growth and colonizationof roots of sunflower and mung bean by *Sclerotium rolfsii*. *Pak. J. Bot.,* **40**(2): 947-953.

Yedidia, I., Benhamou, N. and Chet, I. (1999). Induction of defense responses in cucumber (*Cucumis sativus* L.) by the biocontrol agent *Trichoderma harzianum*. *Appl. Environ. Microbiol.,* **65**: 1061-1070.

Yedidia, I., Shoresh, M., Kareem, Z., Lapulink, Y. and Chet, I. (2003). Concomitant induction of systemic resistance to *Pseudomonas syringae* pv. *lachrymans* in cucumber by *Trichoderma asperellum* (T-203) and accumulation of phytoalexins. *Applied Environmental Microbiology,* **69**: 7343-7353.

2015, Major Constrains and Verdict of Crop Productivity *Pages 303–311*
Editor: Dr. U.N. Bhale
Published by: ASTRAL INTERNATIONAL (P) LIMITED, NEW DELHI

Chapter 21

Weed Diversity of Rice Fields in Cuddalore District, Tamil Nadu

Prabhakaran Jayaraman*, Kavitha Duaraiswamy and Saravanakumar Kaliyaperumal

Botany Wing-DDE, Department of Botany
Annamalai University, Annamalai Nagar – 608 002, Tamil Nadu, India

ABSTRACT

Weeds are at present the major biotic constraint to increased rice production worldwide. Weeds have always been a major disturbing factor in agricultural production systems. If left uncontrolled, weed plants compete with crop plants for resources essential for plant growth, thereby reducing crop yield and quality. Hence, the diversity of weeds an important one to understand the weed dynamics especially in crop fields for effective control of weeds. The survey was conducted to assess the weed diversity of rice fields of some villages of Virudhachalam Taluk, Cuddalore district, Tamil Nadu. Total 45 weed species of which 32dicots and 13 monocots belonging to 21 families and two pteridophytes were recorded, Out of 45species, 26 and 32weeds were recorded in pre and post harvest rice fields respectively. On the basis of IVI, the dominant weed species of pre-harvest rice fields were Cyperus difformis, Echinochloa colona, Cyperus iria, Cyperus rotundus, Ageratum conyzoides, and Croton bonplandianum, Chrozophora rottleri were the dominant species of post harvest rice fields.

Keywords: *Weed diversity, Rice fields, Cuddalore district, IVI.*

Introduction

The rice field ecosystem harbors a rich composition of primary producers, which can be broadly grouped under two categories: the macrophytes and microphytes.

* *Corresponding Author.* E-mail: drprabha2006@gmail.com

Besides the rice plant, the macrophytes include other rooted higher plants, which consist mainly of grasses, sedges and broad-leaved plants. These are generally referred to as weeds, as they compete with the rice plants for growth requirements such as space, nutrients and sunlight (Halwart, 1993). Weeds are the most underestimated crop pests in tropical agriculture although they cause higher reduction/loss in the yields of crops then other pests and disease. Of the total annual loss of agricultural produce from various pests in India, weeds roughly account for 45 per cent, insects for 30 per cent disease for 20 per cent and other pests for 5 per cent (Das, 2008). Without weed management rice yield may be reduced by 16 to 86 per cent or even 100 per cent (Baltazar and De Datta, 1992; Zoschke, 1990).Weeds in simpler terms are plants that interfere with the healthy or normal growth and development of crops. They are now known to limit the production of crops causing serious losses in the output of grains, seeds and fruits etc. (Chaudhri, 1992). Weeds have always been a major disturbing factor in agricultural production systems.To minimize the negative consequences of weeds on crop production, curative actions to remove or kill weed plants shortly after their establishment are often undertaken. Weeds compete with other crops for water, nutrients and space. Weeds also act as alternate hosts for pests and diseases. The abundance or distribution of weed species in a cropped field varies due to the nature of the crop, cultural practices and cropping pattern/system, soil type, moisture availability, location and season. Knowledge of weed flora enables one to use the required herbicide and formulate other suitable management strategies. It is also useful in exploiting abundant weeds as a cover crop or pasture and for other economic uses. Extensive literature is available on weed flora dynamics in field crops (Moyer *et al.*, 1989; Fuente *et al.*, 1999; Derksen *et al.*, 2002).

Weed surveys are useful for determining the occurrence and relative importance of weed species in crop production systems (Frick and Thomas, 1992). A fundamental basis for effective weed management is to know the weed species present and the level of infestation (Labrada and Parker, 1994).Thus the present study has been made to determine the diversity of the weed species in the pre and post harvest rice fields and their occurrence which includes density, frequency, family wise distribution and importance value index of individual weed species.

Materials and Methods

The study was conducted in the rice fields during cultivation period from the date of transplanting rice seedlings to harvest (pre harvest), and after two month from harvest period (post-harvest). The weed population were recorded during 2011-2012 in the pre and post harvest rice fields of nearby villages of Vilakkappadi, Sathappadi and Dharmanallur, Virudhachalam Taluk (11°32'N79° 24' E,)of Cuddalore District, Tamil Nadu (Figure 21.1).

Quadrates of 10sq. m area were laid down in different locations in a random manner.For both pre and post harvest rice field, ten quadrates were included for studying weed flora. Various species and the total number of individuals of each species were noted. Average data of ten replications/quadrate were considered for analysis. The weeds were assessed quantitatively and the information obtained was used to determine the following ecological parameters, such as abundance, density,

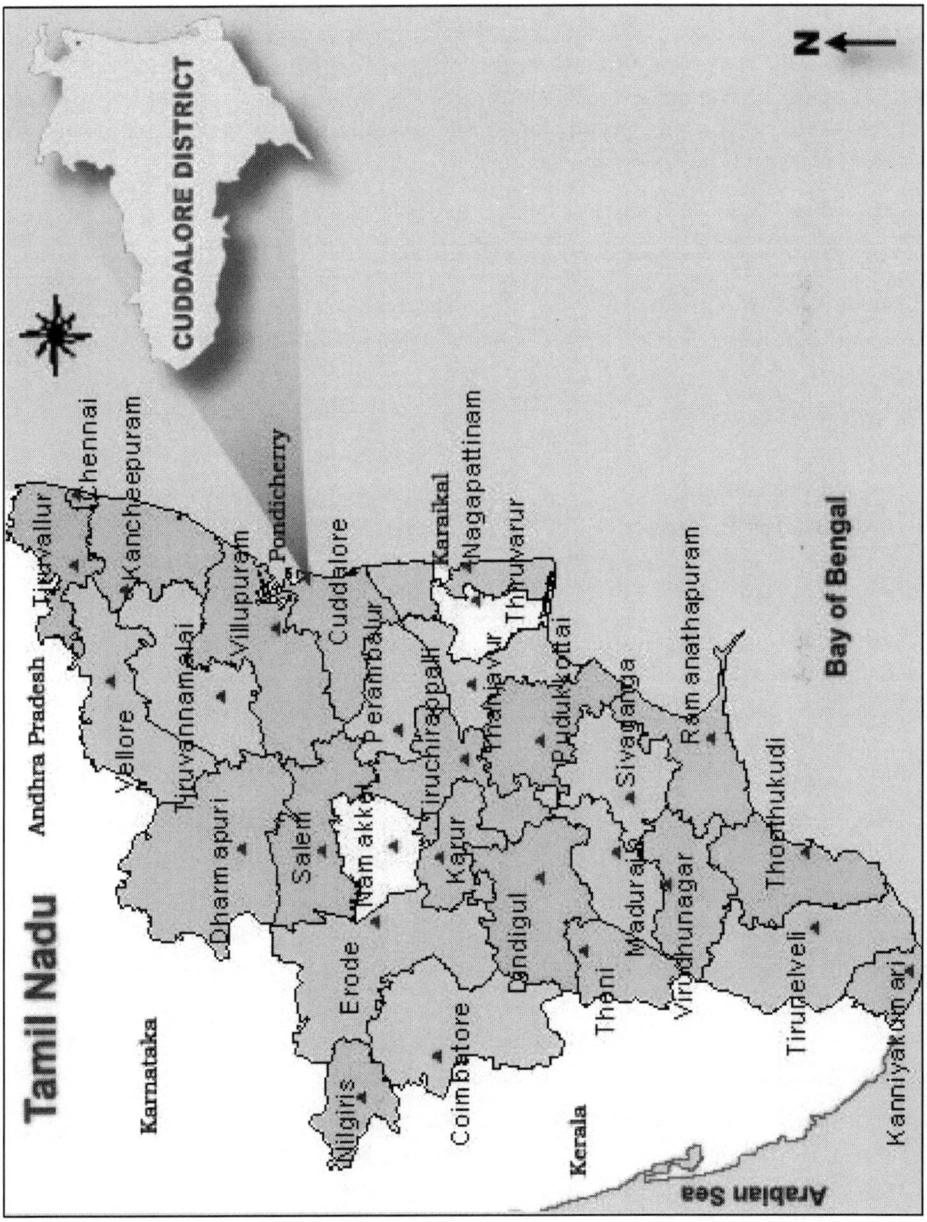

Figure 21.1: Map Showing Cuddalore District of Tamil Nadu.

relative density, frequency, relative frequency, and importance value index(IVI) using the method of Smith and Smith(1998).

Results and Discussion

Forty five weed species were recorded from the pre and post harvest rice fields, of which,11monocotyledons,13 dicotyledons of angiosperms and two pteridophytes (*Salvinea* sp. and *Marsilea* sp.) were in the pre harvest rice fields (Tables 21.1 and 21.2). In the post harvest rice fields, 8 monocotyledons and 24 dicotyledons species were noticed but two pteridophytic plants were completely absent in this field because of the aquatic habitat nature of the species.

Table 21.1: Weed Species Recorded in the Pre- and Post-harvest Rice Fields

Sl.No.	*Name of the Species*	*Family*	*Common Name*
1.	*Abutilon indicum* (L.) Sf.**	Malvaceae	Indian Abutilon
2.	*Achyranthes aspera* L.**	Amaranthaceae	Prickly chaff flower
3.	*Ageratum conyzoides L.****	Asteraceae	Goatweed,
4.	*Ageratum conyzoides* L.***	Asteraceae	Gout weed
5.	*Alternanthera sessilis (L.) R.Br. ex DC.****	Amaranthaceae	Sessile joyweed
6.	*Ammania baccifera* L.*	Lythraceae	Blistering ammania
7.	*Cardiospermum helicacapum* L.**	Sapindaceae	Baloon vine
8.	*Chrozophora rottleri (Geiseler) A.Juss,***	Euphorbiaceae	Suryavarti,
9.	*Cleome chelidoni* L.**	Capparidaceae	Showy cleome
10.	*Cleome gynandra* L.**	Capparidaceae	Butterfly cleome
11.	*Cleome viscosa* L.**	Capparidaceae	Sticky cleome
12.	*Commelina benghalensis* L.*	Commelinaceae	Day flower
13.	*Croton bonplandianum Baill.***	Euphorbiaceae	Ban tulsi
14.	*Cyanotis axillaries* L.***	Commelinaceae	Spreading dayflower.
15.	*Cynodon dactylon (L.) Pers.****	Poaceae	Bermuda Grass
16.	*Cyperus difformis L****	Cyperaceae	Umbrella-sedge
17.	*Cyperus iria* L.***	Cyperaceae	Rice flat sedge
18.	*Cyperus rotundus L.****	Cyperaceae	Purple nut sedge
19.	*Dactylocteniumaegyptiacum*(L.) Willd.***	Poaceae	Durban Crow foot grass
20.	*Desmodium triflorum* (L.).DC.***	Fabaceae	Tick clover
21.	*Dichanthium annulatum* (Forssk) Stafp.**	Poaceae	Kleberg's bluestem
22.	*Echinochloa colona L.**	Poaceae	Panicum grass
23.	*Echinochloa crus-galli (L.) P.Beauv.**	Poaceae	Barnyardgrass
24.	*Eclipta alba* (L.) Hassk.*	Asteraceae	False Daisy
25.	*Eleusine indica* (L.)Gaertn.*	Poaceae	Indian goosegrass
26.	*Enicostemma axillare* (Lam.) Rayanal.*	Gentianaceae	Indian Whitehead
27.	*Euphorbia hirta L.* * *	Euphorbiaceae	Asthma weed

Contd...

Table 21.1–*Contd...*

Sl.No.	*Name of the Species*	*Family*	*Common Name*
28.	*Fimbristylis littoralis* (L.) Vahl.*	Cyperaceae	Fimbry
29.	*Gomphrena celosioides* C.Mart.**	Amaranthaceae	Gomphrena weed
30.	*Heliotropium indicum* L.***	Boraginaceae	India Heliotrope
31.	*Hybanthus enneaspermus* L.) F. Muel.***	Violaceae	Spade Flower
32.	*Hygrophila auriculata*(Sch.)Heine.*	Ac Acanthaceae	Hygrophila
33.	*Kyllinga bulbosa* P.Beauv.***	Cyperaceae	Bulbous bluegrass
34.	*Marsilea quadrifolia* L*.	Marsileaceae	Water clover
35.	*Melilotus indica* L.**	Fabaceae	Alfalfa
36.	*Mollugo oppositifolia* L.**	Aizoaceae	Indian Carpet weed
37.	*Oldenlandia biflora* L.**	Rubiaceae	Two flower mille graines
38.	*Oldenlandia corymbosa* L.***	Rubiaceae	Chay root
39.	*Panicum repens* L.*	Cyperaceae	Victoria grass
40.	*Portulaca oleracea* L.**	Portulacaceae	Common purslane
41.	*Sacciolepis indica* (L.)Chase.**	Poaceae	Glenwood grass
42.	*Salvinia natans* L. (L.) Allioni*	Salviniaceae	Floating water moss
43.	*Stemodia viscosa* Roxb.**	Sc Scrophulariaceae	Sticky Blue Rod
44.	*Trianthema portulacastrum* L.**	Portulacaceae	Meadow purslane
45.	*Tridax procumbens* L.**	Asteraceae	False daisy

Occurrence of weed species *: Only in Pre-harvest; **: Only in Post-harvest; ***: Both in pre-and post-harvest rice fields.

From the view of family wise distribution of the weed species, total 21 plant families were recorded, of which 14 and 16 families of the weeds were occurred in pre and post harvest rice field respectively. Monocotyledonous weed possess maximum frequency, density and IVI in pre harvest rice fields but in the post harvest rice fields the dominant weeds were *Chrozophora rottleri* and *Croton bonplandianum* having high IVI, which were not recorded in pre harvest rice fields (Table 21.3).

Moody (1989) has documented more than 1800 weed species occurring in the rice field ecosystems of Asia and South-East Asia alone. Species of Poaceae (grasses) are the most common, followed by Cyperaceae (sedges) and other broad-leaved families respectively. The barnyard grass (*Echinochloa crus-galli* (L.) Beauv is considered to be the most troublesome weed of rice in the world (Holm *et al.*, 1977), followed by *E. colonum* (L.) Link (Smith, 1983). Other rice field weeds of world importance include: grasses - *Eleusine indica*, sedges - *Cyperus difformis, C.rotundus, C.iria, Fimbristylis miliaceae*, broad-leaved - *Monochoria vaginalis* and *Sphenochlea zeylanica* (Smith, 1983).

Weed poses serious problems to agriculture and environment. Out of the total 826 weed species reported in India, 80 are considered as very serious and nearly 200 as serious weeds. Of the total losses caused by the agricultural pests, weeds contribute to as high as 37per cent unlike other pests and weeds are ubiquitous and affect almost all the crops. The composition and competition by weeds is dynamic and is dependent on soil, climate, cropping and management factors. The biology, ecology and management of weeds cannot remain constant for all the regions and hence, weed management strategies will have to be different for each agro-ecological condition. (NRCWS, 2007).

Table 21.2: Weed Families Recorded in the Rice Fields

Sl.No..	*Family*	*No.of Species*	
		Pre harvest	*Post-harvest*
1	Acanthaceae	1	–
2	Aizoaceae	–	1
3	Amaranthaceae	1	3
4	Asteraceae	3	3
5	Boraginaceae	1	1
6	Capparidaceae	–	3
7	Commelinaceae	2	1
8	Cyperaceae	6	4
9	Euphorbiaceae	–	3
10	Fabaceae	1	2
11	Gentianaceae	1	–
12	Lythraceae	1	–
13	Malvaceae	–	1
14	Marsileaceae	1	–
15	Poaceae	5	3
16	Portulacaceae	–	2
17	Rubiaceae	1	2
18	Salviniaceae	1	–
19	Sapindaceae	–	1
20	Scrophulariaceae	–	1
21	Violaceae	1	1
	Occurrence of Total No. of weed species	26	32

–: Absent.

Gurinder Jit *et al.* (2006) reported that the dominant weed species found in upland, paddled direct seeded and transplanted rice, and rain fed lowland rice fields in Tamil Nadu are, *Chloris barbata, Croton bonplandianum, Croton speciflores, Dinebra arabica, Ammania baccifera, Cyperus difformis, Cyperus iria, Digitaria sanguinalis, Eclipta prostrata, Marsilea minuta, Paspalum distichum, Eichornia colonum, Eclipta alba, Fimbristylis miliacea, Monochoria vaginalis, Scirpus mucronatus* and *Sphaeranthus indicus, Cyanodon dactylon, Cyperus rotundus, Salvania* sp.

Table 21.3: Frequency, Density and IVI of some Major Weeds Occurring in Rice Fields

Sl.No.	*Weed Species*	*Pre-Harvest Rice Fields*			*Post-Harvest Rice Fields*		
		Frequency Per cent	*Relative Density Per cent*	*IVI*	*Frequency Per cent*	*Relative Density Per cent*	*IVI*
1.	*Cyperus difformis* L.	14.2	12.6	15.2	9.5	8.2	7.1
2.	*Echinochloa colona* L.	13.5	9.5	11.2	Ab	Ab	Ab
3.	*Cyperus iria* L.	10.2	6.8	8.3	8.3	7.5	5.7
4.	*Cyperus rotundus* L.	9.4	5.2	6.4	6.4	7.1	5.1
5.	*Ageratum conyzoides* L.	5.4	4.4	3.9	3.2	2.4	1.71
6.	*Eleusine indica* (L.) Gaertn.	5.1	4.0	3.2	Ab	Ab	Ab
7.	*Kyllinga bulbosa* P. Beauv	4.6	3.6	2.1	5.2	4.2	3.2
8.	*Marsilea quadrifolia* L.	4.4	3.5	2.0	Ab	Ab	Ab
9.	*Chrozophora rottleri* (Geiseler) A. Juss	Ab	Ab	Ab	17.6	15.8	19.5
10.	*Croton bonplandianum* Baill	Ab	Ab	Ab	15.5	13.8	17.2

*Ab: Absent.

Various environmental factors *i.e.* light, temperature, water, soil and microorganisms determine the specific weed spectrum, their distribution and growth (Barbour *et al.*, 1980). As the environmental conditions vary from one region to another, geographical differentiation has characterized the weed species in each region(Peng, 1984). According to Rao(1983) persistence of weeds is largely influenced by climatic, edaphic and biotic factors which affect their occurrence, abundance, range and distribution. The distribution and population dynamics of weeds depends on local geographical conditions, cropping pattern and agronomic procedures. The ecological status of weeds varies throughout growing season and also with the time of survey, habit conditions and weed control methods followed by the local farmers.

Table 21.3 shows the frequency, density and IVI of some major weeds recorded in the present observation. Among them *Cyperus iria, Echinochloa colona, Cyperus difformis,* and *Cynodon dactylon* are reported as being the important weeds of paddy fields in the world by Holm *et al.* (1977). Of the total 45 weeds species,14 were common to both pre and post harvest rice fields. Twelve weed species present only in pre harvest and 19species were recorded only in post harvest rice fields. The increasing numbers of species in post harvest fields due to the weeds adapted to neither a particularly dry nor particularly wet environment in the fields.

Hallgren *et al.* (1999) reported that sowing season of crops, geographical region and soil types had an over whelming influence on the weed flora composition.Many factors are responsible for changes and establishment of a weed community and community structure depends upon the fundamental niche of weed species (Sikinder and Zaheer, 2003). Smith and Smith (1998) described that species differ in the range of conditions they tolerate. As environmental conditions change in both time and space the possible distribution and abundance of species will also change.

Among the three villages surveyed that the level of frequency, abundance, family wise distribution and IVI values of weed species were more or less similar. The uniformities on the distribution of weed flora in the villages due to the similar soil types, source of irrigation, agronomic practices during cultivation of rice crops in the surveyed crop fields.

References

Baltazar, A.M. and De Datta, S.K. (1992). Weed management in rice. *Weed Abstracts,* **41**: 495–507

Barbour, M.G., Burk, J.H. and Pitts, W.D. (1980). Terrestrial Plant Ecology. Benjamin/ Cummings, Menlo Park, California. *Biological Agriculture and Horticulture,* **18:** 175-185.

Chaudhri, M.N. (1992).Weeds and their identification. In: Identification and Control of Weeds Manual. National Training Course. *Pak. Agric. Res. Council*. Competition between cereals and their natural weed populations. *Weed Research,* **36:** 461-470.

Das, T.K. (2008). Weed Science, Basics and applications. Jain Brothers, New Delhi.

Derksen, A. D., Andersen, L. R., Blackshaw, E. R. and Maxwell, B. (2002). Weed dynamics and management strategies for cropping systems in the Northern Great Plains. *Agron. J.,* **3**: 174–185.

Frick, B. and Thomas, A.G. (1992). Weed survey in different tillage systems in south Western Ontaria field Crops. *Can.J. Pl.Sci.,* **72:** 1337-1347.

Fuente, E. B., Saurez, S. A., Ghersa, C. M. and Leon, R. J. C., (1999). Weed communities: Relationship with cultural history and crop yield. *Agron. J.,* **91:** 234–241.

Gurinder Jit, R., Shashi Bhalla, V. Celia, C., Vandava T., Verma, D.D. and Manoranjan, H. (2006). Document on Biology of rice (*Oryza sativa* L.) in India. NBPGR – ICAR – New Delhi.

Hallgren, E., Palmer, M.W. and Milberg, P. (1999). Data diving with cross validation and investigation of broadscale gradient in Swedish weed communities. *J. Ecology,* **87**: 1015-1037.

Halwart, M. (1993). Fish as biocontrol agents in rice. Verlag, Weikersheim, Germany, 169 p.

Holm, L.G., Plucknett, D.L., Pancho, J.V., and Herberger, J.P. (1977). The World's Worst Weeds. Distribution and Biology. Honolulu, Hawaii, USA: University Press of Hawaii

Labrada, R. and Parker, C. (1994). Weed control in the context of integrated pest management for developing countries. (Ed. Labrada, *et al.*) FAO, Plant Production and Production Paper 120, Rome.3-8.

Moody, K. (1989). Weeds reported in rice in South and Southeast Asia. IRRI, Los Banos, Philippines. *Memoranda Societas Pro Fauna Flora Fennica,* **62**: 63-67

Moyer, J. R., Owenya, Z. J. and Kibuwa, S. P. (1989). Weed population and agronomic practices at wheat farms on the Hanang plains of Tanzania. *Trop. Pest Manage.,* **35**: 26–29.

NRCWS (National Research Centre for Weed Science) (2007). In: NRCWS Perspective Plan, Vision – 2025. ICAR, New Delhi

Peng, S.Y. (1984). The Biology and control of Weeds in Sugarcane.Elsevier, Amsterdam, p.25.

Rao, V.S. (1983). Principles of Weed Science. Oxford and IBH Publishing Co. New Delhi, India.

Sikinder, S. and Zaheer, A.N.(2003). Dynamics of weed communities in gram fields of Chakwal, Pakistan. *Asian J. Pl.Sci.*, **2**(17-24): 1198-1204.

Smith, R. J. (1983). Weeds of major economic importance in rice and yield losses due to weed competition. In: Proc. of the Conf. on Weed Control in Rice, International Rice Research Institute, Manila, Philippines, International Weed Science Society.

Smith, R.L. and Smith, T.H. (1998). Elements of Ecology. The Benjamin/Cummings Pub.Co.Inc.2725.Sand Hill road, Menlo Park.

Zoschke, A. (1990). Yield loss in tropical rice as influenced by the competition of weed flora and the timing of its elimination. In: Grayson B.T., Green M.B., Copping L.G. (eds): Pest Management in Rice. *Elsevier Science*, London: 301-313.

2015, Major Constrains and Verdict of Crop Productivity *Pages 312–323*
Editor: Dr. U.N. Bhale
Published by: ASTRAL INTERNATIONAL (P) LIMITED, NEW DELHI

Chapter 22

Angiospermic Floral Diversity in Four Lentic Water Bodies of Cuddalore District, Tamil Nadu, India

Saravanakumar Kaliyaperumal[1]*, Prabhakaran Jayaraman[1] and Kavitha Doraiswamy[2]

[1]Botany Wing-DDE
[2]Department of Botany,
Annamalai University, Annamalai Nagar – 608 002, Tamil Nadu, India

ABSTRACT

Aquatic weeds often reduce the effectiveness of water bodies for drinking, irrigation and fish production. Aquatic weeds are those unabated plants which grow and complete their life cycle in water and cause harm to aquatic environment directly and to related eco-environment relatively. Water is one of most important natural resource and in fact basis of all life forms on this planet. Aquatic weeds can assimilate large quantities of nutrients from the water reducing their availability for planktonic algae. They may also cause reduction in oxygen levels and present gaseous exchange with water resulting in adverse fish production. Although excessive weed growth may provide protective cover in water for small fish growth it may also interfere with fish harvesting. In the present study was made to record aquatic weed flora of lending water bodies of four selected village ponds of Cuddalore district. Identified aquatic weeds were classified according to various habitats which form their eco-environment and become conducive for their growth, reproduction and dissemination. The major flora of water bodies were, floating weeds, free floating weeds (Eichhornia crassipes, Salvinia natans, Pistia stratiotes, Azolla pinnata, etc), Rooted floating (Nelumbo nucifera, Nymphaea alba,

* *Corresponding Author.* E-mail: drsaravanak2007@gmail.com

Ipomoea aquatic); Submerged weeds (Vallisneria spiralis, Potamogeton spp. etc.); Deep water rooted submerged weeds (Hydrilla verticillata, Utricularia spp.); Emergent (Typha spp. Polygonum glabrum) etc.

Keywords: *Aquatic flora, Lentic water bodies, Cuddalore district.*

Introduction

The definition of aquatic plants or hydrophytes has been debated for decades. In general, all plants that tolerate or require flooding for a minimum duration of saturation/inundation are considered wetland plants (Gopal and Junk, 2000). Aquatic plants play an important role in aquatic systems worldwide because they provide food and habitat to fish, wildlife and aquatic organisms. Plants stabilize sediments, improve water clarity and add diversity to the shallow areas of lakes. Unfortunately, non-native plants that are introduced to new habitats often become a nuisance by hindering human uses of water and threaten the structure and function of diverse native aquatic ecosystems. Significant resources are often expended to manage infestations of aquatic weeds because unchecked growth of these invasive species often interferes with use of water, increases the risk of flooding and results in conditions that threaten public health.

An aquatic plant can usually be found growing in association with standing water including ponds, shallow lakes, marshes, ditches, reservoirs, swamps, bogs, canals, and sewage lagoons. Less frequently, they occur in the flowing water of streams, rivers, and springs. There are 87 families and 407 genera of aquatic vascular plants (Cook, 1990). Of these it is estimated that up to 2 per cent (or 250 species) of pteridophytes (ferns and allies) and 1 per cent (or 2,500 species) of angiosperms are aquatic (Groombridge and Jenkins, 1998). The geographic patterns of aquatic vascular plant diversity have not been summarized globally, however, a significant amount of information exist for regions where some aquatic plants are considered pests (University of Florida, 2001). There are also a few studies, with limited geographic scope, that identify conditions that correlate with higher species richness or abundance (Bornette *et al.*, 1998; Crow, 1993 and Zedler, 2000). Among angiosperms, 396 genera in 78 families are known to contain aquatic species. At the family level, 12 per cent of dicotyledons and 35 per cent of monocotyledons have aquatic species, indicating the widespread occurrence of aquatic species, especially in more advanced herbaceous groups (Cook, 1990).

Threats to freshwater such as pollution of different kinds, unfavourable climatic changes, eutrophication, acidification and alien invasion lead to reduction in macrophytic diversity which also threatens the faunal diversity of aquatic ecosystem (Chambers *et al.*, 2008). Studies on aquatic flora of India are only limited. Earlier, Subramanian (1962) has described 117 aquatic angiosperms. Lavania *et al.* (1990) has compiled the wetland flora of India; Cook (1996) has published the aquatic and wetland flora of India. Identification and detection of aquatic weed problem on irrigation canals, ponds and lakes will leads to effective weed management strategy, increase water for irrigation as well as to maintain healthy ecosystems. Hence, an attempt has been made to list out the aquatic weed flora in the lentic water bodies of Cuddalore District of Tamil Nadu, India.

Materials and Methods

Floristic survey of aquatic angiosperms was made through repeated field visit during 2011-2012 in four lentic water bodies (Vadavadi, Kammapuram, Arasakuzhi, and Edayanpalachori) in Cuddalore District of Tamil Nadu. The collected plant specimens were identified and confirmed with regional floras (Gamble, 1921); Nair and Henry (1983); Cook, 1996); Henry *et al.* (1987 and 1989), Matthew (1991) and regional checklist for hydrophytes (Udayakumar *et al.* (2010) and Muthulingam *et al.*(2010).

The Cuddalore district (North Latitude between 15° 5'/11° 11′ and 12°, East Longitude between 78° 38′ and 80°35′.) has an area of 3,564 km². It is bounded on the north by Viluppuram District, on the east by the Bay of Bengal, on the south by Nagapattinam District, and on the west by Perambalur District (Figure 22.1).

Results and Discussion

Aquatic plants have been found to severely reduce the flow capacity of irrigation canals thereby reducing the availability of water to the farmers field. Dense growth of aquatic weeds may provide ideal habitat for the development of mosquitoes causing malaria, encephalitis, filarasis. These plants may also serve as vectors for disease causing organisms and can greatly reduce the aesthetic value of water bodies from a recreational point of view. The following aquatic members were listed with binomial name of the species followed by common name and family were classified according to various habitats which form their eco-environment.

International Commission on Irrigation and Drainage (2002) reported that the following weeds are the major aquatic plants in Tami Nadu *i.e., Eicchornea crassipes, Typha latifolia, Cyperus* spp., *Chara* spp., *Hydrilla verticillata, Nymphaea* spp., *Nelumbo lutea, Nymphoides* spp., *Vallisneria americana, Potamogeton* spp., *Najas* spp., *Ipomea aquatica* and *Salvinia molesta.*

Floating Weeds

These are plants which grow and complete their life cycle in water. They vary in size from single cell (algae) and may grow up to large vascular plants. In case of drying of water bodies most of them give their seeds and other vegetative reproductive organs in base ground lands. Some of the weeds in this ecosystem freely float and move long distances, while some of them do float on the water surface but anchor down to soil at the bottom of the water body. These weed species make loss of water through evapotranspiration in addition to impediment caused in flow of water. Therefore, these weeds can be classified in two sub groups *viz.* free floating and rooted floating weeds.

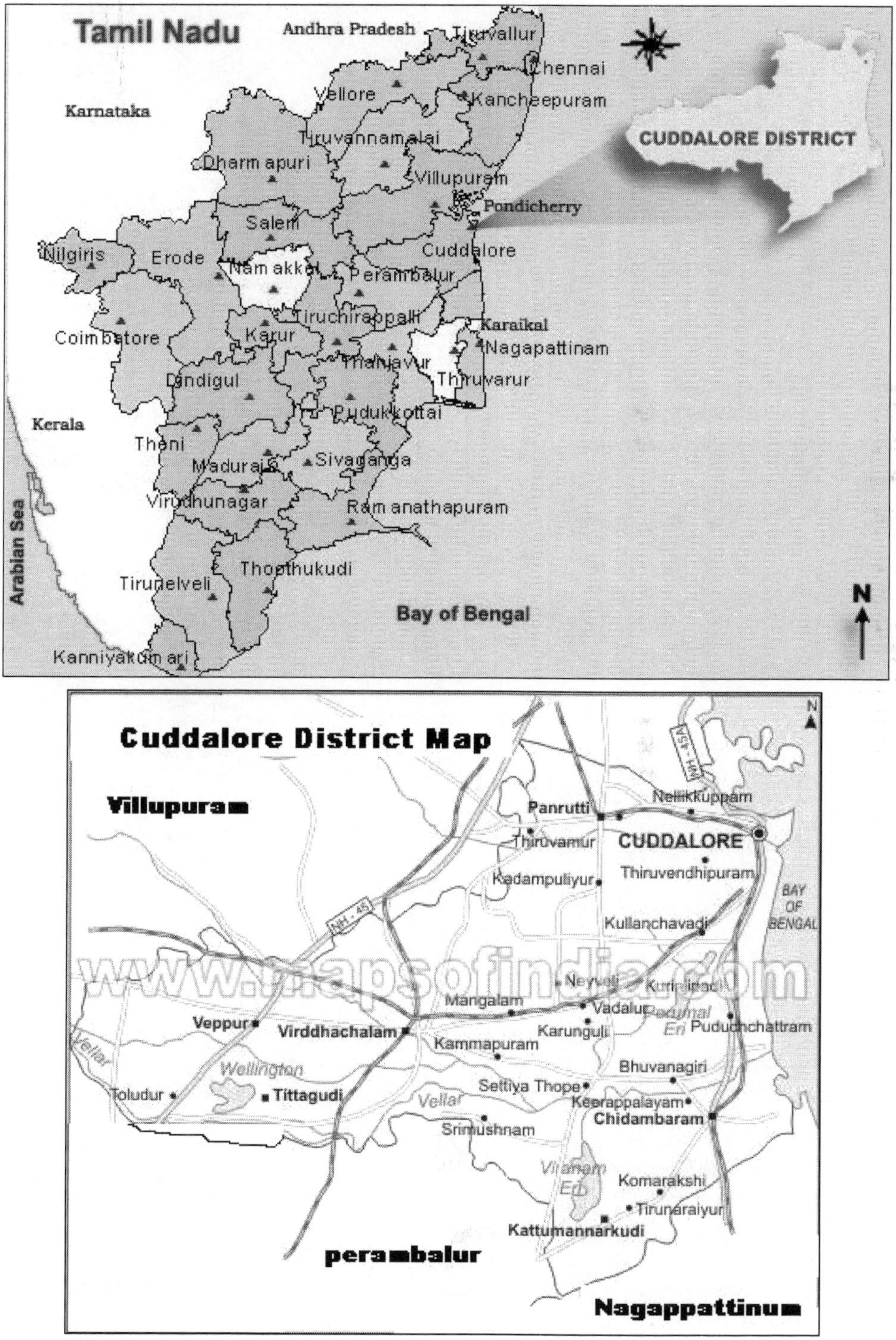

Figure 22.1: Cuddalore District Map.

Vadavadi Lake

Kammapuram Lake

Arasakuzhi Lake

Edyanpalachori Lake

Courtesy: Google map.

Table 22.1

Sl.No.	*Botanical Name*	*Common Name*	*Eco. Group*	*Group*
		Floating		
	Amaranthaceae			
1.	*Alternanthera sessilis* (L.) R.Br. ex. DC.	Sessile joyweed	VV, KP,EP,AK	DICOT
	Araceae			
2.	*Monochoria vaginalis* (Burm. F.) C.Presl	heartleaf	EP, KP	MONO
3.	*Pistia stratiotes* L.	Water lettuce	EP, AK, KP	MONO

Contd...

Table 22.1–*Contd...*

Sl.No.	*Botanical Name*	*Common Name*	*Eco. Group*	*Group*
	Ceratophyllaceae			
4.	*Ceratophyllum demersum* L.	Rigid Hornwort	VV, KP,EP,AK	DICOT
	Lemnaceae			
5.	*Lemna gibba* L.	Duckweed	VV, AK, KP	MONO
6.	*Lemna minor* L.	Common Duck weed	KP, AK	MONO
7.	*Lemna polyrhiza* L.	Duckweed	KP, EP, AK	MONO
8.	*Wolffia globosa* (Roxb.) Hartog and Vander	Asian water Meal	AK, KP	MONO
	Mimosaceae			
9.	*Neptunia prostrata* (Lam.) Baill.	Garden puff	VV, KP,EP,AK	DICOT
	Polygonaceae			
10.	*Polygonum amphibium* (L.) A.Gray	Water smart weed	VV, KP, AK	DICOT
	Pontederiaceae			
11.	*Eichhornia crassipes* (Mart) Solms	Water hyacinth	EP, AK, KP	MONO
	Salviniaceae			
12.	*Azolla pinnata* R.Br	Water velvet	VV, KP,EP,AK	PTERIDO
13.	*Salvinia molesta* D.Mitch.	Giant salvinia	VV, KP,EP	PTERIDO
14.	*Salvinia natans* (L.) All	water spangles	VV, KP,EP,AK	PTERIDO
		Rooted emergent weeds		
	Alismataceae			
1.	*Limnophyton obtusifolium* (L.) Miq.	Forum Flowers	KP,EP,AK	MONO
	Acanthaceae			
2.	*Hygrophila schulli* (Hamilt.) M.R.Almeida and S.M. Almeida	Marsh Barbel	AK, KP	DICOT
	Asteraceae			
3.	*Eclipta alba* Hassk.	False daisy	VV, KP, AK	DICOT
	Brassicaceae			
4.	*Nasturtium indicum* DC.	Orange flower	AK, KP	DICOT
	Convolvulaceae			
5.	*Ipomoea carnea* Jacq.	Pink Morning Glory	VV, KP,EP,AK	DICOT
	Cyperaceae			
6.	*Cyperus bulbosus* Vahl	Wild onion	EP, AK	MONO
7.	*Cyperus rotundus* L.	Purple nut sedge	VV, KP,EP,AK	MONO
8.	*Kyllinga bulbosa.* P. Beauv.	Bulbous blue grass	VV, KP	MONO
9.	*Scirpus articulatus* L.	common reed sedge	VV, KP, AK	MONO

Contd...

Table 22.1–*Contd...*

Sl.No.	*Botanical Name*	*Common Name*	*Eco. Group*	*Group*
	Gentianaceae			
10.	*Enicostemma axillare.* (Lam.) Raynal	Indian Whitehead	KP,EP	DICOT
	Isoetaceae			
11.	*Isoetes coromandelina* L.f.	Indian quillwort	EP, AK, KP	PTERIDO
	Lythraceae			
12.	*Ammania baccifera* L.	Blistering Ammannia	KP, AK	DICOT
	Marsileaceae			
13.	*Marsilea quadrifolia* L.	European water clover	VV, KP,EP,AK	PTERIDO
	Poaceae			
14.	*Chloris barbata* Sw.	Swollen finger grass	VV, KP,AK	MONO
15.	*Cynodon dactylon* (L.) Pers.	Bermuda Grass	VV, KP,EP,AK	MONO
16.	*Dactyloctenium aegyptium* (L.) Willd	Egyptian crowfoot grass	AK, KP	MONO
17.	*Echinochloa colona* (L.) Link	jungle rice	VV, KP,EP,AK	MONO
18.	*Sporobolus coromandelianus* (Retz.) Kunth	Small dropseed	KP,EP	MONO
	Onagraceae			
19.	*Ludwigia adscendens* (L.) Hara	Water-purslane	EP, AK	DICOT
20.	*Ludwigia perennis* L.		KP, EP, AK	DICOT
	Polygalaceae			
21.	*Polygala arvensis* Willd.	Field Milkwort	VV, KP,EP,AK	DICOT
22.	*Polygala chinensis* L.		KP, AK	DICOT
23.	*Polygala erioptera* DC.		VV, AK, KP	DICOT
24.	*Polygala javana* DC.		KP, EP, AK	DICOT
	Polygonaceae			
25.	*Polygonum glabrum* Willd.	Dense flower knotweed	VV, KP, AK	DICOT
	Scrophulariaceae			
26.	*Limnophila aromatica* (Lam.) Merr.		AK, KP	DICOT
27.	*Limnophila heterophylla* (Roxb.) Benth	Rice paddy herb	VV, KP,AK	DICOT
28.	*Striga angustifolia* (D.Don) Saldanha	Asiatic witchweed	KP,EP	DICOT
	Sphenocleaceae			
29.	*Sphenoclea zeylanica* Gaertn.		KP, AK	DICOT

VV: Vadavadi; **KP**: Kammapuram; **AK**: Arasakuzhi; **EP**: Edayanpalachori.

The following weeds grow in shallow waters and situations existing near the water bodies where water recedes and rises with the seasons or regular releases from a large water body or reservoir. Most of such situations are of permanent in nature where minimum and maximum water levels are consistent. Such situations includes banks of canals, rivers, periphery of water bodies, drainage ditches and water ponds near villages

Table 22.2

Sl.No.	*Botanical Name*	*Common Name*	*Eco. Group*	*Group*
	Emergent (near wetlands)			
	Alismataceae			
1.	*Alisma plantago-aquatica* L.	Common water plantain	KP, EP	MONO
	Commelinaceae			
2.	*Commelina benghalensis* L.	Water grass	VV, KP,EP,AK	MONO
	Convolvulaceae			
3.	*Ipomea carnea* Jacq.	Morning grlory	VV, KP,EP,AK	DICOT
	Cyperaceae			
4.	*Cyperus difformis* L.	Umbrella plant	KP, EP, AK	MONO
	Poaceae			
5.	*Phragmites communis* Trin.	Common reed	EP, AK, KP	MONO
	Pondeteriaceae			
6.	*Pontederia cordata* L.	Pickerel weed	KP, EP	MONO
	Typhaceae			
7.	*Typha angustifolia* L.	Lesser Indian reed mace	KP,EP,AK	MONO
	Submerged			
	Lentibulariaceae			
1.	*Utricularia caerulea* L.	Blue bladderwort	VV, KP	DICOT
2.	*Utricularia polygaloides* Edgew.	Annual carnivorous plant	KP, EP, AK	DICOT
	Anchored floating weeds			
	Alismataceae			
1.	*Sagittaria guayanensis* HBK.	Arrowhead	KP, EP	MONO
	Aponogetonaceae			
2.	*Aponogeton natans* (L.) Engler		VV, KP,EP,AK	MONO
	Convolvulaceae			
3.	*Ipomoea aquatica* Forssk.	Water Morning Glory	VV, KP,EP,AK	DICOT
4.	*Ipomea hederacea* Nilkalmi	Ivy-leaved Morning Glory	KP, EP,AK	DICOT
	Gentianaceae			
5.	*Limnanthemum cristatum* Griseb.		VV, KP	DICOT

Contd...

Table 22.2–*Contd...*

Sl.No.	*Botanical Name*	*Common Name*	*Eco. Group*	*Group*
	Nymphaeaceae			
6.	*Nelumbo nucifera* Gaertn.	Indian Lotus	VV, KP,EP,AK	DICOT
7.	*Nymphaea alba* L.	European White Water-lily	KP, EP,	DICOT
8.	*Nymphaea nouchali* Burm. f.	Red and blue water lily	VV, KP,AK	DICOT
9.	*Nymphaea rubra* Roxb. ex. Salisb.	India Red Water-lily	EP, AK, KP	DICOT
10.	*Nymphaea stellata* Willd.	Blue star water lily	KP,EP,AK	DICOT
	Deep water rooted submerged weeds			
	Characeae			
1.	*Chara zeylanica*	Musk grass	KP,EP,	ALGAE
2.	*Nitella hyalina*	Stone wort	EP,AK, KP	ALGAE
	Holoragaceae			
3.	*Myriophyllum spicatum* L.	Eurasian water milfoil	VV, KP,EP,AK	DICOT
	Hydrocharitaceae			
4.	*Hydrilla verticillata* (L.f.) Royle	Esthwaite water weed	KP, AK	MONO
5.	*Elodea canadensis* Michx.	Pond weed	EP,AK, KP	MONO
6.	*Ottelia alismoides* (L.) Pers.	Duck lettuce	VV, KP	MONO
7.	*Vallisneria natans* (Lour.) Hara	Eel grass	VV, KP,EP,AK	MONO
	Lentibulariaceae			
8.	*Utricularia flexuosa* Vahl	Bladderwort	KP,AK	DICOT
	Najadaceae			
9.	*Najas graminea* Del.	Ricefield water nymph	VV, KP,EP	MONO
10.	*Najas indica* (Willd.) Cham.		EP, AK, KP	MONO
11.	*Najas minor* All.	Brittle naiad	VV, KP	MONO

VV: Vadavadi; **KP**: Kammapuram; **AK**: Arasakuzhi; **EP**: Edayanpalachori.

In the marshes or swampy areas the following amphibian species were noticed

Table 22.3

Sl.No.	*Botanical Name*	*Common Name*	*Eco. Group*	*Group*
	Amaranthaceae			
1.	*Alternanthera philoxeroides* (Mart) Griseb.	Alligator weed	VV, KP,EP,AK	DICOT
	Apiaceae			
2.	*Hydrocotyle umbellata* L.	Water pennywort	VV, KP	DICOT

Contd...

Table 22.3–*Contd...*

Sl.No.	*Botanical Name*	*Common Name*	*Eco. Group*	*Group*
	Convolvulaceae			
3.	*Ipomoea aquatica* Forssk.	Floating morining glory	KP,EP	DICOT
	Marsileaceae			
4.	*Marsilea minuta* L.	Pepper west	VV, KP,EP,AK	PTERIDO
	Onagraceae			
5.	*Jussiaea repens* L.	Water primrose	AK, KP	DICOT
6.	*Ludwigia parviflora* Roxb.	Water purslane	VV, KP, AK	DICOT
	Poaceae			
7.	*Panicum purpuracens* Raddi	Paragrass	VV, KP	MONO
8.	*Panicum purpuracens* Kunth	Water paspalum	EP,AK, KP	MONO
	Pondeteriaceae			
9.	*Heteranthera limosa* Sw.	Mud plantain	KP,EP	MONO
10.	*Monochoria vaginalis* C.Presl.	Carpet weed	VV, KP,EP,AK	MONO
	Trapaceae			
11.	*Trapa bispinosa* Roxb.	Water chestnut	EP, KP	DICOT

VV: Vadavadi; **KP**: Kammapuram; **AK**: Arasakuzhi; **EP**: Edayanpalachori.

In the present study there are 84 species from 56 genera belonging to 35 families was noticed. Of which, 14 are floating, 29 are rooted emergents, 7 are emergents, 2 are submersed, 10 are anchored floating, 11 are deep water rooted submersed and 11are marshy species. Among the 84 species listed 21 are found in all the four ponds. Number of species found in pond sites *viz.* 77 in Kammapuram, 55 in Edyanpalchori, 56 in Arasakuzhi and 40 in Vadavadi (Figures 22.2–22.4 and Tables 22.2–22.4).

Further studies required on detailed seasonal survey of frequency, abundance Important value Index, synecology, autecology and phytosociology of all the individual aquatic weed species which interfere on flow of water in irrigation canals.

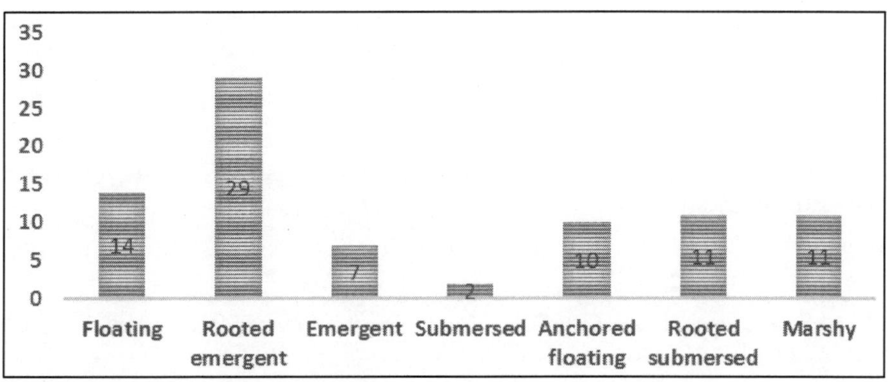

Figure 22.2: Habit-wise Distribution of Macrophytes.

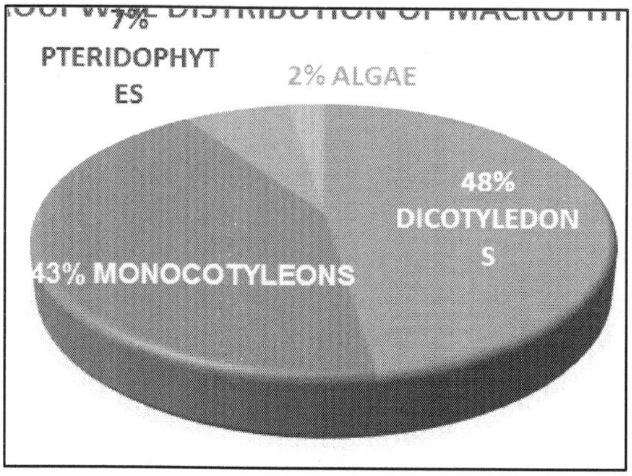

Figure 22.3: Group-wise Distribution of Macrophytes.

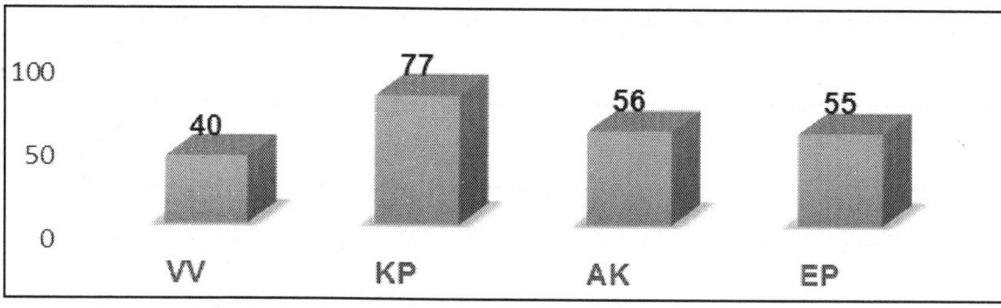

Figure 22.4: No. of Species in different Study Sites.

On the other hand continuous monitoring should be carried out on physico-chemical and biological properties of lentic water bodies to maintain the quality of water for drinking and irrigation purposes and conserve the balanced ecosystem.

References

Bornette, G., Amoros, C and Lamouroux, N. (1998). Aquatic plant diversity in riverine wetlands: the role of connectivity. *Freshwater Biology*, **39**: 267-283.

Chambers, P.A., P. Lacoul., K.J. Murphy and S.M. Thomaz. (2008). Global diversity of aquatic macrophytes in freshwater. *Hydrobiologia*, **595**: 9-26.

Cook, C.D.K. (1990). *Aquatic Plant Book*. The Hague, The Netherlands: SPB Academic Publisher.

Cook, C.D.K. (1996). Aquatic and Wetland Plants of India. Oxford University Press. United Kingdom. 1-385.

Crow, G.E. (1993). Species diversity in aquatic angiosperms: latitudinal patterns. *Aquatic Botany*, **44**: 229-258.

Gamble, J.S, and Fischer, C.E.C. (1921). Flora of Presidency of Madras. Vol. 1-3, Adlard and Sons Ltd., London.

Gopal, B., W. J. Junk, and J.A. Davis (Eds.). (2000). *Biodiversity in Wetlands: Assessment, function, and conservation,* volume 1. Leiden, The Netherlands: Backhuys Publishers.

Groombridge, B. and M. Jenkins. (1998). *Freshwater Biodiversity: a preliminary global assessment*. Cambridge, UK: WCMC-World Conservation Press.

Henry, A.N, Chitra, V, and Balakrishnan, N.P. (1989). Flora of Tamil Nadu, India. **1(3)**: Botanical Survey of India, Southern Circle, Coimbatore; 1-171.

Henry, A.N, Kumari, G.R, and Chitra V. (1987). Flora of Tamil Nadu, India. **1**(2): Botanical Survey of India, Southern Circle, Coimbatore. **1**-258.16.

Lavania, G.S., Paliwal, S.C., and Gopal, B.(1990). Aquatic vegetation of the Indian subcontinent. In Ecology And Management of the aquatic Vegetation of the Indian Subcontinent. Edited by B.Gopal. Academic Publishers, Dordrecht pp.29–78.

Matthew K.M. (1991). An Excursion Flora of Central Tamil Nadu. Tamil Nadu: Thiruchirapalli, Rapinat Herbarium. 1-682.

Muthulingam, U, Narayanasamy D, Kanakashanthi A, and Thangavel, S. (2010). Floristic Study in a Perennial Lake of Thiruvallur District, South India. *Web med Central Ecology*;**1**(10): WMC001037.

Nair, N.C and Henry, A.N. (1983). Flora of Tamil Nadu, India. Series 1, Vol. 1, Botanical Survey of India, Southern Circle, Coimbatore. 1-184.

Subrahmanyam, K. (1962). Aquatic Angiosperms. Botanical Monograph 3. New Delhi: Council of Scientific and Industrial Research, p.190.

Udayakumar M, and Ajithadoss K. (2010). Angiosperms, Hydrophytes of five ephemeral lakes of Thiruvallur District, Tamil Nadu, India. *Checklist*, **6** (2): 270-274.

University of Florida. (2001). Aquatic, Wetland and Invasive Plant Database. Gainesville, FL: Center for Aquatic and Invasive Plants, University of Florida.

Zedler, J.B. (2000). Restoration of biodiversity to coastal and inland wetlands. Pages 311-330 in B. Gopal, W.J. Junk, and J.A. Davis eds. *Biodiversity in wetlands: assessment, function,* and *conservation,* Volume 1. Leiden: The Netherlands: Backhuys Publishers.

2015, Major Constrains and Verdict of Crop Productivity *Pages 324–336*
Editor: Dr. U.N. Bhale
Published by: ASTRAL INTERNATIONAL (P) LIMITED, NEW DELHI

Chapter 23

Lipoxygenase as Biochemical Marker for Plant Disease Resistance

Uma Maheswari Devi*

Department of Applied Microbiology and Biochemistry
Sri Padmavati Mahila Visva Vidyalayam (S.P. Women's University)
Tirupati – 517 502, A.P., India

Introduction

Plants respond to pathogen attack or elicitor treatments by activating a wide variety of protective mechanisms designed to prevent pathogen replication and spreading. It has been proposed that membrane damage produced by Lipoxygenase(LOX) products is responsible for the initiation of the hypersensitive response, an important feature of the resistance response to pathogens (Maccarone *et al.*, 2000). Phytopathogens induce plant phospholipase activity, resulting in the release of free fatty acids (FAs) essential for oxylipin biosynthesis (Dhondt *et al.*, 2000). Breakdown of FAs occurs primarily in the plant peroxisomes and results in the production of cyclic or acyclic oxidation compounds termed oxylipins. LOXs play an important role in the production of oxylipins. Many oxylipins participate in plant defense by serving signaling functions to induce defense responses or by directly functioning as antimicrobial compounds (Blee, 2002).

* *Corresponding Author.* E-mail: umadevi66@yahoo.co.in

Plant Lipoxygenases

Lipoxygenases have been detected in a large number of Plant species (Vick and Zimmerman, 1987). Plant LOXs are monomeric proteins of about 95–105 kDa that consist of two domains. The amino-terminal domain of about 25–30 kDa is a β-barrel domain that is structurally related to C-2 domains (Corbin *et al.*, 2007). The carboxy-terminal domain of about 55–65 kDa consists primarily of a-helices and harbours the catalytic site of the enzyme (Schneider *et al.*, 2007a). LOX enzyme contains one iron atom per protein molecule. In addition, LOXs have been described as originating from fungi that harbor a manganese atom instead, in the active site (Andreou *et al.*, 2009).

Within a given plant, the level of lipoxygenase activity present in any tissue can vary markedly not only among different organs of the plant, but also between different developmental stages of a particular tissue type (Axelrod, 1974). LOXs are present in a wide range of biological organs and tissues, but they are particularly abundant in grain legume seeds. The first LOX enzyme characterized in detail were the isoforms which occur in high amounts in soybean seeds. High levels of LOX are also present in soybean leaves where the 94kDa vegetative storage protein (VSP) was identified as LOX and also in potato tubers. The major bulk of LOXs is localized in the cytoplasm, chloroplasts, lipid bodies, mitochondria and vacuole of the plant cell (Porta and Rocha-Sosa, 2002).

Lipoxygenase Pathway

Plant LOXs (linoleate: oxygen oxidoreductase, EC 1.13.11.12) are a group of non-heme iron containing enzymes that catalyze the addition of molecular oxygen to linolenic (C18:2) and linolenic acid (C18:3) results in the formation of fatty acid hydroperoxides. LOXs are present in a wide range of biological organs and tissues, but they are particularly abundant in grain legume seeds and potato tubers. LOX enzymes can be grouped into two types according to their region specificity: 9-LOX, which specifically form 9-hydroperoxy fatty acid, and 13-LOX, which predominantly catalyze the formation of 13-hydroperoxy fatty acid. These fatty acid hydroperoxides are the substrates for other enzymatic systems that transform these highly reactive molecules into a series of oxylipins via so called "LOX pathway" (Porta *et al.*, 2008). The primary LOX products of LA (Hydroperoxyoctadecadienoic acids (13-HPOD and 9-HPOD) and ALA (Hydroperoxyoctadecatrienoic acids 13-HPOT and 9-HPOT) are very reactive and cause oxidative damage to membranes, leading to cell necrosis and death (Huang and Schwab, 2011). Similarly the reduced products of hydroperoxides, known as Hydroxides (Hydroxyoctadecadienoic and trienoic acids – 13-HOD, 13-HOT, 9-HOD and 9-HOT) also play an important role in defense mechanism.

The 13-LOX Pathway

In plants, most reported LOXs are 13-LOXs, which are important in the biosynthesis of plant hormone, jasmonic acid (JA).Products of 13-LOX enzymatic activity can be utilized by several downstream pathway branches, however, currently the best characterized are allene oxide synthase (AOS) and HPL pathways. AOS mediated or so called octadecanoid pathway produces JA and its precursors collectively known as jasmonates (Wassim *et al.*, 2008). This pathway is ubiquitously found in Plastids, where the 12-oxo-phytodienoic acid (OPDA) is formed and converted in to JA (Feussner and Wasternack 2002). A different set of oxylipins are produced by HPL pathway localized in chloroplast. The best studied HPL-derived oxylipins are c_6 green leafy volatiles (GLV) have been shown to possess both antimicrobial and defense signaling activities (Prost *et al.*, 2005).

JAs are well characterized with respect to their role in regulating defense responses against herbivore attack and infection by some pathogens (Kessler and Baldwin, 2002). Moreover, accumulating evidence suggests that jasmonates have a role in plant resistance against abiotic stimuli such as mechanical stress, salt stress, UV irradiation, low temperatures and ozone exposure.JA pathway plays roles in direct defense against arthropods and pathogens (Li *et al.*, 2002) and is important for resistance to the biotrophic fungal pathogen powdery mildew, *Erysiphe cichoracearum*. The JA pathway also contributes to regulation of the production of many low molecular mass compounds (*e.g.*, arabidopsides and glucosinolates in *Arabidopsis*) that are involved in plant defense responses (Gundlach *et al.*, 1992). Volatile JA derivatives serve as long distance communication signals between other plants and other organisms. MeJA, for example, can act as a signal within and possibly between plants.

HPL was expressed at relatively high levels in mature inflorescence, flower budsand flowers, at low levels in root tissues and at substantially low levels in rosette leaves (Bate *et al.*, 1998). The lyases involved can be roughly classified in to two groups, according to their substrate specificity. The one group cleaves the 13-hydroperoxy derivatives of linoleic or linolenic acid to form 3(Z)-hexenal and 3(Z)-hexanol. The other enzyme cleaves 9-hydroperoxy derivatives yielding 3(Z), 6(Z)-nonadienal and 3(Z)-nonenal respectively. Trans-2-hexenal is formed upon numerous interactions such as pathogen infestation, wounding, pests and disease (Croft *et al.*, 1993). Another fate of these 13-HPOTs is their reduction to corresponding alcohols and further transformation by peroxygenase (Xue *et al.*, 2012). In peroxygenase pathway of 13-HPOT, intramolecular oxygen transfer converts α-linolenic acid to either 9-hydroperoxy linolenic acid or 13-hydrperoxy linolenic acid. These metabolites may serve as substrates for HPL leading to the formation of either (3Z, 6Z)-nonadienal or (3Z)-hexenal. (3Z)-Alkenals may then either be isomerized by a cis-3: trans-2-enal isomerase to the corresponding (2E0-alkenals or also converted by a LOX and or a peroxygenase to 4-hydroxy Alkenals, such as HHE (Kohlmann *et al.*, 1999). The formation of peroxygenase pathway products has been demonstrated in cereal seeds and barley leaf segments (Kohlmann *et al.*, 1999).

The 9-LOX Pathway

The role of the 9-LOX pathway in plant defense is mainly emphasized in Solanaceae plants and 9-LOX pathway might play an important role in defense against diverse pathogens (Fammartino *et al.*, 2007). This is commonly explained by the fact that 9-LOX-derived oxylipins act as signals to regulate programmed cell death (Cacas *et al.*, 2005) or by their potent antimicrobial activity (Prost *et al.*, 2005).The major metabolites from 9-LOX derived hydroperoxides, 9-HPOD and 9-HPOT are Divinyl ethers and epoxy alcohols by the catalytic activity of Divinyl ether synthase (DES) and epoxy alcohol synthase (EAS) respectively. Colneleic acid (CA) and Colnelenic acid (CnA) are Divinyl ethers derived from linoleic acid and linolenic acid by the concerted activities of 9-lipoxygenase (9-LOX) and 9-divinyl ether synthase (9-DES).CA and CnA act in vitro as antimicrobials against a range of pathogenic bacteria, fungi, and oomycetes (Li *et al.*, 2012; Prost *et al.*, 2005). Several 9-LOX-derived oxylipins, including 9-hydro-peroxides, 9-hydroxides, CA and CnAwere reported to accumulate in tobacco or potato in response to elicitor treatment or pathogen ingress (Rusterucci *et al.*, 1999; Gobel *et al.*, 2002).In another pathway, the hydroperoxide intermediates can be converted to aldehydes and oxoacids by hydroperoxidelyase (HPL). The Aldehydes produced by 9-HPL are known to function as defense compounds as well as contributing to special flavors. For example, (2E)-nonenal gives the cardboard flavor of aged beer, and the off flavor of aged milled rice. 9-HPL activity has only been reported from a few plant species including rice and cucumber (Kuroda *et al.*, 2005).The short chain aldehydes are components of the green odor of plants and have antimicrobial effect in vitro. The physiological function of aldehydes is mainly addressed to plant defense against microbes and herbivores (Rosahl *et al.*, 2005). The effect has, for example been demonstrated upon inoculation of French bean with an avirulent variety of *Pseudomonas syringae* pv. *syringae.*

Role of Lipoxygenases in Disease Resistance

LOX pathways are crucial for lipid peroxidation process during plant defense responses to pathogen infection. LOX metabolites act as antimicrobial compounds leading to the development of resistance. The primary products of LOX, the fatty acid hydroperoxides, are very reactive and cause oxidative damage to membranes, leading to cell necrosis and death (Hildebrand, 1995).In plants, increased levels of linoleic acid (18:2) and linolenic acid (18:3) all appear to participate in defense signaling directly or indirectly. For example, increased levels of 18:0 likely inhibit soybean seed colonization by the seed-borne pathogen *Diaporthe phaseolorum* (Xue *et al.*, 2006). *Rhizobacteria*-induced resistance to *Botrytis cinerea* is associated with the accumulation of 18:2 and 18:3 fatty acids in bean (Ongena *et al.*, 2004). Another aspect of LOX response to pathogen infection is the observation that resistant and susceptible plant lines show differential responses. A major difference between the resistant and the susceptible forms in their response to pathogens is in the timing of induction of host defense genes. In *Arabidopsis,* 13-LOX–derived oxylipin biosynthesis is induced upon recognition of the avirulence protein Avr-Rpm1 and is associated with HR induced, during race-specific resistance to *P. syringae* (Andersson *et al.*, 2006). In potato cells, pathogen-derived elicitors preferentially stimulate a second lipoxygenase pathway catalyzed by a 9- LOX enzyme (Gobel *et al.*, 2002). 9-LOX–derived oxylipinsaccumulate in potato during non-host interactions with *P. syringae* pv. *maculicola*and compatible interactions with *Phytophthora infestans* (Goble *et al.*, 2002). Thus, the induced expression of LOX isozymes during pathogenesis would be useful for conferring resistance against pathogen attack.

Antimicrobial Activity of LOX Products

LOX metabolites act as antimicrobial compounds leading to the development of resistance. In order to test such possibility, different chilli LOX products of both LA and ALA were screened for their antimicrobial effects by filter disc assay. As shown in Figure 23.1, Filter discassay showed that both hydroperoxides of LA (9-HPODE) and ALA (9-HPOTrE)were effective against *Pseudomonas, Erwiniasps* and *Xanthomonas campestris* as antibacterial agents.The ALA hydroperoxides were found to be more effectiveantifungal agents than the LA products (Figure 23.2).

Differential Effect of 13-HPODE/13-HPOTE and 9-HPODE/ 9-HPOTE on Aflatoxin Production

Fatty acid derivatives of the plant lipoxygenase (LOX) pathway are implicated in the *Aspergillus* spp.-seed interaction, affecting the growth, development and AF production of the fungus. The LOX specific placement of the hydroperoxy moiety on the ninth or thirteenth carbon atom of linoleic acid and linolenic acid leads to the formation of 9S-HPODE and/or 9S-HPOTE and 13S- HPODE and/or 13S-HPOTE, respectively and this could account for the differential effects on AF biosynthesis by *A. flavus* and *A. parasiticus* (Munikumari *et al.*, 2011). The effect of linoleic acid, LA hydroperoxides (9*S*-HPODE and 13*S*-HPODE) on wet weight of *A.flavus* showed that both 13*S*-hydroperoxides and 9*S*-hydroperoxides except linoleic acid decreased mycelial growth of *A.flavus* (Table 23.1). But the difference in mycelial weight was statistically significant only at concentrations of 10 µM and 100 µM of 13*S*-hydroperoxide. Similar results were presented while studying the effect of lipoxygenase products on *Aspergillus* in maize (Burow *et al.*, 1997). Similarly, aflatoxin production was initiated at 48 h and increased gradually to a maximum of 470 ppb and 650 ppb respectively with the addition of either LA or 9*S*- HPODE. On contrary, 13*S*-HPODE showed consistently reduced aflatoxin levels. With increase in 13*S*-HPODE concentration, aflatoxin content decreased significantly and aflatoxin level reached to zero with the continuous addition of 100µM 13*S*-HPODE. These results inferred that the 13S-HPODE may be responsible for inducing resistance against Aflatoxigenic fungi by blocking the biosynthesis of Aflatoxins. The peanut 13-lipoxygenase pathway is involved in conferring the resistance to peanut against *A. flavus*. Reciprocal "cross talk" between fungus and seed was also demonstrated when plant LOX expression was altered in response to fungal infection in peanut, maize and pigeonpea (Munikumari *et al.*, 2010http://scialert.net/fulltext/?doi=ijpp.2011.153.164 and org=12 - 816940_ja).

Germicidal Properties of LOX Hydroperoxides

Microscopic visualization approach was used to investigate the effect of hydroperoxy fatty acids on germination of sporangia, an important developmental stage in the life cycle of fungi. Usually, sporangia germinate directly in a nutrient medium by producing germ tubes that develop into mycelial masses. The control, in the presence of water, showed the formation of sporangia which are papillate and ovoid with the widest part close to the base and release of zoospores from the sack

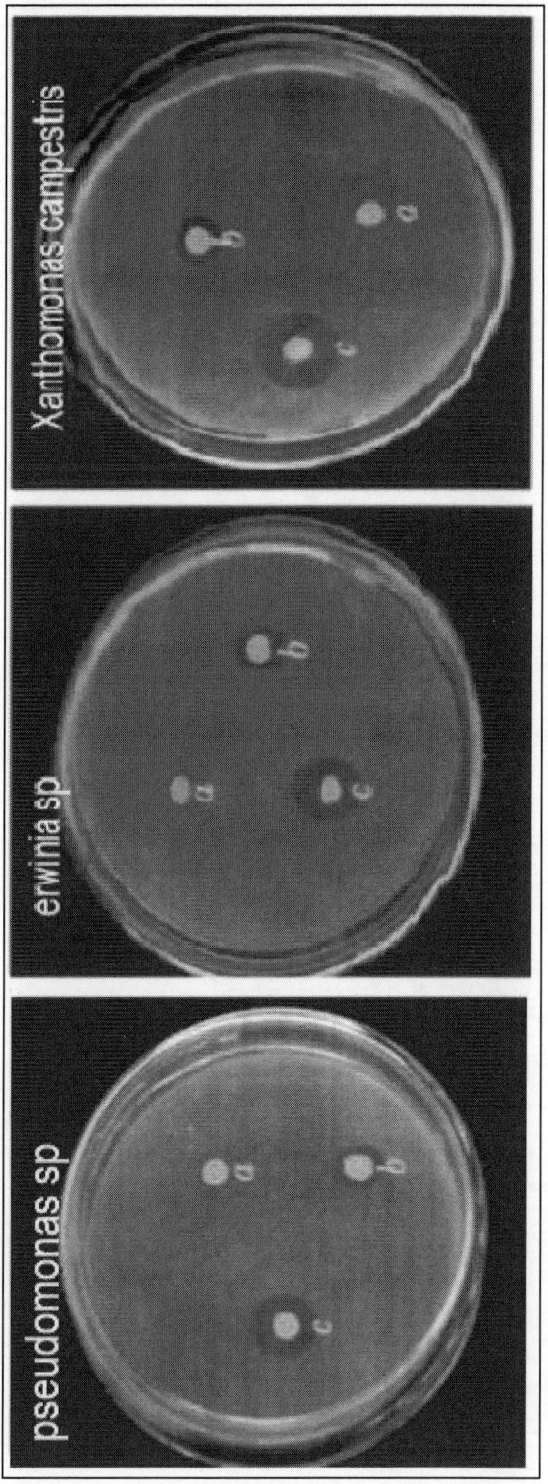

Figure 23.1: Antibacterial Activity of Chilli LOX Products by Filter Disc Assay. Control (10 µl 10 per cent ethanol) (a), 5 µg/ml 9-HPODE(b), 5 µg/ml 9-HPOTrE (c).

Table 23.1: Effect of LA and LA Hydroperoxides on Wet Mycelial Weight of *A. flavus*. Different concentrations (1μm, 10μm and100 μm) of LA and LA hydroperoxides were added to the *A.flavus* culture and mycelia were harvested at different time points. The control was maintained with 0.1 per cent methanol. Means of triplicates were analyzed by Fisher's Protected Least Significant Difference at P=0.05. The aflatoxin levels were detected in the harvested culture filtrates by Indirect Competition ELISA.

Treatment	*Wet Weight of A. flavus Mycelium (mg)*				*Aflatoxm Content in Culture Filtrate of A. flavus ppb/kg*			
	48h	*72h*	*96h*	*120h*	*48h*	*72h*	*96h*	*120h*
Control	11.66±0.21	12.18±0.34	12.47±0.31	13.42±0.35	20	90	230	470
13HPODE 1μM	11.66±0.23	12.07±0.26	12.38±0.43	13.41±037	0	10	50	80
13HPODE 10μM	10.96±0.34	11.85±0.24	12.06±0.47	13.11±0.41	0	5	16	20
13HPODE 100μM	10.63±0.32	10.66±0.23	1129±0.18	13.0 S±0.24	0	0	5	12.5
9HPODE 1μM	11.84±0.41	12.14±0.35	12.49±0.34	13.39±0.32	21	110	360	540
9HPODE 10μM	11.99±0.29	12.16±0.33	12.46±0.23	13.24±0.42	26	140	440	630
9HPODE 100μM	1191±0.27	12.19±0.21	12.59±0.25	13.46±0.41	80.5	240	560	650
LA 1μM	1199±0.21	12.57±0.36	1256±0.4	3.57±0.25	10	95	310	480
LA 10μM	12.03±0.34	12.45±0.54	12.94±0.44	13.86±0.33	22	120	430	540
LA 100μM	12.11±0.31	12.68±0.41	13.97±0.31	13.97±0.26	40	160	450	580

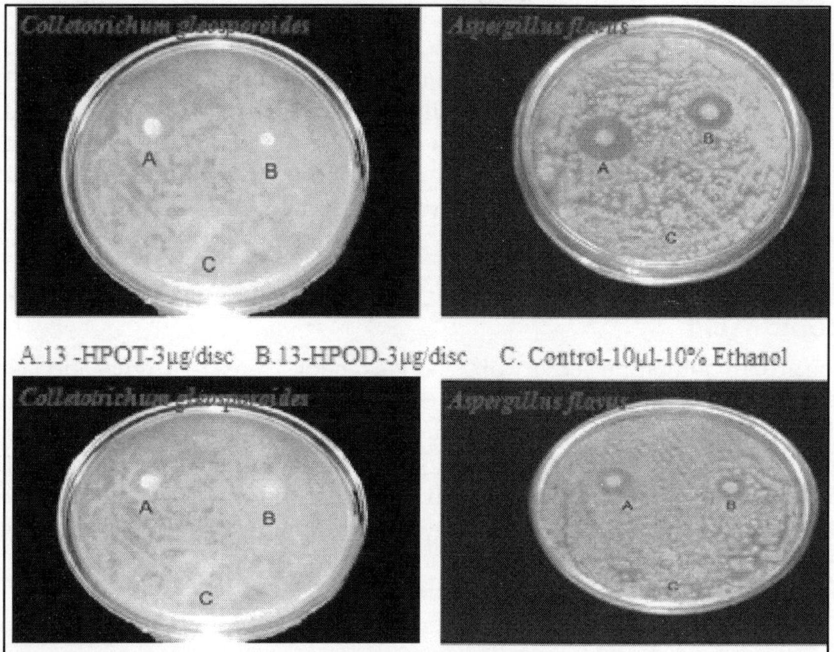

Figure 23.2: Antifungal Activity of LOX Hydroperoxides.

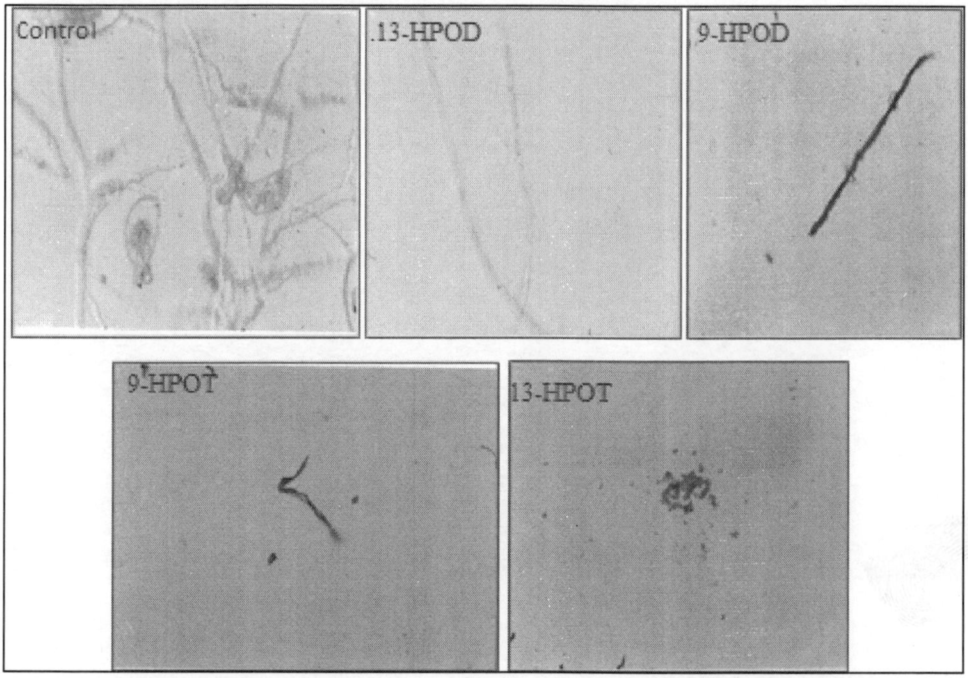

Figure 23.3: Microscopic Visualization of Germination of Sporangia Magnified Under 40X. The Spores of *Phytophthora palmivora* were incubated for 12 hr in Microtitre Plates.

of germinating sporangia and the detached sporangium contains a short pedicel (Figure 23.3). The formation of spore sac is impeded in the presence of 9-HPOD), whereas 9-HPOT restricted the formation of mycelium. The initial fragmentation of mycelium was found in the treatment with 13-HPOD and complete fragmentation of mycelium was found in the presence of 13-HPOT (Figure 23.3). This data clearly indicates that the LOX products are potential not only to inhibit the germination of sporangia but also the growth of mycelium. The results suggest the involvement of LOX in the defense mechanism of papaya against *Phytophthora palmivora* infection.

Bactericidal Activity of Cis-3-Hexenol, Trans-2-Hexenal and Jasmonic Acid

The growth inhibiting activity of Cis-3-Hexenaol, Trans-2-Hexenal and Jasmonic acid was measured on *Erwiniaherbicola, Enterobacter cloacae, Pseudomonas aeruginosa* and *Xanthomonas axonopodis* in liquid medium. The bacteria seeded flasks were incubated at 37°C for 24hr and measured for turbidity at 660nm in a colorimeter.Trans-2-Hexenal, the high antibacterial oxylipin prevented the growth of all bacteria at low concentration. A direct relationship between amount of oxylipins (Cis-3-hexenol, Trans-2-Hexenal and Jasmonic acid) and retardation of bacterial growth (Figure 23.4) was observed in liquid culture as shown in Table 23.2.

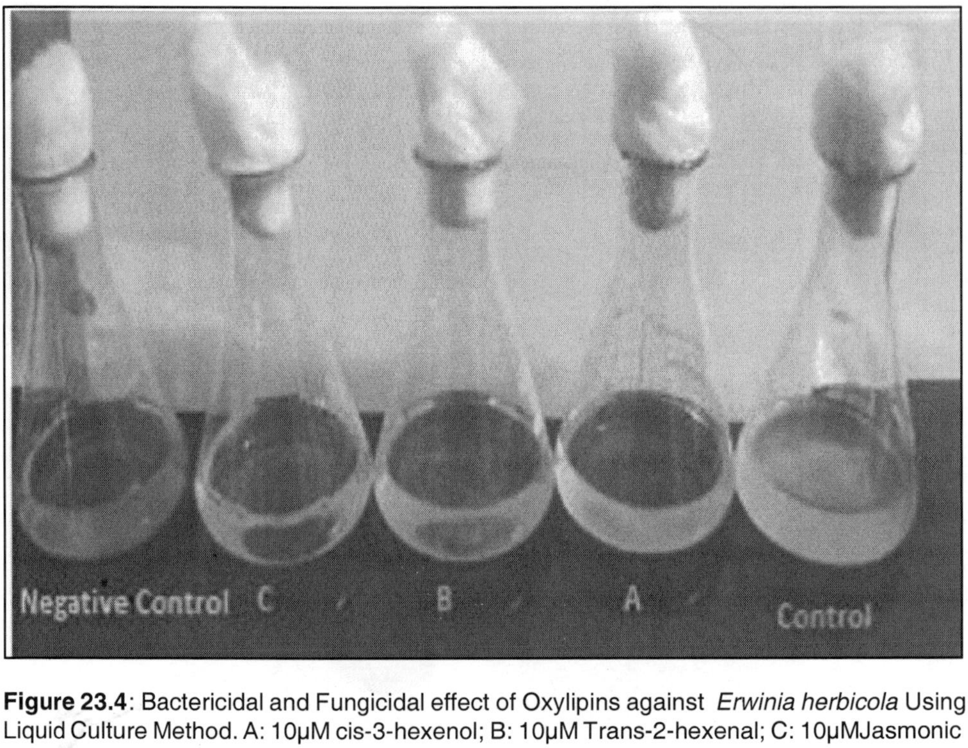

Figure 23.4: Bactericidal and Fungicidal effect of Oxylipins against *Erwinia herbicola* Using Liquid Culture Method. A: 10μM cis-3-hexenol; B: 10μM Trans-2-hexenal; C: 10μMJasmonic acid Control, Negative control.

Table 23.2: Bactericidal Activity of Oxylipins. Data represented as mean± SD of two trials with three replicates. All the values are significant at P<0.05. Values are mean ±SD of two trials with three replicates.

Treatment	*Erwinia herbicola*	*Enterobacter cloacae*	*Pseudomonas aeruginosa*	*Xanthomonasa xonopodis*
Cis-2-Hexenol				
Control	0.20±0.02	0.40 ± 0.02	0.20 ±0.20	0.15±0.02
1μM	0.20±0.04	0.3± 0.03	0.15 ± 0.06	0.20 ±0.08
5μM	0.01±0.03	0.0 ±0.05	0.08 ±0.04	0.0 ±0.02
10μM	0.0 ± 0.01	0.0 ±0.05	0.0 ±0.03	0.0 ±0.03
		Trans-2-Hexenal		
Control	0.20±0.02	0.40 ± 0.02	0.20 ±0.02	0.1±0.02
1μM	0.06±0.05	0.03 ± 0.08	0.04 ± 0.07	0.04±0.01
5μM	0.01±0.04	0.0 ±0.02	0.0 ±0.01	0.0 ±0.06
10μM	0.0 ± 0.02	0.0 ±0.03	0.0 ±0.03	0.0 ±0.04
		Jasmonic acid (JA)		
Control	0.20±0.02	0.40 ± 0.02	0.20 ±0.20	0.1±0.02
1μM	0.04±0.03	0.30 ± 0.03	0.10 ± 0.04	0.07±0.02
5μM	0.01±0.02	0.0 ±0.04	0.05 ±0.05	0.0 ±0.06
10μM	0.0 ± 0.04	0.0 ±0.06	0.0 ±0.03	0.0 ±0.03

Conclusion

Lipoxygenase is one of the key sources to protect the plants from the various types of pathogens. This enzyme provides a number of secondary metabolites by oxygenation which has ubiquitous functions in providing resistance to plant systems against the stress and pathogen attack.LOX can be used as a diagnostic tool for resistance in screening different varieties of seeds against bacterial and fungal pathogens.

References

Andersson, M.X., Hamberg, M., Kourtchenko, O., Brunnstrom, A., McPhail, K.L., *et al.* (2006). Oxylipin profiling of the hypersensitive response in *Arabidopsis thaliana*. Formation of a novel oxo-phytodienoic acid-containing galactolipid, arabidopside. *J. Biol. Chem.*, **281**: 31528-37.

Andreou, A.Z., Hornung, E., Kuznets, S., Rosahl, S., and Feussner, I. (2009). On the substrate binding of linoleate 9-lipoxygenases. *Lipids.*, **44**: 207-215.

Axelrod, B. (1974). Lipoxygenases. *ACS AdvChem Ser.*, **136**: 324-348.

Bate. N.J., and Rothstein, S.J.(1998). C6-volatiles derived from the lipoxygenase pathway induce a subset of defense-related genes. *Plant J.*, **16**: 561-569.

Blee, E. (2002).Impact of phyto-oxylipins in plant defense. *Trends Plant Sci.***7**: 315-322.

Burow, G.B., Nesbitt, T.C., Dunlap, J., and Keller, N.P.(1997). Seed lipoxygenase products modulate *Aspergillus*mycotoxin biosynthesis.*Molecular Plant-Microbe interactions,* **10**: 380-387.

Cacas, J.L., Vailleau, F., Devoine, C., Ennar, N., Agnel, J.P., Tronchet, M., Ponchet, M., Blein, J. P., Roby, D., Triantaphylides, C., and Montillet, J. L. (2005). The combined action of 9- lipoxygenase and galactolipase is sufficient to bring about programmed cell death during tobacco hypersensitive response. *Plant Cell Environ.,* **28**: 1367-1378.

Corbin, J.A., Evans, J.H., Landgraf, K.E., and Falke, J.J. (2007). Mechanism of specific membrane targeting by C2 domains: localized pools of target lipids enhance Ca2+ affinity. *Biochemistry.,* **46**: 4322-4336.

Croft, K.P.C., Jiittner, F., Slusarenko, A.J. (1993). Volatile products of the lipoxygenase pathway evolved from *Phaseolus vulgaris* (L.) leaves inoculated with *Pseudomonas syringae*pv*phaseolicola. Plant Physiol.,* **101:** 13-24.

Dhondt, S., Geoffroy, P., Stelmach, B.A., Legrand, M., and Heitz, T. (2000).Soluble phospholipase A2 activity is induced before oxylipin accumulation in tobacco mosaic virus-infected tobacco leaves and is contributed by patatin like enzymes. *Plant J.,* **23**: 431-440.

Fammartino, A., Cardinale, F., Göbel, C., Mene-Saffrane, L., Fournier, J., Feussner, I., and Esquerre-Tugaye, M. T. (2007). Characterization of a Divinyl ether biosynthetic pathway specifically associated with pathogenesis in tobacco. *Plant Physiol.,* **143**: 378-388.

Feussner, I., Wasternack, C. (2002). The lipoxygenase pathway. *Annu Rev Plant Biol.,* **53**: 275–297.

Gobel, C., Feussner, I., Hamberg, M., and Rosahl, S. (2002).Oxylipin profiling in pathogen-infected potato leaves. *Biochem. Biophys. Acta,* **1584**: 55-64.

Gundlach, H., Muller, M.J., Kutchan, T.M., Zenk, M.H. (1992). Jasmonic acid is a signal transducer in elicitor-induced plant cell cultures. *Proc. Natl. Acad. Sci. USA,* **89**: 2389-93

Hildebrand, D (1995). Effect of volatile methyl jasmonate on the oxylipin pathway in tobacco, cucumber and Arabidopsis. *Plant Physiol.,* **109**: 1227-1230.

Huang, F.C., and Schwab, W. (2011). Cloning and characterization of a 9-lipoxygenase gene induced by pathogen attack from *Nicotiana benthamiana* for biotechnological application. *BMC Biotechnology,* **2011:** 11-30.

Kessler, A., and Baldwin, I.T. (2002). Plant responses to insect herbivory: the emerging molecular analysis. *Annual Review of Plant Biology,* **53**: 299-328.

Kohlmann, M., Bachmann, A., Weichert, H., Kolbe, A., Balkenhohl, T., *et al.* (1999). Formation of lipoxygenase-pathway-derived aldehydes in barley leaves upon methyl jasmonate treatment. *Eur. J. Biochem,*.**260**: 885–95

Kuroda, H., Oshima, T., Kaneda, H., Takashio, M (2005). Identification and functional analyses of two cDNAs that encode fatty acid9-/13-hydroperoxide lyase (CYP74C) in rice. *BiosciBiotechnolBiochem,* **69**: 1545-1554.

Li, C., Williams, M.M., Loh, Y.T., Lee, I.G. and Howe, GA. (2002). Resistance of cultivated tomato to cell content-feeding herbivores is regulated\ by the octadecanoid-signaling pathway. *Plant Physiol,* **130**: 494–503

Li, P., Zhao, P.J., Ma, C.L., and Zeng, Y (2012).A Chitosan Induced 9-Lipoxygenase in *Adelostemma gracillimum* Seedlings. *Int J Mol Sci.,* **13**(1): 540–551.

Maccarrone, M., Van Zadelhoff, G., Veldink, G.A., Vliegenthart, J.F.G., Finazzi-Agro, A(2000). Early activation of lipoxygenase in lentil (*Lens culinaris*)root protoplasts by oxidative stress induces programmed cell death. *Eur. J. Biochem,* **267:** 5078–5084.

Muni kumari, M., Uma maheswari Devi, P., and sucharitha, A (2011).Differential Effect of Lipoxygenase on Aflatoxin Production by *Aspergillus* spp. *International journal of plant Pathology,* **2**(4): 153-164.

Ongena, M., Duby, F., Rossignol, F., Fauconnier, M.L., Dommes, J., Thonart, P (2004). Stimulation of the lipoxygenase pathway is associated with systemic resistance induced in bean by a nonpathogenic *Pseudomonas* strain. *Mol. Plant-Microbe Interact.,* **17**: 1009–18.

Porta, H., Rocha-Sosa, M (2002). Plant lipoxygenase. Physiological and molecular features. *Plant Physiol,* **130**: 15-21.

Porta, H., Figueroa-Balderas, R.E., and Rocha-Sosa, M. (2008). Wounding and pathogen infection induce a chloroplast-targeted lipoxygenase in the common bean (*Phaseolus vulgaris* L.). *Planta,* **227**: 363-373.

Prost, I., Dhondt, S., Rothe, G., Vicente, J., Rodriguez, M.J., *et al.* (2005). Evaluation of the antimicrobial activities of plant oxylipins supports their involvement in the defense against pathogens, *Plant Physiol.,* **139**: 1902-12.

Rosahl, S., Feussner, I (2005). Oxylipins. *In: Murphy D (ed.) Plant lipids.* Oxford: Blackwell Publishing, pp.329–354.

Rusterucci, C., Montillet, J.L., Agnel, J.P., Battesti, C., Alonso, B, Knoll, A., Bessoule, J.J., Etienne, P., Suty, L., Blein, J.P., and Triantaphylides, C. (1999). Involvement of lipoxygenase dependent production of fatty acid hydroperoxides in the development of the hypersensitive cell death induced by cryptogein in tobacco leaves. *J. Biol. Chem.* **274:** 36446–36455.

Schneider, C., Niisuke, K., Boeglin, W.E., Voehler, M., Stec, D.F., Porter, N.A. and Brash, A.R. (2007a). Enzymatic synthesis of a bicyclobutane fatty acid by a hemoproteinlipoxygenase fusion protein from the *Cyanobacterium Anabaena* PCC 7120. *Proc. Natl. Acad. Sci. USA.* **104**: 18941–18945.

Vick, B.A., Zimmerman, D.C. (1987). Oxidative systems for the modification of fatty acids: the lipoxygenase pathway. *In* PK Stumpf, EE Conn, eds, *The Biochemistry of Plants.Academic Press,* New York, **9:** 53-90.

Wassim E.C., Kaspi, R., Savchenko, T., Rowe, H., Negre-Zakharov, F., Kliebenstein, D., and Dehesh, K. (2008). Distinct Roles of Jasmonates and Aldehydes in Plant-Defense Responses. *PLoS ONE* **3**(4): *e1904.*

Xue, H.Q., Upchurch, R.G., Kwanyuen, P (2006). Ergosterol as a quantifiable biomass marker for *Diaporthepha seolorum* and *Cercospora kikuchii*. *Plant Dis.,* **90**: 1395–1415.

Xue, Y.Y., Jiang, W.J., and Yu, H.J (2012).The Expression Profiling of the Lipoxygenase (LOX) Family Genes during Fruit Development, Abiotic Stress and Hormonal Treatments in Cucumber (*Cucumis sativus* L.) *Int. J. Mol. Sci.,* **13**: 2481-2500.

2015, Major Constrains and Verdict of Crop Productivity *Pages 337–341*
Editor: **Dr. U.N. Bhale**
Published by: **ASTRAL INTERNATIONAL (P) LIMITED, NEW DELHI**

Chapter 24

Deterioration of Lipids in the Diseased Groundnut Kernels due to Infection of Fungi

Pawan Shingare[1] and Avinash Ade[2]*

[1]Department of Botany,
Dr. B.N. Purandare Arts and Smt S.G. Gupta Commerce and Science College, Lonavala, Dist. Pune, M.S., India
[2]Department of Botany,
University of Pune, Pune, M.S., India

ABSTRACT

Groundnut is useful to human in many ways. It is especially considered as the oil seed crop. The oil is extracted from the groundnut kernels. Even after harvesting groundnut kernels there is possibility to have infection of the fungi. Due to the infection there is a loss of lipids which directly affects the quality of the groundnut kernels. In the present investigation ten different fungi were isolated from the agricultural field associated with the groundnut rhizosphere and treated with the groundnut kernels belong to five different varieties. The degradation of the lipids was observed in terms of the total content of oil from the affected groundnut kernels. The results showed that the significant lipid loss was observed due to the fungal infection in all the five varieties of groundnut.

Introduction

India is the second largest producer of groundnuts after China. Groundnut is the largest oilseed in India in terms of production. On an average it is accounted for

* *Corresponding Author.* E-mail: avinashade@unipune.ac.in

31.81 per cent of the oilseeds production of the country. The annual production of seed and oil is 5-8 and 1.5 million tons respectively. About 80 per cent of the total groundnut produced in India undergoes processing so that it is utilized as oil or cake. Groundnut is the fifth most important oilseed crop in the world (Ingale and Shrivastava, 2011). It is the rich source of oil as contains 48-50 per cent oil.

The groundnut, being a highly nutritive crop it is parasitized with several pathogens which causes a great loss in the yield. Early and late leaf spots are the most widespread and damaging foliar diseases of groundnut. Leaf spot disease consists primarily of early leaf spot caused by the fungus, *Cercospora arachidicola*. Early leaf spot is most severe where groundnut is grown in humid areas. Late leaf spot, is caused by a closely related fungus, *Cercospora personata*. The disease cycles of early and late leaf spot and the type of damage caused by the two diseases are similar. Late leaf spot often, but not always, occurs towards the end of the growing season and can be more destructive and difficult to control than early leaf. Therefore the effect of different fungi on the groundnut was studied in this investigation.

Materials and Methods

The material used for comparative study of Groundnut Germplasm was of 10-different varieties of Groundnut. The Groundnut varieties like LGN-2, AK-320, TKG-19A, PHULE UNAP, K-411 were collected from "Agricultural College, Latur.

Five varieties of groundnut were surface sterilized with 0.1 per cent mercuric chloride solution and subsequently washed and soaked in sterile distilled water for two hours. Excess of water was decanted from the seeds. The seeds of each variety were distributed into eleven flasks (100 g per flask) and were inoculated separately with 2 ml spore suspension of the ten test fungi and one flask without spore suspension was kept as a control. The flasks were incubated at room temperature for 14 days. After this incubation period mycelium mat was removed with the help of blotter paper. Powder of each sample was prepared in electrical grinder. These powders were further used for the estimation of physical and chemical parameters.

The crude fat in the seed sample was estimated by the standard Soxhlet method (A.O.A.C., 1970). The fat present in the seed material was extracted in the solvent mixture of chloroform ($CHCl_3$) and methanol (CH_3OH) in Soxhlet extraction assembly. 2g dry seed powder was placed in a thimble of Whatman filter paper No. 1. The mouth of thimble was plugged with fat free absorbent cotton. Solvent was added in dry 250 ml receiver flask from the soxhlet assembly just to reach the level of the neck. The thimble with sample was introduced into the soxhlet. The apparatus was assembled and placed on heating mantle with temperature controlling device. The fat was extracted for 8 hours at 60°C. When the extraction was over, thimble was removed from soxhlet. Apparatus was again assembled and heated to recover most of the solvent from the receiver flask. About 25ml solvent along with the extracted fat was left in the receiver flask, the receiver flask was disconnected. The solvent was then transferred in a clean, previously weighed beaker. After drying in a hot air oven at 95°C, it was then cooled in desiccators and weighed. The amount of fat was measured from extracted per 2g of the sample and amount of crude fat per per cent of dry matter (DM) was calculated.

Results and Discussion

The groundnut seeds were inoculated with ten fungal spore suspensions for fifteen days. After fifteen days lipids were analyzed. All ten fungi reduced the lipid content as compare to the control (Table 24.1). In the variety, LGN-2 the highest degradation was found in *Aspergillus niger* (0.271g) and in *Macrophomina phaseolina* it was 0.226 g. In control, lipid was 0.910 g/2 g. In *Fusarium oxysporum* it was 0.075 g, *Rhizopus stolonifer* 0.091 g, *Penicillium digitatum* 0.037 g, *Aspergillus flavus* 0.103 g was observed which is quite low as compared to above fungi.

Table 24.1: Lipid Content of Diseased Kernels of Groundnut Varieties (g/2g)

Sl.No.	*Name of Fungi*	*LGN-2*	*AK-320*	*TKG-19 A*	*PHULE UNAP*	*K-411*
1.	Control	0.910	0.900	1.000	1.000	0.972
2.	*Curvularialunata*	0.758	0.829	0.920	0.937	0.811
3.	*Fusariumoxysporum*	0.835	0.823	0.890	0.910	0.919
4.	*Aspergillusniger*	0.639	0.749	0.829	0.864	0.755
5.	*Rhizopusstolonifer*	0.819	0.838	0.803	0.911	0.925
6.	*Alternariaalternata*	0.732	0.849	0.749	0.882	0.900
7.	*Aspergillusterreus*	0.762	0.886	0.900	0.921	0.865
8.	*Penicilliumdigitatum*	0.873	0.765	0.870	0.855	0.933
9.	*Aspergillus fumigatus*	0.761	0.784	0.789	0.906	0.882
10.	*Macrophominaphaseolina*	0.684	0.843	0.840	0.886	0.907
11.	*Aspergillus flavus*	0.807	0.732	0.900	0.780	0.843
	S.D.	0.67	0.05	0.05	0.05	0.07
	S.E.	0.20	0.02	0.02	0.02	0.02
	C.D.	0.52	0.04	0.04	0.04	0.05

In variety, AK-320 *Aspergillus niger* showed highest (0.151 g) degradation and next is *Aspergillus flavus* which showed 0. 135 g degradation, *Penicillium digitatum* 0.135 g and 0.116 g lipid was degraded in case of *Aspergillus fumigates*. *Curvularia lunata* showed 0.071 g, *Fusarium oxysporum* showed 0.077 g, *Rhizopus stolonifer* could degrade 0.062 g lipids, *Macrophomina phaseolina* showed 0.057 g lipid degradation. Very low degradation was found in *Aspergillus terrus* that is 0.014 g, and 0.051 g in *Alternaria alternata*.

In variety, TKG 19A highest degradation was found in *Alternaria alternata* that is 0.251 g, *Aspergillus fumigates* degraded 0.211 g, *Aspergillus niger* degraded 0.171 g, *Rhizopus stolonifer* degraded 0.197 g, *Macrophomina phaseolina* degraded 0.160 gm which is also considerable. Average degradation was observed in *Aspergillus flavus* (0.100 g), *Aspergillus terrus* (0.100 g), *Penicellium digitatum* (0.130 g) and *Curvularia lunata* (0.080 g). In this variety very low degradation of lipid was observed in *Fusarium oxysporum* that is 0.010 g.

In the variety, PHULE UNAP five fungi showed below the 0.1 g degradation of lipid and five fungi showed above the 0.1 g. The above degradation was observed in *Aspergillus flavus* (0.220 g), *Penicellium digitatum* (0.145 g), *Aspergillus niger* (0.136 g), *Alternaria alternata* (0.118 g), *Macrophomina phaseolina* (0.114 g). Below the 0.1 g observed in *Aspergillus fumigates* (0.094 g), *Fusarium oxysporum* (0.090 g), *Rhizopus stolonifer* (0.089 g), *Aspergillus terrus* (0.079 g) *and Curvularia lunata* (0.063 g). The highest degradation found in *Aspergillus flavus i.e.*0.220 g and lowest degradation found in *Curvularia lunata.*

In the variety, K-411 *Aspergillus niger* showed lipid degradation as 0.217 g and it is highest in this variety. *Aspergillus flavus* showed 0.129 g, *Curvularia lunata* showed 0.161 g, *Aspergillus terries* showed 0.107 g degradation of lipid. *Alternaria alternata* showed 0.072 g, *Macrophomina phaseolina* 0.065 g, *Fusarium oxysporum* showed 0.053 g and *Rhizopus stolonifer* 0.047 g degradation which is somewhere equal to each other. The lowest degradation found in *Aspergillus fumigates* which is 0.09g. All these lipid content is shown in Table 24.1.

The groundnut seeds inoculated with ten fungal spores for fifteen days. After fifteen days lipid content was analyzed. All ten fungi reduced the lipids as compare to the control. In the variety, LGN 2 the highest degradation was found in *Aspergillus niger* (0.271g) and in *Macrophomina phaseolina* it was 0.226 g. In control lipid was 0.910 g/2g. In *Fusarium oxysporum* 0.075 g, *Rhizopus stolonifer* 0.091 g, *Penicellium digitatum* 0.037 g, In *Aspergillus flavus* 0.103 g was observed which was quite low as compared to the above fungi, and that type of results were found in other four varieties not exactly similar but quite similar. Umechuruba *et al.* (1992) also analyzed oil content of healthy and infected groundnut seeds with *A niger*, *A. flavus* and *Macrophomina phaseolina* and he also found that 2 to 6 per cent of oil loss was done. Maximum reduction of nitrogen in groundnut, as compare to other oil seeds were caused due to *Aspergillus sp.* (Hatti, 2010). Kesare, (2008) found that *Alternaria alternata* and *Fusarium oxysporum* were mainly responsible fungi for deteriorating nitrogen content in soybean. Protein degradation was also found maximum in all varieties; Chavan (2011) reported loss of protein content in sunflower due to associated fungi. Similarly change in protein content was observed by Singh and Prasad (1988) in *Sesamum* and Kumar and Prasad (1993) observed in mustard.

References

AOAC (1970) Official Methods of Analysis. *Association of Official Agricultural Chemists* (11 th ed.) Washington. D.C.

Chavan, A. M. (2011) Nutritional changes in oilseeds due to *Aspergillus* spp. *Journal of Experimental Sciences*, **2**(4): 29-31.

Hatti, A. D. (2010) Impact of *Aspergillus* on biodegradation of oil seeds. Ph.D. Thesis, Dr. Babasaheb Ambedkar Marathwada University Aurangabad.

Ingale, S. and Shrivastava, S.K. (2011) Chemical, nutritional and anti-nutritional study of new varieties of oil seeds from sunflower, safflower and groundnut, *International Journal of Biotechnology Applications*, **3**(4): 118-129.

Kesare, U. T. (2008) Studies on seed mycoflora of soyabean. Ph.D. Thesis, Dr. Babasaheb Ambedkar Marathwada University, Aurangabad.

Kumar, S. and Prasad, B. K. (1993) Total free amino acid content of mustard seed due to storage fungus *Aspergillus flavus*. *Indian Phytopath*, **43**(3): 329.

Singh, B. K. and Prasad, T. (1988) Effect of seed quality on germination and seed mycoflora in sunflower. *Journal of Indian Botanical Society,* **67**(3-4): 316-317.

Umechuruba, C. I., Out, K. A. and Ataga, A. E. (1992) The role of seed-borne *Aspergillus flavus* link ex Fr., *Aspergillus niger* Van Tiegh and *Macrophomina phaseolina* (Tassi) Goid on deterioration of groundnut (*Arachis hypogaea* L.) seeds. *International Biodeterioration and Biodegradation*, **30**(1): 57–633.

2015, Major Constrains and Verdict of Crop Productivity Pages 342–348
Editor: Dr. U.N. Bhale
Published by: ASTRAL INTERNATIONAL (P) LIMITED, NEW DELHI

Chapter 25

In vitro Efficacy of Different Phytoextracts and Bioagents against *Colletotrichum falcatum* Went. Causing Red Rot of Sugarcane (*Saccharum officinarum* L.)

M.D. Jadhav, J.J. Kadam*, M.S. Joshi and P.G. Borkar

Department of Plant Pathology,
College of Agriculture, Dapoli – 415 712, Maharashtra, India

ABSTRACT

The severe incidence of red rot disease incited by Colletotrichum falcatumWent.was observed on sugarcane crop cultivated at the Department of Agronomy and Central Experiment Station, Wakawali (Asond Block), Dr. Balasaheb SawantKonkan Krishi Vidyapeeth, Dapoli, Dist. Ratnagiri (M.S.) in the range of 11.62 to 29.14 per cent during 2012 to 2014. Among different plant extracts tested against C. falcatum the maximum mycelial inhibition (88.15 per cent) was achieved due to fruit extract of Soapnut followed Wild brinjal (87.44 per cent). For biological control, complete inhibition of the C. falcatumwas achieved due to Trichoderma harzianum followed by Saccharomyces cerevisiae (NCIM - 3315), T. viride and Aspergillus niger.

Keywords: *Red rot, Colletotrichum falcatum, sugarcane, Trichoderma, Aspergillus, phytoextracts, bio-agents.*

* *Corresponding Author.* E-mail: jjkdm1234@gmail.com

Introduction

Sugarcane, a potential crop for commercial cultivation, is subjected to be attacked by many pathogens leading to cause serious diseases. In India, production of sugarcane is 3,47,870 (000′ tonnes) covering area 5081 (000′ ha) and productivity 68.46 t/ha. In India, Uttar Pradesh is the major sugarcane growing state. Maharashtra is the second largest sugarcane growing state in the country. It contributed 13.53 per cent hectare to total area and 15.06 per cent to total production of sugarcane in the country. In Maharashtra, sugarcane production is 81991 (000′ tonnes) covering area 1022 (000′ ha.) with productivity of 80.10 (t/ha.) (Deokate, 2013). More than hundred diseases of sugarcane have been reported from different parts of the world which are obstacle to get maximum production of this crop. Red rot, the most dreaded disease of sugarcane, is of great economic importance and responsible for the failure of many popular varieties in different countries. The present investigations was undertaken to test *in vitro*the efficacy of phytoextracts and bio-agents against *C. falcatum.*

Materials and Methods

Efficacy of Phytoextracts against *C. falcatum*

The plant extracts of soapnut, nilgiri, rui, sarpagandha, tulsi, neem, garlic, jatropha, wild brinjal and glyricidia were used during the studies. Aqueous phytoextracts were obtained as per the method described by Bhatti (1988). Hundred gram fresh plant materials were washed thoroughly with sterile distilled water and ground well in 100 ml distilled sterile water. The macerate was filtered through sterilized double layered muslin cloth and centrifuged at 4000 rpm for 5 minutes. After centrifuging, the supernatant was then filtered through filter paper. Extracts thus obtained were passed through Sintered glass filter separately to avoid bacterial contamination. This formed the standard plant extract solution (100 per cent).

The effect of plant extracts on mycelial growth was studied by 'Poisoned Food Technique' (Nene and Thapliyal, 1997). All the plant extracts were tested at 10 per cent concentration against the test pathogenusing oat meal agar as a basal medium. The culture discs of 5 mm diameter were cut from 7 day old culture of the test fungus with the help of sterile cork borer and transferred aseptically to the centre of the plate already poured with poisoned medium. The plates without plant extract were also inoculated with fungal culture disc which served as control. The plates were then incubated at 27 ± 2ºC in the BOD incubator. Three replications of each treatment were maintained. The observations on colony diameter of the fungus were recorded when Petri plate in control treatment was fully covered with mycelial growth.

Per cent inhibition of growth of the test fungus was calculated by following formula (Horsfall, 1956).

$$X = \frac{Y - Z}{Y} \times 100$$

where,

X = Per cent inhibition

Y = Growth of fungus in control (mm)

Z = Growth of fungus in treatment (mm)

Efficacy of Bioagents against *C. falcatum*

The laboratory experiment was conducted with five bio-agents *viz; Trichoderma harzianum, T. viride* and *Aspergillus niger* and two strains of yeast (*Saccharomyces cerevisiae* Culture No. NCIM-3095 and NCIM-3315) and *Pseudomonas fluorescens* (bacterial bio-agent) against causal fungus *C. falcatum*.

The trial was conducted by using dual culture technique (Padder *et al.*, 2010) in which the fungal discs of 5 mm diameter were placed in a petri plate in such a way that both the organisms in the plate will get equal opportunity for growth. For testing of *P. fluorescens* and yeast cultures, Bengal method and streaking were followed.

Each treatment was replicated three times. The plates were then incubated at room temperature (27 ± 2°C) for seven days. The observations on colony diameter of the test fungus were recorded seven days after inoculation when Petri plate in control treatment was fully covered with mycelial growth of the test pathogen.Per cent inhibition of growth was calculated by following formula (Horsfall, 1956).

Results and Discussion

Efficacy of Phytoextracts against *C. falcatum*

All the plant extracts tested at 10 per cent concentration were found significantly effective in inhibiting the mycelial growth of *C. falcatum*. Maximum per cent inhibition of *C. falcatum* over control was achieved due to fruit extract of Soapnut (88.15 per cent)and was significantly superior over rest of the treatments and was at par with fruit extract of Wild brinjal with 87.44 per cent inhibition of test pathogen over control.These were followed by extracts of Garlic, Neem, Nilgiri and Tulsi which showed 66.66, 64.82, 61.11 and 55.55 per cent inhibition, respectively over control. Leaf extracts of *Glyricidia*, Sarpagandha and Rui showed moderate effect with 52.60, 50.00 and 48.15 per cent inhibition over control, respectively. Jatropha was found to be least effective phytoextract in inhibiting the mycelial growth of the test pathogen which recorded only 40.74 per cent inhibition over control. These results are in line with the earlier findings of Chauhan and Joshi (1990) and Chavan (1996) who concluded that garlic, neem, nilgiri and tulsi extracts gave good inhibition of mycelial growth of *C. gloeosporioides*. Similarly, Mala *et al.* (1998) and Radha *et al.* (1998) reported that phytoextract of *Allium sativum* significantly inhibited mycelial growth of *C. falcatum*. Raheja and Thakore (2002) observed that *Allium sativum* (cloves) and *Azadirachta indica* (leaves) were found most effective to check mycelial growth of *C. gloeosporioides*. The present findings are also in close conformity with Imtiaj *et al.* (2007) who reported that leaf extracts of *Curcuma domestica, Ocimum sanctum*, seed extract of *Azadirachtaindica* and bark extract of *Swietenia mahogani* were inhibitor for conidia and mycelial growth of red rot pathogen. Watve *et al.* (2009), Mishra and Behera (2011b) and Prince and Prabakaran (2011) reported that *A. sativum, A. indica* and *O. sanctum* extracts showed antifungal activity against *Colletotrichum* species including *falcatum*.

Table 25.1: Effect of Plant Extracts on Growth of *C. falcatum*

Tr.No.	*Treatments*	*Conc. (per cent)*	*Mean Colony Diameter (mm)*	*Per cent Inhibition Over Control*
T_1	Soapnut (*Sapendus tripholiarus*)	10	10.66	88.15
T_2	Nilgiri (*Eucalyptus globulus*)	10	35.00	61.11
T_3	Rui (*Calotropis gigantea*)	10	46.66	48.15
T_4	Sarpagandha (*Rauvolfia serpentina*)	10	45.00	50.00
T_5	Tulsi (*Ocimum sanctum*)	10	42.66	52.60
T_6	Neem (*Azadirachta indica*)	10	31.66	64.82
T_7	Garlic (*Allium sativum*)	10	30.00	66.66
T_8	Jatropha (*Jatropha curcas*)	10	53.33	40.74
T_9	*Glyricidia* (*Glyricidia maculata*)	10	40.00	55.55
T_{10}	Wild brinjal (*Solanum nigrum*)	10	11.30	87.44
T_{11}	Control	–	90.00	–
	S.Em. ±C.D. at 1 per cent		0.27	
	1.07			

Efficacy of Bio-agents against *C. falcatum*

The present study revealed that *Trichoderma harzianum* showed 100 per cent inhibition of *C. falcatum* over control which was followed by *Saccharomyces cerevisiae* – NCIM-3315 (96.11 per cent) when tested by Bengal method. *T. viride* showed 87.22 per cent inhibition of the test pathogen over control which was followed by *Aspergillus niger* and *S. cerevisiae* – NCIM-3315 with 86.11 and 85.55 per cent inhibition of the pathogen over control, respectively and were found statistically at par with each other. Moderate inhibition of test pathogenwas recorded due to *S. cerevisiae* – NCIM-3095 when tested by both streaking (53.22) and Bengal method (42.77). Least inhibition of *C. falcatum* was noticed due to *Pseudomonas fluorescens* which showed27.41 per cent inhibition when tested by Bengal method and 25.00 per cent when streaked on two sides. The above findings are in close conformity with the results of Narendra Singh (1992) and Kumar and Satyavir (1998) who recorded maximum inhibition of *C. falcatum*by *T. harzianum*. Similarly, Begum *et al.* (2007) reported that *T. harzianum, T. viride* and *A. niger* prominently inhibited the mycelial growth of *C. falcatum* under laboratory conditions following the dual culture technique. Singh *et al.* (2008b) found that *T. harzianum* was most useful for effective management of red rot. Krishnamma and Reddy (2009) recorded 84.1 per cent inhibition of *C. falcatum* in dual culture by *T. viride*. Mishra and Behera (2011a) stated that *T. viride*was proved superior in reducing radial growth of *C. falcatum*upto 45.53 per cent. The present findings are also in line with the results of Prince *et al.* (2011) and Shailbala *et al.* (2013).

Table 25.2: Effect of Bio-agents on Growth of *C. falcatum*

Tr.No.	*Placement details*	*Mean Colony Diameter (mm)*	*Per cent Inhibition Over Control*
T_1	*Cf Th*	0.00	100
T_2	*Cf Tv*	11.50	87.22
T_3	*Cf* *An*	12.50	86.11
T_4	Sc-3315 *Cf* O Sc-3315 Sc-3315	3.50	96.11
T_5	Sc-3315 *Cf* Sc-3315	13.00	85.55
T_6	Sc-3095 *Cf* O Sc-3095 Sc-3095	42.10	53.22
T_7	Sc-3095 *Cf* Sc-3095	51.50	42.77
T_8	*Pf* *Cf* O *Pf* *Pf*	65.33	27.41
T_9	Pf Cf Pf	67.50	25.00
T_{10}	Control	90.00	–
	S.Em. ±		0.18
	C.D. at 1 per cent		0.74

Cf = *Colletotrichum falcatum*,Tv; *Trichoderma viride*,Th: *T. harzianum*; An: *Aspergillus niger*, Sc: *Saccharomyces cerevisiae* (NCIM- 3315, NCIM- 3095); Pf: *Pseudomonas fluorescens.*

Conclusion

From the present investigation it is concluded that fruit extracts of soapnut and wild brinjal are the most effective plant extracts against red rot causing pathogen. Extracts of Garlic, Neem, Nilgiri and Tulsi are also inhibitory to *C. falcatum*. Similarly, *Trichoderma harzianum*, *T. viride*, *Aspergillus niger* and *Saccharomyces cerevisiae* – NCIM-3315 are the potential bio-control agents for the red rot pathogen.

References

Begum, F., Talukder, M. I. and Hoque, M. Z. (2007). Comparative *in vitro* antagonistic potential of some bio-agents against *C. falcatum* Went., the cause of red rot of sugarcane. *Planter.*, **83** (978) : 607-616.

Bhatti, B. S. (1988). Utilization of toxic plants for the control of nematode pest of economic crop. Final Technical Report April 1, 1983 to March 31, 1988, HAU Hissar, India, p. 56.

Chauhan, H. L. and Joshi, H. U. (1990). Evaluation of phyto-extracts for control of mango fruit anthracnose. Paper presented in National Symposium on *Botanical Pesticides inIntegrated Pest Management* held on January 21-22, 1990 at CTRI, Rajahumundry. pp. 455-459.

Chavan, M. (1996). Effect of plant extracts on some important pathogenic fungi of Konkan region. *M.Sc.* (*Agri.*) *Thesis*, KKV, Dapoli, Maharashtra, India.

Deokate, T. B. (2013). Quantification of yield gaps in different planting types of sugarcane in Maharashtra. Website: *http.//www.igidr.ac.in/pdf/publication/WP-2013-011.pdf* (Source: Ministry of Agriculture, Govt. of India).

Horsfall, J. G. (1956). Principles of fungicidal action. *Cronical Botanica.* p. 207.

Imtiaj, A., Shah Alam M., Rafiul Islam, A.K.M., ShahidulAlam and Tae-Soo Lee (2007). *In vitro* studies on *C. falcatum* the causal of red rot disease of sugarcane. *American - Eurasian J. Agric. and Environ. Sci.*, **2** (5) : 511-517.

Kumar, A. and Satyavir(1998).Evaluation of biological control agents against redrot (*Colletotrichum falcatum*) of sugarcane. *Tests of Agrochemicals and Cultivars*. **(19)** : 72-73.

Krishnamma, K. and Reddy, P. N. (2009). Efficacy of *Trichoderma viride* against *Colletotrichum falcatum* in sugarcane. *Indian Journal of Plant Protection*, **37** (1/2) : 111-115.

Mala, S. R., Radha, M., Revathy, G. and Solayappan, A. R. (1998).Garlic as a biocide for *Colletotrichum falcatum. Indian Journal of Plant Protection,* **26** (1) : 49-51.

Mishra, M. K. and Behera, B. (2011a). Antagonistic effect of various bioagents against different isolates of *Colletotrichum falcatum* Went. causing red rot of sugarcane. *Journal of Plant Protection and Environment*, **8** (1) : 71-77.

Mishra, M. K. and Behera, B. (2011b). Bioefficacy of phytoextracts against sugarcaneredrot pathogen *Colletotrichum falcatum* Went. *Journal of Plant Protection and Environment*, **8** (2) : 57-60.

Narendra Singh, (1992). Biological control of red rot disease of sugarcane. *Indian Phytopath.*, **43** : 64.

Nene, Y. L. and Thapliyal, P. N.(1997). *Fungicides in plant disease control*. Oxford and IBH Publishing Co., New Delhi. pp. 531 - 532.

Padder, B. A., Sharma P. N., Kapil, R., Pathania, A. and Sharma, O. P. (2010).Evaluation of Bioagents and Biopesticides against *Colletotrichum lindemuthianum* and its Integrated Management in Common Bean.*Not Sci Bio.*, **2** (3) : 72-76.

Prince, L. and Prabakaran, P. (2011).Antifungal activity of medicinal plants against plant pathogenic fungus *C. falcatum*. *Asian Journal of Plant Science and Research,* 1 (1) : 84-87.

Prince, L., Raja, A. and Prabakaran, P. (2011).Antagonistic potentiality of some soil mycoflora against *Colletotrichum falcatum*.*World Journal of Science and Tech.*, **1** (4): 39-42.

Radha, M., Mala, S. R., Revathy, G. and Solayappan, A. R. (1998). *In vitro* antifungal properties of garlic (*Allium sativum* L.) extracts against *Colletotrichum falcatum* Went. *Journal of Biological Control*, **12** (1) : 81-83.

Raheja, S. and Thakore, B.B.L. (2002). Effect of physical factors, plant extracts and bioagents on *Colletotrichum gloeosporioides*Penz, the causal organism of anthracnose of Yam. *Journal of Mycol. and Plant Patho.*, **32** : 293-294.

Singh, V. Srivastava, S. N. Lal, R. J. Awasthi, S. K. and Joshi, B. B. (2008b).Biological control of red rot disease of sugarcane through *Trichoderma harzianum* and *Trichoderma viride*. *Indian Phytopath*. **61** (4) : 486-491.

Shailbala, Singh, V. K. and Kashyap, S. (2013). Field response of *Trichoderma harzianum* on control of red rot and quantitative attributes of sugarcane. *J. Research on Crops*, **14**(2) : 561-566.

Watve, Y. G., Diwakar M. P., Kadam J. J., Sawant U. K. and Mundhe V. G. (2009). Effect of Different chemicals, plant extracts and bioagents against *C. gloeosporioides* causing leaf spot disease of Jatropha. *J. Pl. Dis. Sci.*, **4** (2) : 199-203.

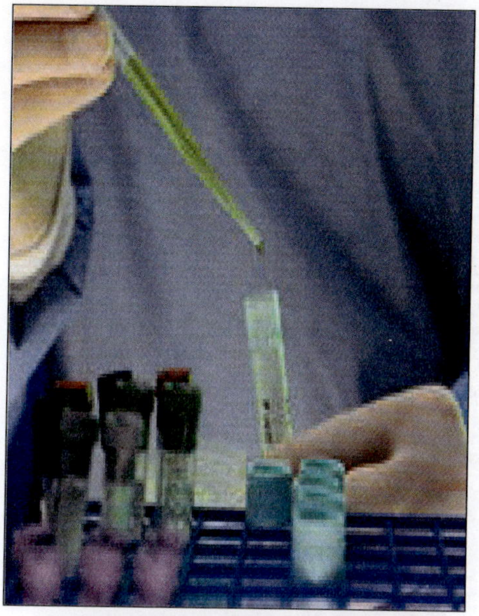

GENETIC MARKERS

"The green section indicates the presence of a desirable gene in an organisms' genetic code that is associated with two genetic markers (red flags)."

(P. 79)

Initial Symptoms

Green Ear Symptoms
Plate 6.1 (P. 101)

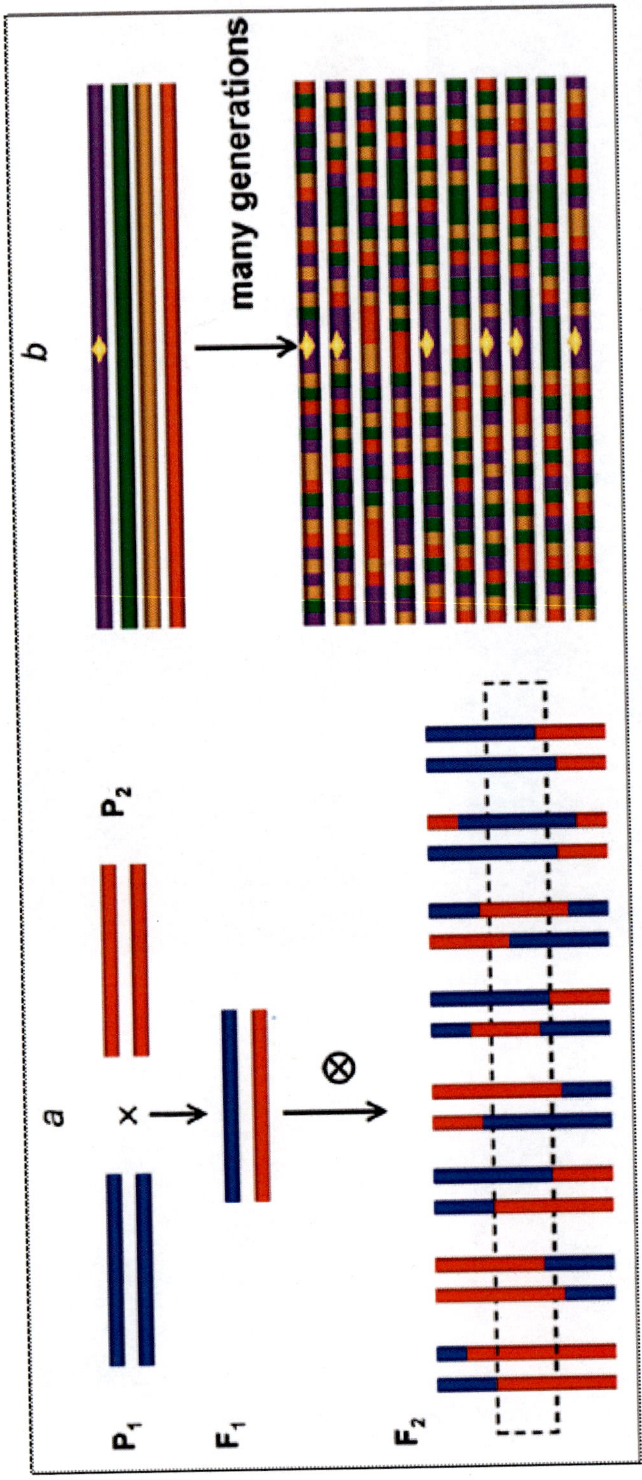

Figure 7.1: Schematic Comparison of Linkage Analysis with Designed Mapping Populations and Association Mapping with Diverse Collections (*Source:* Zhu *et al.*, 2008). (P. 121)

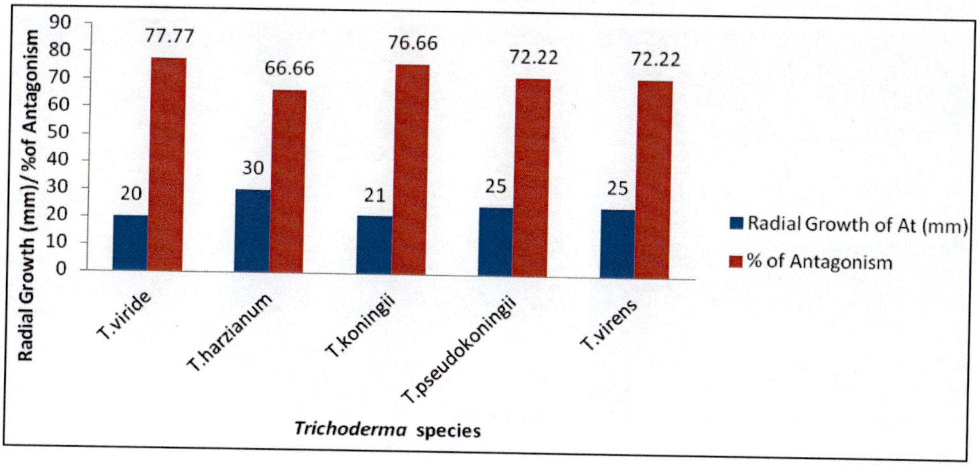

Figure 9.1 (P. 157)

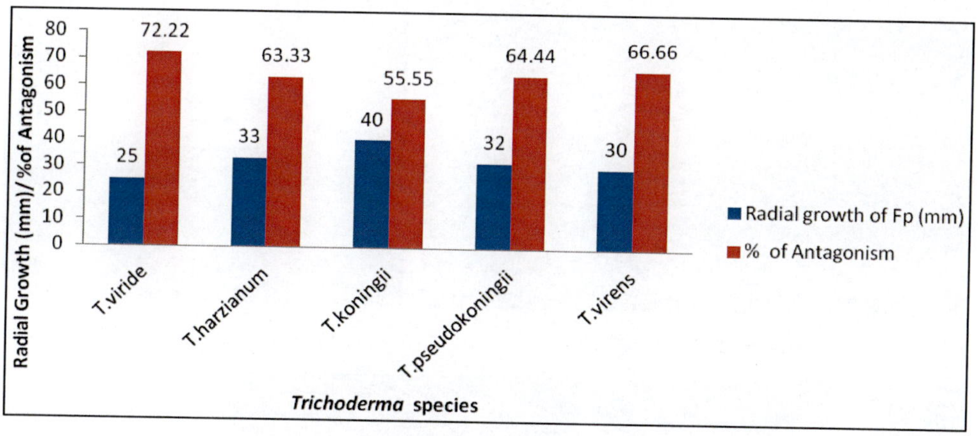

Figure 9.2 (P. 157)

Plate 13.1a (P. 203)

Plate 13.1b (P. 203)

Plate 13.2a (P. 204)

Plate 13.2b (P. 204)

Plate 12.3a (P. 206)

Plate 12.3b (P. 206)

Figure 16.1 (P. 238)

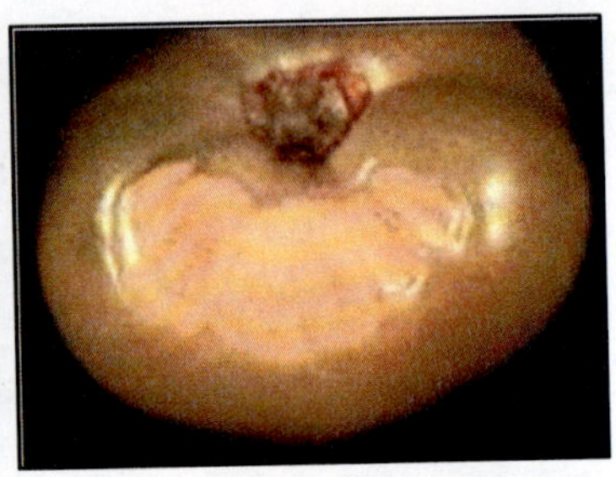

Figure 16.2 (P. 238)

Figure 16.3 (P. 239)

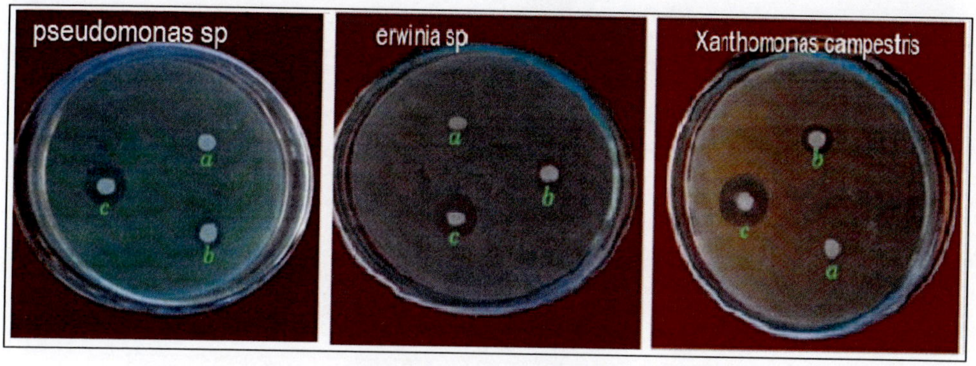

Figure 23.1 (P. 329)

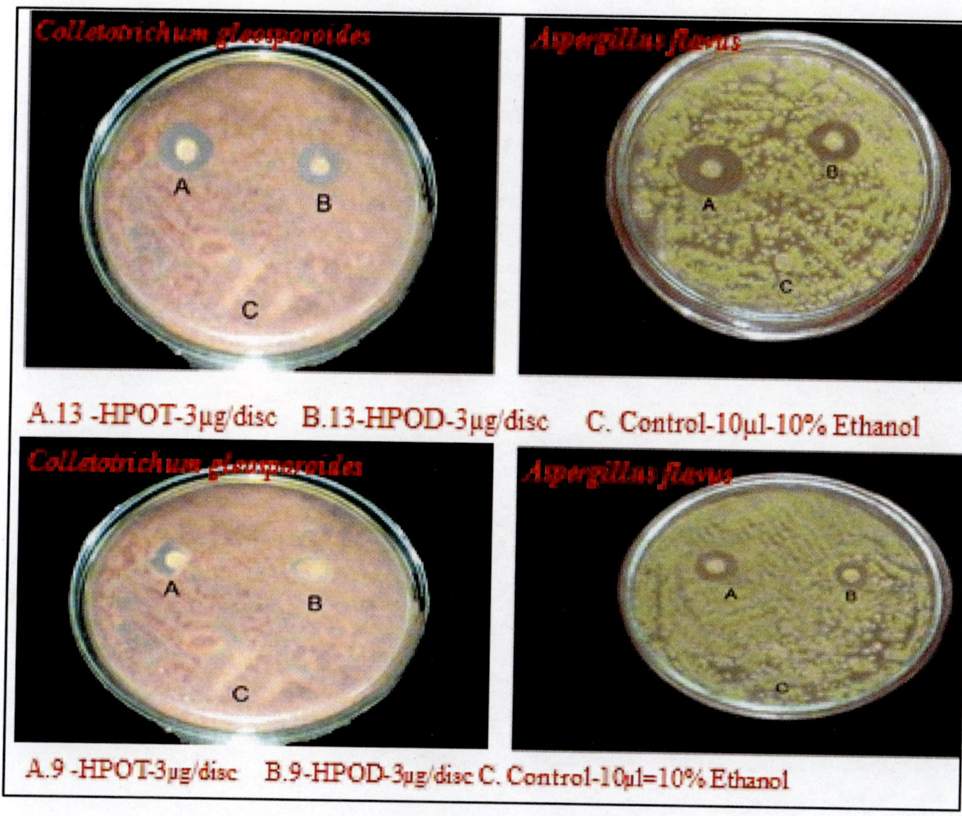

Figure 23.2 (P. 331)

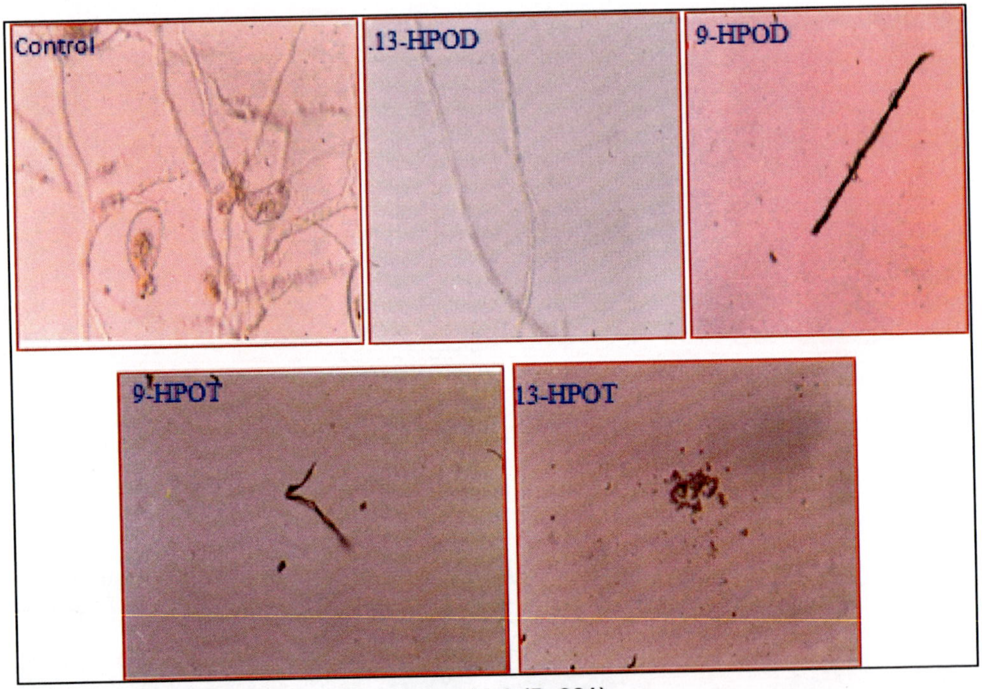

Figure 23.3 (P. 331)

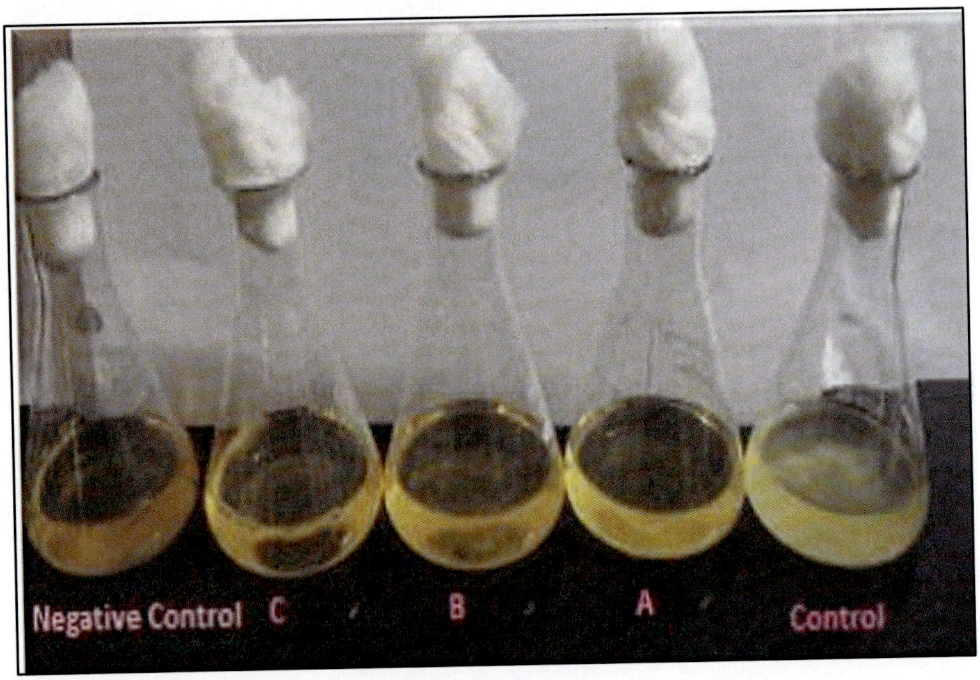

Figure 23.4 (P. 332)